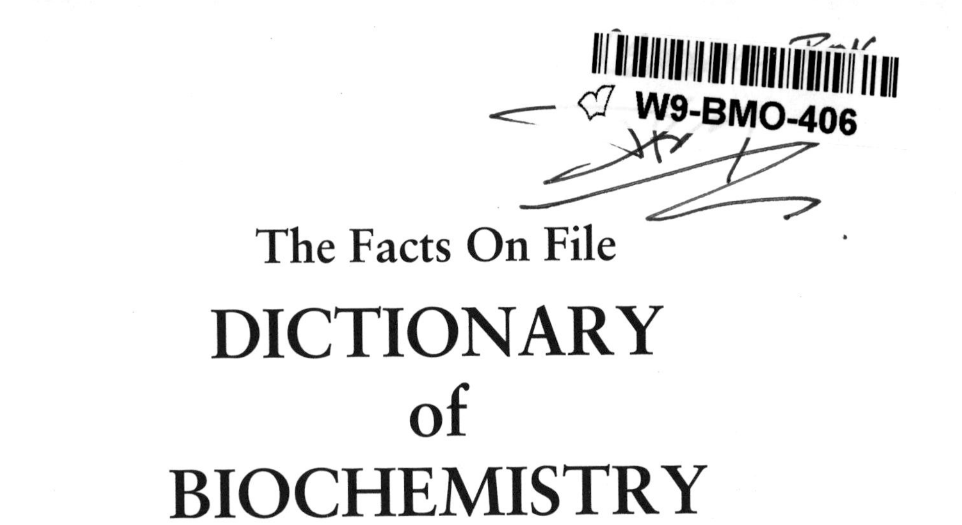

The Facts On File

DICTIONARY of BIOCHEMISTRY

The Facts On File

DICTIONARY of BIOCHEMISTRY

Edited by
John Daintith

The Facts On File Dictionary of Biochemistry

Checkmark Books
An imprint of Facts On File, Inc.
132 West 31st Street
New York NY 10001

Library of Congress Cataloging-in-Publication Data

The Facts on File dictionary of biochemistry / edited by John Daintith.
p. cm.
Includes bibliographical references.
ISBN 0-8160-4914-9 (hc : alk. paper). — ISBN 0-8160-4915-7 (pbk. : alk. paper)
1. Biochemistry—Dictionaries. I. Title: Dictionary of biochemistry. II. Daintith, John.

QP512.F33 2002
572'.03—dc21 2002035203

Compiled and typeset by Market House Books Ltd, Aylesbury, UK

Printed in the United States of America

MP 10 9 8 7 6 5 4 3 2 1

This book is printed on acid-free paper

CONTENTS

PREFACE

This dictionary is one of a series covering the terminology and concepts used in important branches of science. *The Facts On File Dictionary of Biochemistry* is planned as an additional source of information for students taking Advanced Placement (AP) Science courses in high schools, and will also be helpful to older students taking introductory college courses.

This volume covers the whole area of modern biochemistry, including basic organic and physical chemistry, classes of compound, basic cytology and histology, nutrition and metabolism, cell metabolism and signaling, and natural-product chemistry. The definitions are intended to be clear and informative and, where possible, we have provided helpful diagrams and examples. The book also has a selection of short biographical entries for people who have made important contributions to the field. There are a number of useful appendices including a chronology of main advances in the subject, together with structures of amino acids, the genetic code, and chemical elements. We have also added lists of webpages and an informative bibliography.

The book will be a helpful additional source of information for anyone studying the AP Biology course, especially the first part on Molecules and Cells. It will also be useful to students of AP Chemistry. However, we have not restricted the content to these syllabuses. Biochemistry is a dynamic and ever-expanding field of research with relevance to many other branches of bioscience and to medicine. We have tried to cover it in an interesting and informative way.

ACKNOWLEDGMENTS

Contributors

Robert Hine B.Sc.
Elizabeth Owen B.A.

ABA *See* abscisic acid.

abiogenesis The development of living from nonliving matter, as in the origin of life. The term is sometimes used for the now discredited doctrine of spontaneous generation.

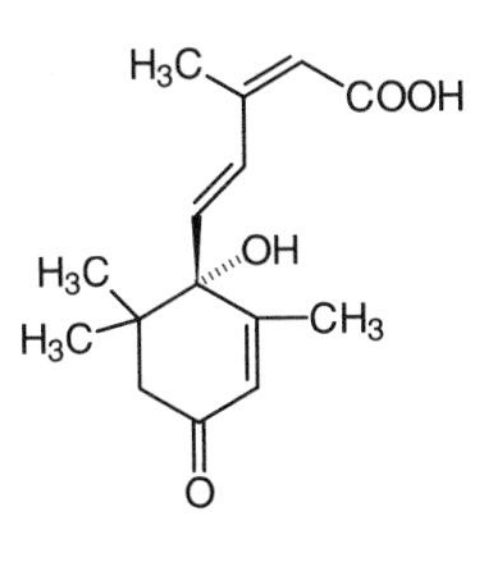

Abscisic acid

abscisic acid (**ABA**) A PLANT HORMONE once thought to be responsible for the shedding (abscission) of flowers and fruit and for the onset of dormancy in buds (hence its early name, *dormin*). The compound is associated with the closing of pores (stoma) in the leaves of plants deprived of water.

absorption **1.** The uptake of liquid by cells. Most digested food is absorbed in the small intestine; the liquid products of digestion are absorbed into the blood and lymphatic systems. In plants, water and mineral salts are absorbed mainly by the root hairs, just behind the root tips.
2. The conversion of electromagnetic radiation into other forms of energy in a medium.

absorption spectrum A plot of the amount of absorption by a substance of radiation at different wavelengths, usually of ultraviolet, visible, or infrared radiation. An absorption spectrum can give information about the identity or quantity of a substance. It can also give information about the way photosensitive compounds interact with radiation. Chlorophylls, for example, have absorption peaks in the red and blue (and therefore reflect green light).

abzyme An antibody that has enzyme activity.

accessory pigment *See* photosynthetic pigments.

acentric Denoting a chromosome or fragment of a chromosome that lacks a centromere.

acetal A type of organic compound formed by addition of an alcohol to an aldehyde. Addition of one alcohol molecule gives a *hemiacetal*. Further addition yields the full acetal. Similar reactions occur with ketones to produce hemiketals and ketals.

acetaldehyde *See* ethanal.

acetate (**ethanoate**) A salt or ester of acetic acid. Acetates contain the ion CH_3COO^- or the group $CH_3COO–$.

acetate–CoA ligase An ENZYME that catalyzes the reaction between acetate, CoA and ATP to form AMP and ACETYL COA. It is important in organisms that use acetate as a carbon source (e.g. *E. coli*, algae, and many fungi).

acetazolamide A compound used as an

inhibitor of carbonic anhydrase. It has been used as a diuretic but its main use is in treating glaucoma (by reducing the formation of fluid in the eye).

acetic acid (**ethanoic acid**) A carboxylic acid, CH_3COOH, obtained by the oxidation of ethyl alcohol. Acetic acid is a component of vinegar (which is obtained by bacterial oxidation of wine waste).

acetic anhydride The ACID ANHYDRIDE of acetic acid, $CH_3COOCOH_3$; commonly used in acetylation reactions.

acetone (**dimethylketone**) A KETONE, CH_3COCH_3, commonly used as a solvent.

acetone body *See* ketone body.

acetylation A reaction in which an acetyl group, $CH_3CO–$, is introduced into a compound. *See also* acylation.

acetylcholine (**ACh**) A NEUROTRANSMITTER found at the majority of synapses, which occur where one nerve cell meets another. Nerves that produce acetylcholine are called *cholinergic nerves*; they form the parasympathetic nervous system and also supply the voluntary muscles.

acetylcholinesterase An ENZYME that catalyzes the hydrolysis of ACETYLCHOLINE to give choline (and acetate). It occurs in spinal fluid and cholinergic neurons. The ester is inhibited by a number of compounds. A class of drugs (*acetylcholinesterase inhibitors*) act by inhibiting the action of the enzyme, so increasing the amount of acetylcholine available to transmit nerve impulses. These are used in the treatment of Alzheimer's disease.

acetyl CoA (**acetyl coenzyme A**) An important intermediate in cell metabolism, particularly in the oxidation of sugars, fatty acids, and amino acids, and in certain biosynthetic pathways. It is formed by the reaction between pyruvate (from GLYCOLYSIS) and COENZYME A, catalyzed by the enzyme pyruvate dehydrogenase. The acetyl group of acetyl CoA is subsequently oxidized in the KREBS CYCLE, to yield reduced coenzymes and carbon dioxide. Acetyl CoA is also produced in the initial oxidation of fatty acids and some amino acids. Other key roles for acetyl CoA include the provision of acetyl groups in biosynthesis of fatty acids, terpenoids, and other substances.

ACh *See* acetylcholine.

acid A substance that gives rise to hydrogen ions (or H_3O^+) when dissolved in water. An acid in solution will have a pH below 7. This definition does not take into account the competitive behavior of acids in solvents and it refers only to aqueous

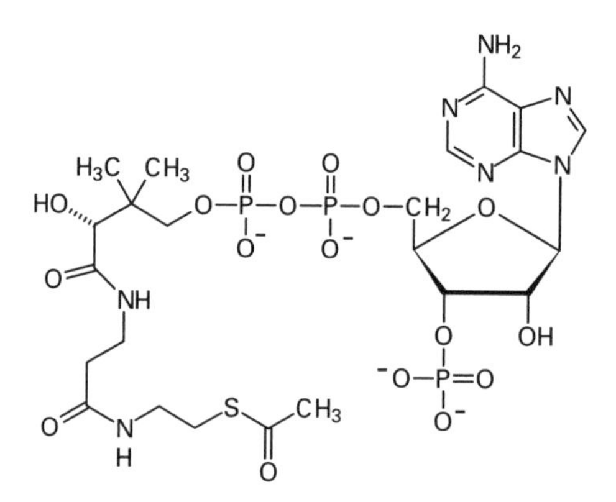

Acetyl CoA

systems. The *Lowry–Brønsted theory* defines an acid as a substance that exhibits a tendency to release a proton, and a base as a substance that tends to accept a proton. Thus, when an acid releases a proton, the ion formed is the *conjugate base* of the acid. Strong acids (e.g. HNO_3) react completely with water to give H_3O^+, i.e. HNO_3 is stronger than H_3O^+ and the conjugate base NO_3^- is weak. Weak acids (e.g. CH_3COOH and C_6H_5COOH) are only partly dissociated because H_3O^+ is a stronger acid than the free acids and the ions CH_3COO^- and $C_6H_5COO^-$ are moderately strong bases. The *Lewis theory* defines an acid as a substance capable of accepting a pair of electrons from a base.

acid anhydride A chemical compound that produces an acid when added to water. For example, sulfur trioxide is the acid anhydride of sulfuric acid:

$$SO_3 + H_2 \rightarrow H_2SO_4$$

Organic acid anhydrides are derived by removing the elements of water (i.e. two hydrogen atoms and one oxygen) from two acidic groups. A simple example is acetic anhydride from acetic acid:

$$CH_3COOH + HOCOCH_3 \rightarrow CH_3COOCOCH_3 + H_2O.$$

acid–base balance Maintenance of the optimum pH of body fluids by regulating the acid:base ratio. This usually involves a buffer system (*see* buffer). For example, mammalian blood must be maintained at a pH of 7.4, which requires a ratio of carbonic acid to bicarbonate of 1:20; any serious deviation would result in the conditions of acidosis or alkalosis. The optimum pH for higher plants is around 6.7, and they tend to be somewhat more tolerant to pH changes.

acidic hydrogen A hydrogen atom in a molecule that enters into a dissociation equilibrium when the molecule is dissolved in a solvent. For example, in acetic acid (CH_3COOH) the acidic hydrogen is the one on the carboxyl group, –COOH.

acidic stain *See* staining.

acid value (**acid number**) The number of milligrams of potassium hydroxide (KOH) required to neutralize the free fatty acids in one gram of a given fat. It is used as a measure of free acid.

acquired immune deficiency syndrome *See* AIDS.

acridine An aromatic heterocyclic compound found in coal tar. It is the parent compound of a number of effective mutagens.

ACTH *See* corticotropin.

actin A protein that is a major constituent of muscle and is heavily involved with the process of muscle contraction. It is also found in the microfilaments (part of the CYTOSKELETON) present in almost all eukaryote cells, where it accounts for 1–5 % of the total protein. Actin has two forms. *G-actin* consists of globular monomers. At higher ionic strengths it polymerizes into *F-actin*, which is a fibrous form with a double-stranded helix structure. In muscles, there are overlapping alternating sets of thin actin and thick myosin filaments. When muscles contract the actin and myosin produce a complex, called ACTOMYOSIN.

actin-binding protein Any of a number of proteins that bind to ACTIN and modify its properties. Examples are dystrophin, filamin, caldesmon, and the myosins.

actinomycetes (**filamentous bacteria**) A group of bacteria characterized by a mycelial fungus-like growth habit and belonging to the phylum Actinobacteria. They are numerous in the topsoil and are important in soil fertility. Some may cause infections in animals and humans. Most of the antibiotics (e.g. streptomycin, actinomycin, and tetracycline) are obtained from actinomycetes.

actinomycin Any of a number of antibiotics produced by certain bacteria. The main one, *actinomycin D* (or *dactinomycin*), can bind between neighbouring base pairs in DNA, preventing RNA synthesis. It is used in the treatment of some cancers.

action potential The transitory change in electrical potential that occurs across the membrane of a nerve or muscle fiber during the passage of a nervous impulse. The degree of change is independent of the strength of the impulse; it either occurs or does not. In the absence of an impulse a *resting potential* of about –70 mV exists across the membrane (inside negative) as a result of the unequal distribution of ions between the intracellular and extracellular media (*see* sodium). However, during the passage of an impulse the resistance of the membrane falls and allows an inward current of sodium ions, which makes the inside less negative and eventually alters the membrane potential to about +30 mV (inside positive). An outward flow of potassium ions follows shortly afterwards and restores the resting potential. Local currents flow ahead of the action potential that in turn give rise to further action potentials. In this way a wave of activity is propagated along the membrane to produce the impulse. *See also* ion channel.

action spectrum A graph showing the effect of different wavelengths of radiation, usually light, on a given process. It is often similar to the ABSORPTION SPECTRUM of the substance that absorbs the radiation and can therefore be helpful in identifying that substance. For example, the action spectrum of photosynthesis is similar to the absorption spectrum of chlorophyll.

activated complex The partially bonded system of atoms in the transition state of a chemical reaction.

activation energy Symbol: E_a The minimum energy a particle, molecule, etc., must acquire before it can react; i.e. the energy required to *initiate* a reaction regardless of whether the reaction is exothermic or endothermic. Activation energy corresponds to an energy barrier that must be overcome if a reaction is to take place. *See* Arrhenius equation.

active site The region of an ENZYME molecule that combines with and acts on the substrate. It consists of catalytic amino acids arranged in a configuration specific to a particular substrate or type of substrate. The ones that are in direct combination are the *contact amino acids*. Other amino acids may be further away but still play a role in the action of the enzyme. These are *auxilliary amino acids*. Binding of a regulatory compound to a separate site, known as the ALLOSTERIC SITE, on the enzyme molecule may change this configuration and hence the efficiency of the enzyme activity.

active transport The transport of molecules or ions across a cell membrane against a concentration gradient, with the expenditure of energy: it is probably an attribute of all cells. Anything that interferes with the provision of energy will interfere with active transport. The mechanism typically involves a carrier protein (a transporter) that spans the cell membrane and transfers substances in or out of the cell by changing shape. *Compare* facilitated diffusion.

actomyosin A protein complex found in muscle, formed between molecules of ACTIN and MYOSIN present in adjacent thick and thin muscle filaments. These actomyosin complexes are involved in the process of muscle contraction; their formation is important in the concerted mechanism that pulls the thin filaments past the thick ones. The complex can also be formed by mixing actin with myosin *in vitro*.

acylation A chemical reaction in which an acyl group is introduced into a compound.

acyl group An organic group with the general formula R–CO–. The acetyl group, CH_3–CO– is a simple example.

adaptive enzyme (**inducible enzyme**) An enzyme that is produced by a cell only in the presence of its substrate. *Compare* constitutive enzyme.

addition reaction A reaction in which additional atoms or groups of atoms are introduced into an unsaturated compound. A simple example is the addition of bromine across the double bond in ethene:

$$H_2C{:}CH_2 + Br_2 \rightarrow BrH_2CCH_2Br$$

Addition reactions can be induced either by electrophiles or by nucleophiles.

adduct A compound formed by direct addition of two chemical species, especially one produced by forming a coordinate bond between a donor and an acceptor.

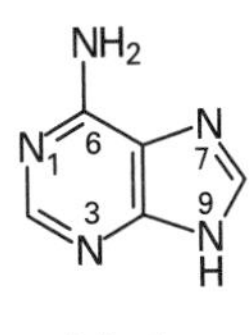

Adenine

adenine A nitrogenous base found in DNA and RNA. It is also a constituent of certain coenzymes, e.g. NAD and FAD, and when combined with the sugar ribose it forms the nucleoside adenosine found in AMP, cAMP, ADP, and ATP. Adenine has a purine ring structure. *See also* DNA.

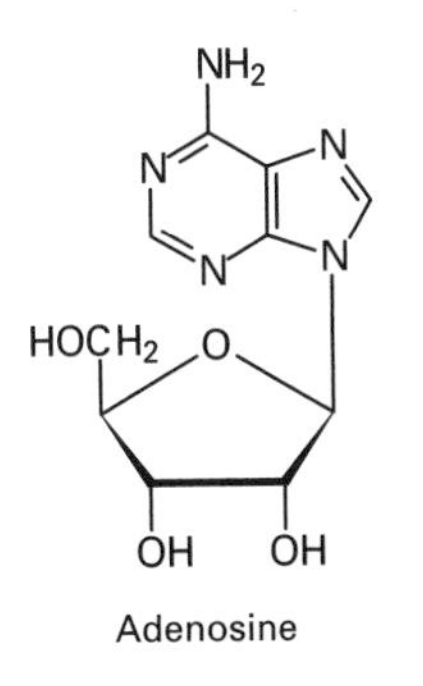

Adenosine

adenosine (**adenine nucleoside**) A NUCLEOSIDE formed from adenine linked to D-ribose with a β-glycosidic bond. It is widely found in all types of cell, either as the free nucleoside or in combination in nucleic acids. Phosphate esters of adenosine, such as ATP, are important carriers of energy in biochemical reactions.

adenosine diphosphate *See* ADP.

adenosine monophosphate *See* AMP.

adenosine triphosphate *See* ATP.

ADH (**antidiuretic hormone**) *See* vasopressin.

adhesion molecule Molecules that protrude from the cell surface and mediate the adhesion of the cell to other cells or the extracellular matrix. Some play a role in morphogenesis and others are involved in wound healing.

adipose tissue (**fatty tissue**) A type of connective tissue consisting of closely packed cells (*adipocytes*) containing fat. Adipose tissue is found in varying amounts in the dermis of the skin and around the kidneys, heart, and blood vessels. It provides an energy store, heat insulation, and mechanical protection.

Most adipose tissue is in the form of *white fat*, but hibernating and newborn animals have deposits of darker colored *brown fat*, rich in unsaturated fatty acids, which is well supplied with blood vessels and acts as a readily mobilizable source of heat energy. Some theories on the cause of obesity in humans postulate a lack of brown fat in those affected.

ADP (**adenosine diphosphate**) A nucleotide consisting of adenine and ribose with two phosphate groups attached. *See also* ATP.

adrenal glands A pair of glands that are situated above each kidney and produce various hormones, notably epinephrine. Secretion is controlled by the nervous system. The adrenal glands have a central medulla and an outer cortex, each behaving independently. The former produces norepinephrine and epinephrine. The adrenal cortex is rich in vitamin C and cholesterol, and produces three types of hormones: aldosterone, cortisol, and sex hormones.

adrenaline *See* epinephrine.

adrenergic Designating the type of nerve fiber that releases epinephrine or norepinephrine from its ending when stimulated by a nerve impulse. Vertebrate sympathetic motor nerve fibers are adrenergic. *Compare* cholinergic.

adrenocorticotropic hormone *See* corticotropin.

aerobe An organism that can live and grow only in the presence of free oxygen, i.e. it respires aerobically (*see* aerobic respiration). *Compare* anaerobe.

aerobic respiration (**oxidative metabolism**) Respiration in which free oxygen is used to oxidize organic substrates to carbon dioxide and water, with a high yield of energy. Carbohydrates, fatty acids, and excess amino acids are broken down yielding acetyl CoA and the reduced coenzymes NADH and $FADH_2$. The acetyl coenzyme A enters a cyclic series of reactions, the KREBS CYCLE, with the production of carbon dioxide and further molecules of NADH and $FADH_2$. NADH and $FADH_2$ are passed to the ELECTRON-TRANSPORT CHAIN (involving cytochromes and flavoproteins), where they combine with atoms of free oxygen to form water. Energy released at each stage of the chain is used to form ATP during a coupling process. The substrate is completely oxidized and there is a high energy yield. There is a net production of 38 ATPs per molecule of glucose during aerobic respiration, a yield of about 19 times that of anaerobic respiration. Aerobic respiration is therefore the preferred mechanism of the majority of organisms. *See also* oxidative phosphorylation; respiration.

aerotaxis (**aerotactic movement**) A taxis in response to an oxygen concentration gradient. For instance, motile aerobic bacteria are positively aerotactic, whereas motile obligate anaerobic bacteria are negatively aerotactic. *See* taxis.

aflatoxin *See Aspergillus.*

agent orange A herbicide consisting of

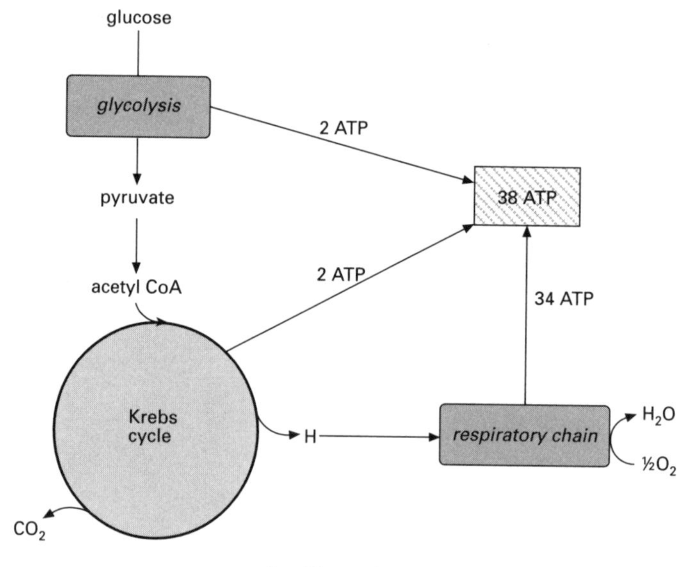

Aerobic respiration

a mixture of two weedkillers (2,4-D and 2,4,5-T), used in warfare to defoliate trees or to destroy crops. It also contains traces of the highly toxic chemical dioxin, which may cause cancers and birth defects.

agglutination The clumping together of red blood cells or bacteria as a result of the action of antibodies. Agglutination may occur in transfusion if blood of the wrong group is given. The surfaces of the donor's red blood cells contain antigen molecules that are attacked by antibody molecules in the serum of the recipient, which causes the red cells to clump together. These clumps may block capillaries, causing fatal damage to the heart or brain. Agglutination of bacteria by antibodies causes them to disintegrate. *See also* antibody; blood groups.

agonist A substance, such as a drug or hormone, that binds to a cell's receptors and elicits a response.

Agrobacterium A genus of soil bacteria, the species *A. tumefaciens* being the causative agent of crown gall, a type of tumor in plants. A segment of DNA (transferred DNA, T-DNA) from a plasmid in the bacterium is transferred into the host DNA and induces tumor formation. Since the plasmid is capable of independent replication in host cells of many dicotyledonous plants, it has been used as a cloning vector in genetic engineering. Once the desired segment of DNA, for example a gene, has been spliced into the T-DNA, the plasmid can be introduced into certain plant cultures and entire plants with the desired characteristic can be produced. Unfortunately, the bacterium does not infect monocotyledonous plants, which include important cereal crops. *See also* gene cloning; tissue culture.

AIDS (acquired immunodeficiency syndrome) A viral disease of humans characterized by destruction of the lymphocytes responsible for cell-mediated immunity (*see* T-cell), the patient consequently succumbing to opportunistic fatal infections, cancers, etc. The causative agent – HIV (human immunodeficiency virus) – is transmitted in blood, semen, or vaginal fluid; it is a retrovirus and can remain inactive in its host for several years without causing serious symptoms. Such carriers, described as *HIV-positive*, can nevertheless transmit the virus. *See also* HIV.

alanine *See* amino acids.

albinism The absence of pigmentation in the eyes, skin, feathers, hair, etc., of some animals. It is a hereditary condition in vertebrates, resulting from abnormalities in production or function of the pigment cells. Albinism in humans is due to a recessive gene and results in the absence of a dark brown pigment (melanin). The complete albino has milk-white skin and hair and the irises of the eyes appear pink.

albumen *See* albumin.

albumin One of a group of simple proteins, which are usually deficient in glycine. They are water-soluble and when heated coagulate. Albumins are the products of plants and animals; e.g. legumelin in peas and the serum albumin of blood. The protein in egg white (*albumen*) is a mixture of albumins. These also contain carbohydrate groups and consequently are classified as conjugated proteins.

alcohol A type of organic compound of the general formula ROH, where R is a hydrocarbon group. Examples of simple alcohols are methanol (CH_3OH) and ethanol (C_2H_5OH). If the compound is aromatic, the –OH group is not directly attached to the ring.

Alcohols can have more than one –OH group; those containing two, three or more such groups are described as *dihydric*, *trihydric*, and *polyhydric* respectively (as opposed to those containing one –OH group, which are *monohydric*). Alcohols are further classified according to the environment of the –C–OH grouping. If the carbon atom is attached to two hydrogen atoms, the compound is a *primary alcohol*. If the carbon atom is attached to one hydrogen atom and two other groups, it is a *secondary alcohol*. If the carbon atom is attached to three other groups, it is a *tertiary alcohol*.

aldaric acid *See* sugar acid.

aldehyde A type of organic compound with the general formula RCHO, where the –CHO group (the aldehyde group) consists of a carbonyl group attached to a hydrogen atom. Simple examples of aldehydes are methanal (formaldehyde, HCHO) and ethanal (acetaldehyde, CH_3CHO). Aldehydes are formed by oxidizing a primary alcohol. They can be further oxidized to carboxylic acids. Reduction produces the parent alcohol. *See also* ketone.

alditol *See* sugar alcohol.

aldohexose An aldose SUGAR with six carbon atoms.

aldonic acid *See* sugar acid.

aldopentose An aldose sugar with five carbon atoms. *See* sugar.

aldose A sugar containing an aldehyde (CHO) or potential aldehyde group. *See* sugar.

aldosterone A hormone secreted by the cortex of the adrenal glands. It is a steroid, and is important in the control of sodium and potassium ion concentrations in mammals. The hormone has an important effect on the handling of sodium and potassium by the kidney tubules, favoring the reabsorption of sodium ions and the excretion of potassium ions. It also has the effect of increasing the uptake of sodium ions by the gut. It acts by stimulating the transcription of sodium transport protein genes. The overall result is that sodium ion concentration in the blood rises, whereas potassium falls.

alginic acid (**algin**) A yellow-white organic solid that is found in brown algae. It is a complex polysaccharide and produces, in even very dilute solutions, a viscous liquid. Alginic acid has various uses, especially in the food industry as a stabilizer and texture agent.

aliphatic compound An organic compound with properties similar to those of the alkanes, alkenes, and alkynes and their derivatives. Most aliphatic compounds have an open chain structure but some, such as cyclohexane and sucrose, have rings. The term is used in distinction to aromatic compounds, which are similar to benzene. *Compare* aromatic compound.

alkali A water-soluble base.

alkaloid One of a group of organic compounds found in plants, that are poisonous insoluble crystalline compounds. They contain nitrogen and usually occur as salts of acids such as citric, malic, and succinic acids. Their function in plants remains obscure, but it is suggested that they may be nitrogenous end-products of metabolism. Some may have a protective function against herbivores. Many have pharmacological activities. Important examples are quinine, nicotine, atropine, opium, morphine, codeine, and strychnine. They occur mainly in the poppy family, the buttercup family, and the nightshade family of plants.

alkane A type of hydrocarbon with general formula C_nH_{2n+2}. Alkanes are saturated compounds, containing no double or triple bonds. Methane (CH_4) and ethane (C_2H_6) are typical examples. The alkanes are fairly unreactive (their former name, the *paraffins*, means 'small affinity').

alkene A type of hydrocarbon with the general formula C_nH_{2n}. The alkenes (formerly called *olefins*) are unsaturated compounds containing double carbon–carbon bonds. They can be obtained from crude oil by cracking alkanes. Important examples are ethene (C_2H_4) and propene (C_3H_6).

alkyl group A group obtained by removing a hydrogen atom from an alkane or other aliphatic hydrocarbon.

alkyne A type of hydrocarbon with the general formula C_nH_{2n-2}. The alkynes are unsaturated compounds containing triple carbon–carbon bonds. The simplest member of the series is ethyne (C_2H_2).

allele (**allelomorph**) One of the possible forms of a given gene. The alleles of a particular gene occupy the same positions (*loci*) on homologous chromosomes. A gene is said to be *homozygous* if the two loci have identical alleles and *heterozygous* when the alleles are different. When two different alleles are present, one (the *dominant* allele) usually masks the effect of the other (the *recessive* allele). The allele determining the normal form of the gene is usually dominant while mutant alleles are usually recessive. Thus most mutations only show in the phenotype when they are homozygous. In some cases one allele is not completely dominant or recessive to another allele. Thus an intermediate phenotype will be produced in the heterozygote. *See also* co-dominance; multiple allelism.

allelomorph *See* allele.

allelopathy Inhibition of the germination, growth, or reproduction of an organism effected by a chemical substance released from another organism. This is a common anticompetition mechanism in plants; for example, barley secretes an alkaloid substance from its roots to inhibit competing weeds. *See also* phytoalexin.

allergy A type of abnormal immune response in which the body produces ANTIBODIES against substances, such as dust or pollen, that are usually not harmful and are normally removed or destroyed in other individuals. The allergy-triggering substances (*allergens*) stimulate certain lymphocytes (B-cells) to secrete antibodies, termed *reagins*, which bind to mast cells. When an allergen encounters and binds to the mast cell the latter degranulates, discharging histamine and other substances. Histamine dilates blood capillaries in the region and increases their permeability. This results in inflammation and increased mucus secretion. It also stimulates contraction of smooth muscle, leading to bronchial constriction. Certain drugs that block histamine (*antihistamines*) are used to relieve the symptoms of hay fever, asthma, and similar allergies.

allograft *See* graft.

allosteric site A part of an enzyme separate from the active site to which a specific effector or modulator can be attached. This attachment is reversible and alters the activity of the enzyme. Allosteric enzymes possess an allosteric site in addition to their ACTIVE SITE. This site is as specific in its relationship to modulators as active sites are to substrates. *See* active site. Some iron-enzymatic proteins e.g. hemoglobin also undergo allosteric effects.

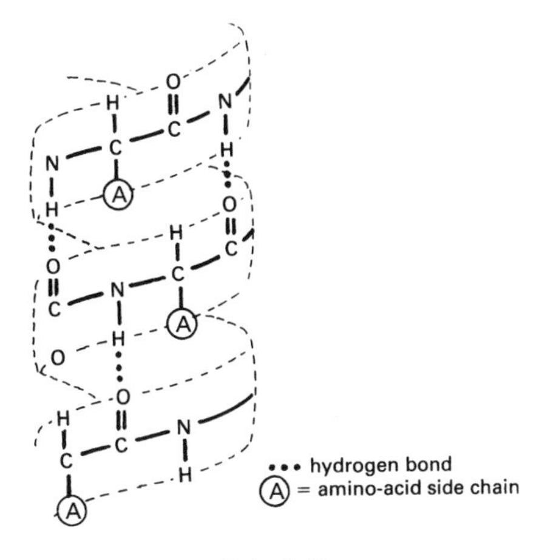

Alpha helix

alpha helix A highly stable structure in which peptide chains are coiled to form a spiral. Each turn of the spiral contains approximately 3.6 amino-acid residues. The R group of these amino-acids extends outward from the helix and the helix is held together by hydrogen bonding between successive coils. If the alpha helix is stretched the hydrogen bonds are broken but reform on relaxation. The alpha helix is found in muscle protein and keratin. It is one of the two basic secondary structures of proteins. *See* protein.

alpha-naphthol test (**Molisch's test**) A standard test for carbohydrates in solution. Molisch's reagent, alpha-naphthol in alcohol, is mixed with the test solution. Concentrated sulfuric acid is added and a violet ring at the junction of the two liquids indicates the presence of carbohydrates.

alternative splicing *See* intron.

Altman, Sidney (1939–) Canadian-born American biochemist who discovered the catalytic properties of RNA. He was awarded the Nobel Prize for chemistry in 1989 jointly with T. R. Cech.

amide A type of organic compound of general formula $RCONH_2$ (primary), $(RCO)_2NH$ (secondary), and $(RCO)_3N$ (tertiary). Amides are white, crystalline solids and are basic in nature, some being soluble in water.

amination The introduction of an amino group ($-NH_2$) into an organic compound. An example is the conversion of an aldehyde or ketone into an amide by reaction with hydrogen and ammonia in the presence of a catalyst:

$$RCHO + NH_3 + N_2 \rightarrow RCH_2NH_2 + H_2O$$

amine A compound containing a nitrogen atom bound to hydrogen atoms or hydrocarbon groups. Amines have the general formula R_3N, where R can be hydrogen or an alkyl or aryl group. Amines can be prepared by reduction of amides or nitro compounds. An amine is classified according to the number of organic groups bonded to the nitrogen atom: one, primary; two, secondary; three, tertiary. Since amines are basic in nature they can form the quaternium ion, R_3NH^+.

amino acids Derivatives of carboxylic acids in which a hydrogen atom has been replaced by an amino group ($-NH_2$). Thus, from acetic acid, the amino acid, glycine, is formed. All are white, crystalline, soluble in water, and with the sole exception of the simplest member, glycine, all are optically active. In solution at neutral pH they are predominantly dipolar ions with the amino group protonated ($-NH_3^+$) and the carboxyl group dissociated ($-COO^-$). The ionization state varies with pH. In the body the various proteins are assembled from the necessary amino acids and it is important therefore that all the amino acids should be present in sufficient quantities. All proteins, in all species are composed of the same twenty amino acids. In humans ten of the twenty amino acids can be synthesized by the body itself. Since these are not required in the diet they are known as *nonessential amino acids*. The remaining ten cannot be synthesized by the body and have to be supplied in the diet. They are known as *essential amino acids*. *See Appendix for structures*. Each of the twenty amino acids has a single letter and three-letter symbol. The single letter designation is more common.

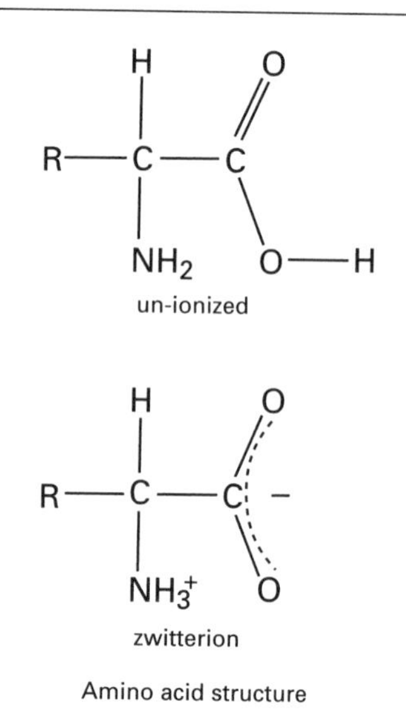

Amino acid structure

Various other amino acids fulfill important roles in metabolic processes other than as constituents of proteins. For example, ornithine ($H_2N(CH_2)_3CH(NH_2)COOH$) and citrulline ($H_2N.CO.NH.(CH_2)_3CH(NH_2)COOH$) are intermediates in the production of urea. Amino acids in excess of those used in protein synthesis are not stored or excreted. The amino group is removed and converted into urea and the carbon part converted into a major meta-

amino acid	*symbol*	*R group*
alanine	Ala	$-CH_3$
valine	Val	$-CH(CH_3)_2$
leucine	Leu	$-CH_2-CH(CH_3)_2$
isoleucine	Ile	$-CH(CH_3)-CH_2-CH_3$
proline	Pro	(complete structure: COO^-, H, C ring with H_2C, CH_2, CH_2, N–H)
methionine	Met	$-CH_2-CH_2-S-CH_3$
phenylalanine	Phe	$-CH_2-C_6H_5$
tryptophan	Trp	$-CH_2-C$ (indole ring: C=CH–NH fused benzene)

amino acids with nonpolar R groups

The **amino acids** commonly found in proteins are alpha amino acids. In all except proline, where the complete structure is given, the remainder of the molecule has the structure

$$HOOC-\underset{NH_2}{\overset{H}{C}}-$$

Amino acids

amino acid	symbol	R group
glycine	Gly	$—H$
serine	Ser	$—CH_2—OH$
threonine	Thr	$—CH(OH)—CH_3$
cysteine	Cys	$—CH_2—SH$
asparagine	Asn	$—CH_2—C(=O)NH_2$
glutamine	Gln	$—CH_2—CH_2—C(=O)NH_2$
tyrosine	Tyr	$—CH_2—C_6H_4—OH$

amino acids with uncharged polar R groups

amino acid	symbol	R group	
aspartic acid	Asp	$—CH_2—C(=O)O^-$	acidic amino acids
glutamic acid	Glu	$CH_2—CH_2—C(=O)O^-$	acidic amino acids
lysine	Lys	$—CH_2—CH_2—CH_2—CH_2—\overset{+}{N}H_3$	basic amino acids
arginine	Arg	$—CH_2—CH_2—CH_2—NH—C(=\overset{+}{N}H_2)—NH_2$	basic amino acids
histidine	His	$—CH_2—C—CH$, HN, NH, C(+), H (imidazolium ring)	basic amino acids

amino acids with charged polar R groups

Amino acids: The essential amino acids are Val, Leu, Ile, Met, Phe, Trp, Thr, Tyr, Lys, and His.

bolic intermediate: acetyl CoA, acyl CoA, pyruvate, or an intermediate of the KREBS CYCLE.

amino group The group $-NH_2$.

amino sugar A sugar in which a hydroxyl group (OH) has been replaced by an amino group (NH_2). Glucosamine (from glucose) occurs in many polysaccharides of vertebrates and is a major component of chitin. Galactosamine or chondrosamine (from galactose) is a major component of cartilage and glycolipids. Amino sugars are important components of bacterial cell walls.

amount of substance Symbol: n A measure of the number of entities present in a substance. *See* mole.

AMP (**adenosine monophosphate**) A nucleotide consisting of adenine, ribose, and phosphate. *See* ATP; cyclic AMP.

amphipathic Describing a molecule that contains both hydrophobic and hydrophilic parts. Phospholipids and glycolipids are examples of amphipathic compounds.

amphiprotic *See* solvent.

ampholyte ion *See* zwitterion.

amphoteric A material that can display both acidic and basic properties. The term is most commonly applied to the oxides and hydroxides of metals that can form both cations and complex anions. For example, zinc oxide dissolves in acids to form zinc salts and also dissolves in alkalis to form zincates, $[Zn(OH)_4]^{2-}$.

amu *See* atomic mass unit.

amylase A member of a group of closely related ENZYMES, found widely in plants, animals, and microorganisms, that hydrolyzes starch or glycogen to the sugars maltose, glucose, or dextrin. Both α- and β-amylases occur in plants, the latter particularly in malt (being used in the brewing industry). Only α-amylase is found in animals, in the pancreatic juices and in saliva (*see* ptyalin), having an important role in digestion.

amylopectin The water-insoluble fraction of starch. *See* starch.

amylose The water-soluble fraction of starch. *See* starch.

anabolic steroid Any STEROID hormone or synthetic steroid that promotes growth and formation of new tissue. Anabolic steroids are used in the treatment of wasting diseases. They are also sometimes used in agriculture to boost livestock production. People also use them to build up muscles, although this is now generally outlawed in sporting activities. Most androgens have anabolic activity, but when used by women they lead to masculinization.

anabolism Metabolic reactions in which molecules are linked together to form more complex compounds. Thus, anabolic reactions are concerned with building up structures, storage compounds, and complex metabolites in the cell. Starch, glycogen, fats, and proteins are all products of anabolic pathways. Anabolic reactions generally require energy provided by ATP produced by CATABOLISM. *See also* metabolism.

anaerobe An organism that can live and grow in the absence of free oxygen, i.e. it respires anaerobically (*see* anaerobic respiration). Anaerobes can be *facultative*, in that they usually respire aerobically but can switch to anaerobic respiration when free oxygen is in short supply, or *obligate*, in that they never respire aerobically and may even be poisoned by free oxygen. *Compare* aerobe.

anaerobic respiration Respiration in which oxygen is not involved. It is found in yeasts, bacteria, and occasionally in muscle tissue. In this type of respiration the organic substrate is not completely oxidized and the energy yield is low. In the absence of oxygen in animal muscle tissue, glucose is degraded to pyruvate by GLYCOLYSIS, with the production of a small amount of energy and also lactic acid, which may be oxidized later when oxygen becomes available (*see* oxygen debt). FERMENTATION is an example of anaerobic respiration, in which certain yeasts produce ethanol and carbon dioxide as end products. Only two molecules of ATP are produced by this process. *See* anaerobe. *Compare* aerobic respiration.

anaphase The stage in mitosis or meiosis when chromatids are pulled towards opposite poles of the nuclear spindle. In mitosis the chromatids moving towards the poles represent a single complete chromosome. During anaphase I of meiosis a pair of chromatids still connected at their centromere move to the spindle poles. During anaphase II the centromeres divide and single chromatids are drawn towards the poles.

anaphylaxis The severe reaction of an animal to an antigen to which it has previously been exposed. In humans it may rarely occur after injection of antiserum or antibiotics, after bee or wasp stings, etc. It is caused by the release of histamine and other substances from mast cells, following antigen–antibody interaction.

androgen A male sex hormone (a STEROID) that controls the development, function, and maintenance of secondary male characteristics (e.g. facial hair and deepening of the voice), male accessory sex organs, and spermatogenesis. Androgens are produced chiefly by the testis (smaller amounts are produced by the ovary, adrenal cortex, and placenta); the most important is testosterone. Castration (removal of the testes) leads to atrophy of accessory sexual organs; this effect can be prevented by androgen replacement therapy. Androgens are also used in the treatment of certain diseases in which androgen secretion is reduced or absent (e.g. hypogonadism, hypopituitarism) and in the treatment of certain breast cancers. They also have anabolic activity, promoting growth and formation of new tissue. *See also* anabolic steroid.

androsterone A steroid hormone formed in the liver from the metabolism of testosterone. It has only weak androgenic activity. *See* androgen.

Anfinsen, Christian Boehmer (1916–95) American protein chemist. He was awarded the Nobel Prize for chemistry in 1972 for his work on ribonuclease. The prize was shared with S. Moore and W. H. Stein.

angiotensin A polypeptide that is associated with high blood pressure. It exists in two forms in the blood. Angiotensin I is an inactive decapeptide produced by the action of renin, a kidney enzyme released when blood pressure is low, on a globulin in the blood. It has no physiological action. The active form, angiotensin II, is formed by the enzyme-catalyzed removal of two terminal amino acids from angiotensin I. Angiotensin II raises blood pressure by stimulating the constriction of arterioles and the secretion of the hormone aldosterone. It is destroyed rapidly by proteinase enzymes. High blood pressure may be treated by inhibiting the enzyme responsible for converting inactive angiotensin I to active angiotensin II.

angstrom Symbol: Å A unit of length equal to 10^{-10} meters. It is still used occasionally for measurements of wavelength or interatomic distance.

anhydride A compound formed by removing water from an acid or, less commonly, a base. Many non-metal oxides are anhydrides of acids: for example CO_2 is the anhydride of H_2CO_3 and SO_3 is the anhydride of H_2SO_4. Organic anhydrides are formed by removing H_2O from two carboxylic-acid groups, giving compounds with the functional group –CO.O.CO–. These form a class of organic compounds called *acid anhydrides* (or *acyl anhydrides*).

aniline stains *See* staining.

animal starch *See* glycogen.

anion A negatively charged ion, formed by addition of electrons to atoms or molecules. In electrolysis anions are attracted to the positive electrode (the anode). *Compare* cation.

anionic resin An ion-exchange material that can exchange anions, such as Cl^- and OH^-, for anions in the surrounding medium. Such resins are used for a wide range of analytical and purification purposes.

anomer A stereoisomeric form of a sugar.

ANS *See* autonomic nervous system.

antagonism 1. The interaction of two substances, e.g. drugs or hormones, which have opposing effects in a given system, such that one partially or completely inhibits the effect of the other.
2. The interaction of two types of organisms, such that the growth of one is partially or completely inhibited by the other.
3. The opposing action of two muscles, such that the contraction of one is accompanied by relaxation of the other.

anthocyanin One of a group of water-soluble pigments found dissolved in higher plant cell vacuoles. Anthocyanins are red, purple, and blue and are widely distributed, particularly in flowers and fruits, where they are important in attracting insects, birds, etc. They also occur in buds and sometimes contribute to the autumn colors of leaves. They are natural pH indicators, often changing from red to blue as pH increases, i.e. acidity decreases. Color may also be modified by traces of iron and other metal salts and organic substances, for example cyanin is red in roses but blue in the cornflower. *See* flavonoid.

antibiotic One of a group of organic compounds, varying in structure, that are produced by microorganisms and can kill or inhibit the activities of other microorganisms. One of the best-known examples is PENICILLIN, which was discovered by Sir Alexander Fleming. Another example is STREPTOMYCIN. Antibiotics are widely used in medicine to combat bacterial infections. Over-use can lead to microorganisms becoming resistant to a particular antibiotic. Antiobiotics can also be produced synthetically.

antibody A glycoprotein molecule of the immunoglobulin family formed within the body of an animal in order to neutralize the effect of a foreign invading protein (called an *antigen*). Antibodies are produced by LYMPHOCYTES in response to the presence of antigens and are the recognition element of the immune response. Each antibody has a molecule structure that exactly fits the structure of one particular antigen molecule, like a lock and key (i.e. they are specific). Antibody molecules attach themselves to invading antigen molecules (on or from bacteria, transfused red blood cells, or grafted tissue of another animal) and so render them inactive. Some cause agglutination (clumping) of invading cells, so that they disintegrate; others, called *opsonins*, make bacteria more easily engulfed by phagocytic leukocytes. Antibodies are important in defense against infectious diseases and in developing immunities. *See also* agglutination; immunoglobulin; immunity.

anticoagulant A chemical that can prevent blood clotting. An example is HEPARIN, which occurs in the saliva of leeches, mosquitoes, and other blood-sucking animals. Traces are found normally in blood and may help to prevent clotting in the blood. Heparin acts by inhibiting the formation of THROMBIN. Another anticoagulant is acid sodium citrate, which is added to blood taken for transfusion and prevents clotting by removing calcium ions, which are necessary for clotting. *See also* blood clotting.

anticodon A nucleotide triplet on TRANSFER RNA that is complementary to and bonds with the corresponding codon of MESSENGER RNA in the ribosomes.

antidiuretic hormone (ADH) *See* vasopressin.

antigen A substance that induces the production of antibodies that are able to bind to the antigen.

antioxidant A substance that slows or inhibits oxidation reactions, especially in biological materials or within cells, thereby reducing spoilage or preventing damage. Natural antioxidants include vitamin E and β-carotene. These work by reacting with peroxides or oxygen free radicals, effectively mopping them up. Antioxidants are added to a range of foods, e.g. margarines, and industrial materials, such as plastics.

antisense RNA A single-stranded RNA molecule that has a complementary sequence to a specific mRNA. It can therefore bind to the mRNA and inhibit translation. *See* RNA interference.

antiserum A serum containing ANTIBODIES against a particular ANTIGEN, obtained from a human or animal which has been exposed to the antigen, either naturally (through disease) or through immunization. Antisera are injected to give passive immunity against specific diseases and also used in the laboratory to identify unknown pathogens.

apoenzyme The protein part of a conjugate enzyme. It is an enzyme whose cofactor has been removed (e.g. via dialysis) rendering it catalytically inactive. When combined with its PROSTHETIC GROUP or coenzyme it forms a complete enzyme (HOLOENZYME).

apoptosis Programmed cell death that is regulated by an intracellular pathway involving caspase enzymes. It is essential to normal development and ensures damaged or dysfunctional cells destroy themselves in a controlled fashion. Many key regulators of apoptosis are potential oncogenes and failure of apoptosis leads to cells with damaged DNA surviving to cancer cells.

Arber, Werner (1929–) Swiss microbiologist who discovered restriction enzymes. He was awarded the Nobel Prize for physiology or medicine in 1978 jointly with D. Nathans and H. O. Smith.

Archaea One of the three cellular kingdoms – the other two are Eukarya and Bacteria. They can be distinguished from the bacteria on the basis of biochemical differences. For example, they differ in the nature of their lipid constituents, and lack a peptidoglycan layer in the cell wall. Included in the Archaea are the methanogic (methane-producing), thermophilic (heat-loving), and halophilic (salt-loving) species and therefore many live in harsh environments, such as hot springs, salt flats, or sea vents, thought to resemble the early earth environment. However, they are not restricted to such extreme lifestyles, and are widespread in more congenial settings.

arene An organic compound containing a benzene ring; i.e. an aromatic hydrocarbon or derivative.

arginine *See* amino acids.

aromatic compound An organic compound containing benzene rings in its structure. Aromatic compounds, such as benzene, have a planar ring of atoms linked by alternate single and double bonds. The characteristic of aromatic compounds is that their chemical properties are not those expected for an unsaturated compound as the electrons in the double bonds are delocalized over the ring, so that the six bonds are actually all identical and intermediate between single bonds and double bonds.

Arrhenius, Svante August (1859–1927) Swedish chemist and physicist who worked on electrochemistry and on reaction rates. He was awarded the Nobel Prize for chemistry in 1903.

Arrhenius equation An equation relating the rate constant of a chemical reaction and the temperature at which the reaction is taking place:

$$k = A\exp(-E_a/RT)$$

where A is a constant, k the rate constant, T the thermodynamic temperature in kelvin, R the gas constant, and E_a the activation energy of the reaction.

Reactions proceed at different rates at different temperatures, i.e. the magnitude of the rate constant is temperature dependent. The Arrhenius equation is often written in a logarithmic form, i.e.

$$\log_e k = \log_e A - E/2.3RT$$

This equation enables the activation energy for a reaction to be determined.

ascorbic acid *See* vitamin C.

asexual reproduction The formation of new individuals from a single parent without the production of gametes or special reproductive structures. It occurs in many plants, usually by vegetative propagation or spore formation; in unicellular organisms usually by fission; and in multicellular invertebrates by fission, budding, fragmentation, etc.

asparagine *See* amino acids.

aspartic acid *See* amino acids.

Aspergillus A genus of ascomycete fungi including mostly saprophytes and opportunist pathogens. Inhalation of the spores of *A. fumigatus* causes the lung disease *aspergillosis*, while other *Aspergillus* species (including *A. flavus*) produce the poisonous metabolite *aflatoxin*, which may lead to liver diseases when consumed in *Aspergillus*-contaminated nuts and cereals.

aspirin (**acetylsalicylic acid**) A white crystalline powder, widely used for its analgesic, anti-inflammatory, and antipyretic (fever-relieving) properties, and to prevent thrombosis. It acts by inhibiting the synthesis of prostaglandin, which is released from damaged tissue during inflammation.

assimilation The process of incorporation of simple molecules of food that has been digested and absorbed into living cells of an animal and conversion into the complex molecules making up the organism.

association The combination of molecules of a substance with those of another to form more complex species. An example is a mixture of water and ethanol (which are termed *associated liquids*), the molecules of which combine via hydrogen bonding.

asymmetric atom *See* chirality; isomerism; optical activity.

atomic mass unit (**amu**) Symbol: u A unit of mass used for atoms and molecules, equal to 1/12 of the mass of an atom of carbon-12. It is equal to $1.660\ 33 \times 10^{-27}$ kg.

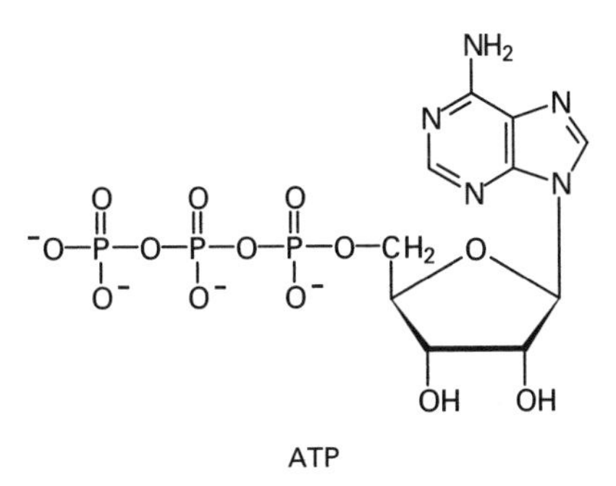

ATP

ATP (**adenosine triphosphate**) The universal energy carrier of living cells. Energy from respiration or, in photosynthesis, from sunlight is used to make ATP from ADP. It is then reconverted to ADP in various parts of the cell by enzymes known as *ATPases*, the energy released being used to drive three main cellular processes: mechanical work (muscle contraction and cellular movement); the active transport of molecules and ions; and the biosynthesis of biomolecules. It can also be converted to light, electricity, and heat.

ATP is a nucleotide consisting of adenine and ribose with three phosphate groups attached. Hydrolysis of the terminal phosphate bond releases energy (30.6 kJ mol^{-1}) and is coupled to an energy-requiring process. Further hydrolysis of ADP to AMP sometimes occurs, releasing more energy. The pool of ATP is small, but the faster it is used, the faster it is replenished. ATP is not transported round the body, but is synthesized where it is needed. *See also* oxidative phosphorylation; electron-transport chain.

ATP synthase (**H^+ transporting ATP synthase**) The enzyme that synthesises ATP from ADP and inorganic phosphate during oxidative phosphorylation. It is a complex enzyme consisting of two units, F_0 and F_1, linked by a stalk with all three components made of several polypeptide chains. F_0 is hydrophobic and is situated in the inner mitochondrial membrane, forming a proton channel through the membrane. F_1 is hydrophilic and is situated in the mitochondrial matrix where it synthesises ATP. The formation and release of ATP by F_1 is driven by the energy released from the flow of protons through F_0. The synthesis of one ATP molecule requires three protons to flow through ATP synthase. The structure of ATP synthase is the same in all eukaryotic cells. Prokaryotic ATP synthase has the same basic structure but the F_0 and F_1 units are joined directly. *See* oxidative phosphorylation.

atropine A poisonous alkaloid found in deadly nightshade. It is used to treat colic and dilate the pupil of the eye.

attenuation **1.** A 'back-up' mechanism that helps to regulate gene expression in bacterial OPERONS. An operon has a leader sequence, which lies downstream of the operon's promotor and operator sites, but before the start of the structural genes. An attenuator site is a segment of DNA within the leader sequence. Even if transcription of the leader sequence by RNA polymerase has commenced, transcription can abort when the polymerase complex reaches the attenuator site, according to the cell's requirements for the gene products. For example, if the amino acid tryptophan is readily available in a cell, the *trp* operon, which encodes genes involved in tryptophan synthesis, is inhibited by the *trp* repressor. However, this inhibition is often incomplete, so regulation is augmented by an attenuation mechanism. The high concentration of tryptophan causes translation of the *trp* operon leader sequence by a ribosome to stall. This in turn prompts termination of transcription, and release of the messenger RNA chain from the transcription complex.
2. The loss of virulence of a pathogenic microorganism after several generations of culture *in vitro*. Attenuated microorganisms are commonly used in vaccines.

atto- Symbol: a A prefix denoting 10^{-18}. For example, 1 attometer (am) = 10^{-18} meter (m).

autoimmune disease *See* immunity.

autolysis The self-destruction of cells by digestive enzyme activity. It is the final stage of cell senescence resulting in complete digestion of all cell components. *See* lysosome.

autonomic nervous system The division of the vertebrate nervous system that supplies motor nerves to the smooth muscles of the gut and internal organs and to heart muscle. It comprises the *sympathetic nervous system*, which (when stimulated) increases heart rate, breathing rate, and blood pressure and slows down digestive processes, and the *parasympathetic nervous system*, which slows heart rate and promotes digestion. Each organ is innervated by both systems and their relative rates of stimulation determine the net effect on the organ concerned. Many functions of the autonomic nervous system, such as the control of heart rate and blood pressure, are regulated by centers in the medulla oblongata of the brain. *See also* parasympathetic nervous system; sympathetic nervous system.

autoradiography A technique in which a thin slice of tissue containing a radioactive isotope is placed in contact with a photographic plate. The image obtained when the plate is developed shows the distribution of the isotope in the tissue.

autosomes Paired somatic chromosomes that play no part in sex determination. *Compare* sex chromosomes.

autotrophism A type of nutrition in which the principal source of carbon is inorganic (carbon dioxide or carbonate). Organic materials are synthesized from inorganic starting materials. The process may occur by use of light energy (*photoautotrophism*) or chemical energy (*chemoautotrophism*). Autotrophic organisms (*autotrophs*) are important ecologically as primary producers, their activities ultimately supplying the carbon requirements of all heterotrophic organisms. *Compare* heterotrophism. *See* chemotrophism; phototrophism.

auxin Any of a group of plant hormones, the most common naturally occurring one being indole acetic acid, IAA. Auxins are made continually in growing shoot and root tips. They are actively moved away from the tip, having various effects along their route before being inactivated. They regulate the rate of extension of cells in the growing region behind the shoot tip and are involved in phototropic and geotropic curvature responses of shoot tips (but probably not root tips) moving laterally away from light and toward gravity.

Auxins stimulate cell enlargement, probably by stimulating excretion of protons leading to acid-induced wall loosening and thus wall extension. They help maintain apical dominance by inhibiting lateral bud development. Root initiation may be stimulated by auxin from the shoot, and auxins have been shown to move towards the root tips. Pollen-tube growth is stimulated by auxin and its production by developing seeds stimulates fruit set and pericarp growth in fleshy fruits. It interacts synergistically with gibberellins and cytokinins in stimulating cell division and differentiation in the cambium. A high auxin:cytokinin ratio stimulates root growth but inhibits regeneration of buds in tobacco pith callus. It is antagonistic to abscisic acid in abscission.

Synthetic auxins, cheaper and more stable than IAA, are employed in agriculture, horticulture, and research. These include indoles and naphthyls: e.g. NAA (naphthalene acetic acid) used mainly as a rooting and fruit setting hormone; phenoxyacetic acids, e.g. 2,4-D (2,4-dichlorophenoxyacetic acid) used as weed-killers and modifiers of fruit development; and more toxic and persistent benzoic auxins, e.g. 2,4,5-trichlorobenzoic acid, also formerly used as herbicides but now widely restricted.

axon (nerve fiber) The part of a neurone that conveys impulses from the cell body towards a synapse. It is an extension of the cell body and consists of an axis cylinder (*axoplasm*), surrounded in most vertebrates by a fatty (myelin) sheath, outside which is a thin membrane (*neurilemma*). *See* neurone.

Azotobacter A genus of free-living aerobic nitrogen-fixing bacteria found in limestone soils and water. The cells are plump rods or cocci, surrounded by slime. A multilayered wall may be synthesized around the cell to produce a microcyst resistant to desiccation.

AZT An inhibitor of reverse transcriptase in retroviruses. It is used to treat patients with HIV infection and AIDS.

Bacteria Bacteria is one of the three cellular kingdoms (Archaea and Eukarya are the other two). A large and diverse group of organisms, which, in terms of numbers and variety of habitats, includes the most successful life forms. In nature, bacteria are important in the nitrogen and carbon cycles, and some are useful to man in various industrial processes, especially in the food industry, and in techniques of genetic engineering (*see also* biotechnology). However, there are also many harmful parasitic bacteria that cause diseases such as botulism and tetanus.

Bacterial cells are simpler than those of animals and plants. They are unicellular and lack well-defined membrane-bound nuclei (i.e. are prokaryotes), and do not contain complex organelles such as chloroplasts and mitochondria. They may divide (by binary fission) every 20 minutes and can thus reproduce very rapidly. They also form resistant spores. *See also* prokaryote.

bactericidal Describing a compound that has a lethal effect on bacteria. A bactericidal compound may act by interfering with a vital biochemical pathway, or by destroying the molecular structure of the cell.

bacteriochlorophyll Any of several types of chlorophyll found in photosynthetic bacteria, such as the purple bacteria. There are seven forms, designated bacteriochlorophylls *a–g*. All are structurally similar to chlorophyll *a* of plants. The bacteriochlorophylls absorb light at longer wavelengths than chlorophyll *a* enabling far-red and infrared light to be used in photosynthesis. *See* photosynthetic pigments.

bacteriophage *See* phage.

bacteriostatic Used to describe a compound that prevents reproduction of bacteria, but does not kill them.

Baltimore, David (1938–) American microbiologist and molecular biologist noted for his work on viral nucleic-acid biosynthesis and his discovery of reverse transcription by the enzyme RNA-directed DNA polymerase in the virions of tumor-producing retroviruses. A classification of animal viruses is named for him. He was awarded the Nobel Prize for physiology or medicine in 1975 jointly with R. Dulbecco and H. M. Temin for their discoveries concerning the interaction between tumor viruses and the genetic material of the cell.

Banting, (Sir) Frederick Grant (1891–1941) Canadian surgeon, physiologist, and endocrinologist distinguished for his discovery (with C. H. BEST) of a way of extracting the hormone insulin from pancreatic tissue. He was awarded the Nobel Prize for physiology or medicine in 1923 jointly with J. J. R. Macleod.

baroreceptor (**baroceptor**) A sensory receptor in the walls of the blood vascular system that responds to changes in blood pressure. Its function is to maintain a steady blood pressure. Any increase in pressure distends the wall and stimulates the receptor to fire nervous impulses that result in reflex slowing of the heart and vasodilation.

Barton, (Sir) Derek Harold Richard (1918–98) British organic chemist who developed the concept of conformation and its application in chemistry. He was awarded the Nobel Prize for chemistry in

1969, the prize being shared with O. Hassel.

basal metabolic rate (**BMR**) The minimum rate of energy expenditure by an animal necessary to maintain the vital processes, e.g. circulation, respiration, etc. It is expressed as the output of heat in joules (or kilojoules) per square meter of body surface area per hour and is measured either directly from heat production or indirectly from oxygen consumption. The thyroid hormones are the prime regulators of the BMR.

base **1.** (*Chemistry*) A compound that reacts with an acid to produce water plus a salt.
2. (*Biochemistry*) A nitrogenous molecule, either a PYRIMIDINE or a PURINE, that combines with a pentose sugar and phosphoric acid to form a nucleotide, the fundamental unit of nucleic acids. The most abundant bases are cytosine, thymine, and uracil (pyrimidines) and adenine and guanine (purines).

base analog An unnatural purine or pyrimidine that can be incorporated into DNA, causing altered base pairing. Some base analogs are used therapeutically as anticancer drugs.

base pairing The linking together of the two helical strands of DNA by bonds between complementary bases, adenine pairing with thymine and guanine pairing with cytosine. The specific nature of base pairing enables accurate replication of the chromosomes and thus maintains the constant composition of the genetic material. In pairing between DNA and RNA the uracil of RNA pairs with adenine.

base ratio The ratio of adenine (A) plus thymine (T) to guanine (G) plus cytosine (C). In DNA the amount of A is equal to the amount of T, and the amount of G equals the amount of C, but the amount of A + T does not equal the amount of C + G. The A + T : G + C ratio is constant within a species but varies between species.

basic Having a tendency to release OH^- ions. Thus any solution in which the concentration of OH^- ions is greater than that in pure water at the same temperature is described as basic; i.e. the pH is greater than 7.

basic stain *See* staining.

B-cell (**B-lymphocyte**) A type of lymphocyte (*see* white blood cell) that originates in the bone marrow and serves as the main instrument of the specific immune response, particularly the production of antibodies. Lymphoid tissue contains vast numbers of B-cells, each with a different type of immunoglobulin receptor on its surface. When an antigen binds to the one or several B-cells carrying its specific receptor, it triggers them to undergo repeated division forming a large clone of cells dedicated to producing antibody specific to that antigen. Hence, according to the *clonal selection theory* of antibody specificity, the antigen selects the few appropriate B-cells from among the huge number present in the body. These B-cells then enter the bloodstream as antibody-secreting *plasma cells*.

Another type of lymphocyte, known as a T-helper cell, is required to trigger clonal expansion and antibody secretion by B-cells. This recognizes the antigen bound to the surface of the B-cell in association with class II MHC proteins (*see* major histocompatibility complex). The T-helper cell binds to the antigen-MHC complex and releases substances (lymphokines) that act as the stimulus for B-cell growth.

After an infection has been dealt with and all antigens removed by the antibodies, a small set of B-cells from the clone, called *memory cells*, remain in the circulation. These carry receptors that will bind avidly to the same antigen in any subsequent infection, prompting a much more rapid response by the immune system. *See* immunity; immunoglobulin; T-cell.

Beadle, George Wells (1903–89) American geneticist who formulated the one gene–one reaction hypothesis. He was awarded the Nobel Prize for physiology or medicine in 1958 jointly with E. L. Tatum for their discovery that genes act by regulating definite chemical events. The prize was shared with J. Lederberg.

Benacerraf, Baruj (1920–) Venezuelan-born American immunologist notable for his discovery of histocompatability genes and their role in regulation of the immune response. He was awarded the Nobel Prize for physiology or medicine in 1980 jointly with J. Dausset and G. D. Snell.

Benson–Calvin–Bassham cycle *See* photosynthesis.

benzaldehyde *See* benzenecarbaldehyde.

benzene (C_6H_6) A colorless liquid hydrocarbon with a characteristic odor. Benzene is a highly toxic compound and continued inhalation of the vapor is harmful. Benzene is the simplest aromatic hydrocarbon. It shows characteristic electrophilic substitution reactions, which are difficult to explain assuming a simple unsaturated structure. This anomolous behavior can now be explained by assuming that the six pi electrons are delocalized.

benzenecarbaldehyde (**benzaldehyde**) A yellow organic oil, C_6H_5CHO, with a distinct almondlike odor. It is used as a food flavoring and in the manufacture of dyes and antibiotics.

benzenecarbonyl group (benzoyl group) The group $C_6H_5.CO–$.

benzenecarboxylic acid (**benzoic acid**) A white crystalline carboxylic acid, C_6H_5COOH. It is used as a food preservative.

benzene ring The cyclic hexagonal arrangement of six carbon atoms that are characteristic of aromatic compounds, i.e. of benzene and its derivatives. *See* aromatic compound, benzene.

benzoic acid *See* benzenecarboxylic acid.

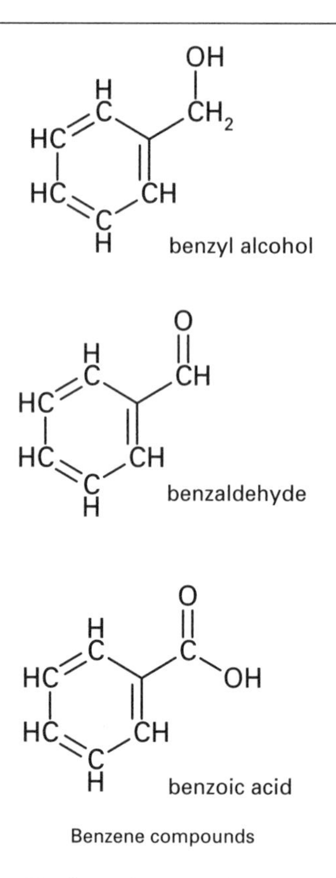

Benzene compounds

Berg, Paul (1926–) American biochemist and molecular geneticist who discovered the method of introducing foreign DNA into bacteria by the use of viruses as vectors. He was awarded the Nobel Prize for chemistry in 1980 for his work on the biochemistry of nucleic acids, with particular regard to recombinant-DNA. The prize was shared with W. Gilbert and F. Sanger.

Bergström, Sune K. (1916–) Swedish biochemist who isolated and purified prostaglandins and certain related substances. He was awarded the Nobel Prize for physiology or medicine in 1982 jointly with B. I. Samuelsson and J. R. Vane.

Best, Charles Herbert (1899–1978) Canadian physician and physiologist. He discovered the enzyme histaminase and the lipotropic action of choline but is best known for the discovery (with F. G. BANTING) of a way of extracting the hor-

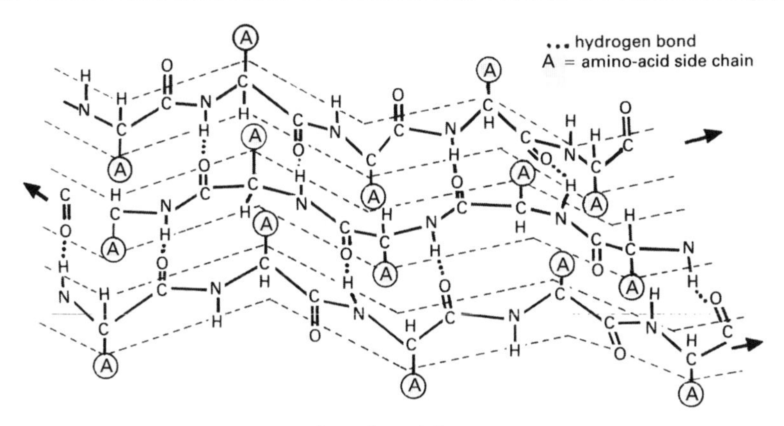

Beta-pleated sheet

mone insulin from pancreatic tissue. Banting divided his share of the Nobel Prize with Best.

beta-pleated sheet A type of PROTEIN structure in which polypeptide chains run close to each other and are held together by hydrogen bonds at right angles to the main chain. The structure is folded in regular 'pleats'. Fibers having this type of structure are usually composed of amino acids with short side chains. The chains may run in the same direction (parallel) or opposite directions (antiparallel). It is one of the two basic secondary structures of proteins.

bile A secretion of the liver that enters the duodenum via the bile duct. It is a mixture of bile salts, bile pigments (bilirubin and biliverdin), cholesterol, and traces of other substances. The bile salts aid digestion by facilitating the emulsification of fats, while the bile pigments and cholesterol are merely excretory products, playing no part in digestion.

bile pigments Pigments excreted in the bile as the products of the degradation of hemoglobin. When hemoglobin is destroyed in the body the protein portion, globin, is degraded to amino acids, while the porphyrin or heme portion gives rise to the bile pigments. The green pigment *biliverdin* is the first of the bile pigments; it is easily reduced to the red-brown pigment *bilirubin*. Bilirubin is the major pigment in human bile, there being only slight traces of biliverdin, which is the chief pigment in the bile of birds. A specific enzyme, biliverdin reductase, catalyzes the reduction of biliverdin to bilirubin. The formation of biliverdin and bilirubin from heme takes place in the reticuloendothelial cells of the liver, spleen, and bone marrow. The pigments are then transported to the liver where they are excreted in the bile into the duodenum. In the intestine, bilirubin undergoes further chemical modification and is secreted in the feces.

bile salts Components of bile that aid digestion in the duodenal region of the gut. Sodium taurocholate and sodium glycocholate emulsify fat droplets by lowering their surface tension, causing them to break up into numerous smaller droplets and facilitating the digestive action of lipase.

bilirubin *See* bile pigments.

biliverdin *See* bile pigments.

binding site A discrete region on a macromolecule (usually a protein) that directly binds a specific ligand.

bioassay An experimental technique for measuring quantitatively the strength of a biologically active chemical by its effect on a living organism. For example, the vitamin activity of certain substances can be measured using bacterial cultures. The increase in bacterial numbers is compared against that achieved with known standards for vitamins.

Biochemical Oxygen Demand (BOD) The standard measurement for determining the level of organic pollution in a sample of water. It is the amount of oxygen used by microorganisms feeding on the organic material over a given period of time. Sewage effluent must be diluted to comply with the statutory BOD before it can be disposed of into clean rivers.

biochemical taxonomy The use of chemical characteristics to help classify organisms; for example, the Asteroideae and Cichorioideae, which are the two main divisions of the plant family Compositae, are separated by the presence or absence of latex. This area of taxonomy has increased in importance with the development of chromatography, electrophoresis, and other analytical techniques.

biochemistry The study of chemical reactions occurring in living organisms.

biodegradable Describing organic compounds that are able to be decomposed by bacteria and other microorganisms, such as the constituents of sewage, as compared with nonbiodegradable compounds, such as most plastics. *See also* pollution.

bioengineering 1. The design, manufacture, and use of replacements or aids for body parts or organs that have been removed or are defective, e.g. artificial limbs, hearing aids, etc.
2. The design, manufacture, and use of equipment for industrial biological processes, such as fermentation.

bioinformatics The use of computing power to analyze biological information. Algorithms are used to compare the sequences of DNA and proteins to detect structural, functional, and evolutionary relationships between the sequences. The sequencing of various genomes produced much DNA sequence data and bioinformatics has been used to great effect to study gene function and predict the function of previously unknown genes. Attention is now being focused on proteins and the prediction of structure and function from only the amino acid sequence.

biological clock The internal mechanism of an organism that regulates circadian rhythms and various other periodic cycles. *See* circadian rhythm.

bioluminescence The production of light by living organisms. Bioluminescence is found in many marine organisms, especially deep-sea organisms. It is also a property of some insects, e.g. fireflies, and certain bacteria. The light is produced as a result of a chemical reaction whereby the compound *luciferin* is oxidized. An enzyme, *luciferase*, catalyzes the reaction in which ATP supplies the energy. There are several quite different types of luciferin.

biopoiesis The origin of organisms from replicating molecules. Biopoiesis is a cornerstone of abiogenesis. Deoxyribonucleic acid (DNA) is the best example of a self-replicating molecule, and is found in the chromosomes of all higher organisms. In some bacterial viruses (bacteriophages) ribonucleic acid (RNA) is self-replicating. Various chemical and physical conditions must be met before either DNA or RNA is able to replicate. *See also* abiogenesis.

biorhythm A periodic physiological or behavioral change that is controlled by a biological clock. Circadian rhythms, hibernation, and migration are examples.

biosynthesis Chemical reactions in which a living cell builds up its necessary molecules from other molecules present. *See* anabolism.

biosystematics The area of systematics in which experimental taxonomic techniques are applied to investigate the relationships between taxa. Such techniques include serological methods, biochemical analysis, breeding experiments, and cytological examination, in addition to the more established procedures of comparative anatomy. Evidence from ecological studies may also be brought to bear. *See also* molecular systematics.

biotechnology The application of technology to biological processes for industrial, agricultural, and medical purposes. For example, bacteria such as *Penicillium* and *Streptomycin* are used to produce antibiotics and fermenting yeasts produce alcohol in beer and wine manufacture. Recent developments in genetic engineering have enabled the large-scale production of hormones, blood serum proteins, and other medically important products. Genetic modification of farm crops, and even livestock, offers the prospect of improved protection against pests, or products with novel characteristics, such as new flavors or extended storage properties. *See also* enzyme technology; genetic engineering.

biotin A water-soluble vitamin generally found, together with vitamins in the B group, in the VITAMIN B COMPLEX. It is widely distributed in natural foods, egg yolk, kidney, liver, and yeast being good sources. Biotin is required as a coenzyme for carboxylation reactions in cellular metabolism.

Bishop, John Michael (1936–) American microbiologist. He was awarded the Nobel Prize for physiology or medicine in 1989 jointly with H. E. Varmus for their discovery of the cellular origin of retroviral oncogenes.

biuret A colorless crystalline organic compound, $H_2NCONHCONH_2$, made by heating urea (carbamide). It is used in a chemical test for proteins.

biuret test A standard colorimetric test for proteins and their derivatives, e.g. peptides and peptones. Sodium hydroxide is first mixed with the test solution and then copper(II) sulfate solution is added drop by drop. A violet color indicates a positive result. The reaction is due to the presence in the molecule of –NH–CO–, the color not appearing with free amino acids. The compound biuret also readily forms from urea around 150°–170°C and thus the biuret test can be used to identify urea.

bivalent A term used for any pair of homologous chromosomes when they pair up during meiosis. Pairing of homologous chromosomes (*synapsis*) commences at one or several points on the chromosome and is clearly seen during pachytene of meiosis I.

Black, (Sir) James Whyte (1924–) British pharmacologist who discovered beta blockers in the 1950s. He also discovered the drug cimetidine for the control of gastric ulcers. He was awarded the Nobel Prize for physiology or medicine in 1988 jointly with G. B. Elion and G. H. Hitchings.

Blöbel, Gunter (1936–) American biochemist who discovered that proteins have intrinsic signals that govern their transport and localization in the cell. He was awarded the 1999 Nobel Prize for physiology or medicine.

Bloch, Konrad Emil (1912–2000) German-born American biochemist who used isotope techniques to study the biosynthesis of cholesterol, creatine, and protoporphyrin. He was awarded the Nobel Prize for physiology or medicine in 1964 jointly with F. Lynen for their work on the mechanism and regulation of the cholesterol and fatty acid metabolism.

blood The transport medium of an animal's body. It is a fluid tissue that circulates by muscular contractions of the heart (in vertebrates) or other blood vessels (in invertebrates). It usually carries oxygen and food to the tissues and carbon dioxide and nitrogenous waste from the tissues, to be excreted. It also conveys hormones and circulates heat throughout the body. Blood consists of liquid plasma in which are suspended white cells (leukocytes), which devour bacteria and produce antibodies. In most animals (except insects) the blood carries oxygen combined with a pigment (HEMOGLOBIN in vertebrates; HEMOCYANIN in some invertebrates). In some invertebrates the pigment is dissolved in the plasma, but in vertebrates it is contained in the red cells (erythrocytes). *See also* blood plasma; erythrocyte; leukocyte; platelet.

blood cell (**blood corpuscle**) Any of the cells contained within the fluid plasma of blood. 45% of blood volume is made up of red cells (erythrocytes) and 1% of white cells (leukocytes). *See* erythrocyte; leukocyte.

blood clotting (**blood coagulation**) The conversion of blood from a liquid to a solid state, which occurs when an injury to the blood vessels exposes blood to air. The clot closes the wound and prevents further blood loss. Blood clotting is brought about in a series of changes that will occur only when at least 14 different *clotting factors* are all present. When platelets in the blood encounter a damaged blood vessel they adhere to the site, forming a plug. They also activate various clotting factors, culminating in the formation of the enzyme thrombokinase (Factor X). This converts the precursor molecule prothrombin (Factor II) into the enzyme thrombin. This then converts fibrinogen, a soluble protein in the plasma, into insoluble fibrin, which forms a network of fibers in which the blood cells become entangled to form a blood clot. *See also* platelet.

blood groups Types into which the blood is classified. Since 1900 it has been known that human blood can be divided into four groups, A (42% of the population), B (9%), AB (3%), and O (46%), based on the presence or absence of certain molecular groups (antigens), called A and B, on the surface of the red blood cells. In group AB, for example, both antigens are present, while group O has neither antigen. Knowledge of a patient's blood group is essential when a blood transfusion is to be given. If blood from a group A donor is given to a group B recipient, the recipient's anti-A antibodies will attack the donor's A antigens, causing the red cells to clump together. Group O blood, having no antigens, can be given to patients of any blood group since it will not provoke an antibody reaction. Group AB, having both antigens and therefore neither antibody, can receive blood from any group. There are various other blood group systems, including that based on the presence or absence of the rhesus factor. *See also* rhesus factor.

blood plasma The liquid that remains when all the cells are removed from blood. It consists of 91% water and 7% proteins, which are albumins, globulins (mainly antibodies), and prothrombin and fibrinogen (concerned with clotting). Plasma also contains the ions of dissolved salts, especially sodium, potassium, chloride, bicarbonate, sulfate, and phosphate. Plasma is slightly alkaline (pH 7.3) and the proteins and bicarbonate act as buffers to keep this constant. It transports dissolved food (as glucose, amino acids, fat, and fatty acids), excretory products, (urea and uric acid), dissolved gases (about 40 mm^3 oxygen, 19 mm^3 carbon dioxide, and 1 mm^3 nitrogen in 100 mm^3 plasma), hormones, and vitamins. Most of the body's physiological activities are concerned with maintaining the correct concentration and pH of all these solutes (this is homeostasis), since plasma supplies the extracellular fluid that is the environment of all the cells.

blood serum The pale fluid that remains after blood has clotted. It consists of plasma without any of the substances involved in clotting. *See also* blood plasma.

blue-green algae *See* Cyanobacteria.

blue-green bacteria *See* Cyanobacteria.

BMR *See* basal metabolic rate.

boat conformation *See* conformation.

BOD *See* Biochemical Oxygen Demand.

Bohr effect The phenomenon in which the affinity of the respiratory pigment of the blood (hemoglobin in vertebrates) for oxygen is reduced as the level of carbon dioxide is increased. This facilitates gaseous exchange, because more oxygen is released in the tissues where the amount of carbon dioxide is rising due to metabolic activity. At the same time, more oxygen is taken up at the lungs or gills where the amount of carbon dioxide is low.

bond The interaction between atoms, molecules, or ions holding these entities together. The forces giving rise to bonding may vary from extremely weak intermolecular forces, about 0.1 kJ mol^{-1}, to very strong chemical bonding, about 10^3 kJ mol^{-1}.

bone A hard connective tissue that makes up most of the skeleton of vertebrates. It consists of fine-branched cells (OSTEOBLASTS) embedded in a matrix, which they have secreted. The matrix is 30% protein (collagen) and 70% inorganic matter, mainly calcium phosphate.

In *compact bone* the matrix is laid down in concentric cylinders called *lamellae*, which surround *Haversian canals* containing blood vessels; the osteoblasts lie in spaces (*lacunae*) between the lamellae and are linked by fine canals (*canaliculi*), which contain the cells' branches. *Spongy bone* has lamellae that form an interlacing network with red marrow in the spaces. Most bones are first formed in cartilage, which is then replaced by bone. A few, called *dermal* or *membrane bones*, are formed by the ossification of connective tissue. *See also* ossification; osteoclast.

boron *See* trace element.

Bowman's capsule The cup-shaped end of a uriniferous tubule of the vertebrate kidney. It surrounds a knot of blood capillaries (glomerulus) and together they form a MALPIGHIAN BODY. *See also* nephron.

Boyer, Paul D. (1918–) American biochemist who worked on the enzyme mechanism underlying the synthesis of ATP. He was awarded the 1997 Nobel Prize for chemistry jointly with J. E. Walker. The prize was shared with J. C. Skou.

bradykinin A peptide formed from a blood protein (kininogen) in certain inflammatory conditions. It has a strong dilating action on blood vessels and causes a fall in blood pressure. *See also* kinin.

Brady's reagent *See* 2,4-dinitrophenylhydrazine.

Bragg, (Sir) William Henry (1862–1942) British physicist who, along with his son W. L. Bragg, pioneered x-ray crystallography. He was awarded the Nobel Prize for physics in 1915 jointly with W. L. Bragg for their work on the analysis of crystal structure by means of x-rays.

Bragg, (Sir) William Lawrence (1890–1971) British experimental physicist, born in Australia, noted for his x-ray crystallographic studies in collaboration with his father, W. H. Bragg, and his subsequent application of x-ray crystallography to biological macromolecules. He was awarded the Nobel Prize for physics in 1915 jointly with W. H. Bragg.

brain The most highly developed part of the nervous system, which is located at the anterior end of the body in close association with the major sense organs and is the main site of nervous control within the animal. Except during prolonged starvation it is solely reliant on glucose as a source of fuel and requires a continuous supply as it has no storage capacity. During starvation, ketone bodies partially replace glucose as fuel for the brain.

The human brain has two greatly folded cerebral hemispheres, which cover most of its surface. The outer layer of these forms the cerebral cortex, which is the principal site of integration for the entire nervous system and is also concerned with memory and learning. The cerebellum, lying beneath the cerebral hemisphere at the rear of the brain, makes fine adjustments to muscle actions initiated by the cortex. Involuntary muscle actions, such as those involved in breathing and swallowing, are governed by the medulla oblongata, located where the spinal cord enters the brain. The hypothalamus, which lies deep within the brain, controls various metabolic functions and also influences the activity of the pituitary gland. The brain has four interconnected internal cavities, the ventricles, which are filled with cerebrospinal fluid. External protection is provided by three membranes, the meninges.

Brown, Michael Stewart (1941–) American physician and molecular geneticist noted for his discovery of cellular receptors for low-density lipoproteins and their role in the removal of cholesterol from the bloodstream. He was awarded the Nobel Prize for physiology or medicine in 1985 jointly with J. L. Goldstein.

brown fat *See* adipose tissue.

Brownian motion The random motion of microscopic particles due to their continuous bombardment by the much smaller and invisible molecules in the surrounding liquid or gas. Particles in Brownian motion can often be seen in colloids under special conditions of illumination.

brush border The outer surface of columnar epithelial cells lining the intestine, kidney tubules, etc. The electron microscope shows it to be made of fine hairlike processes called *microvilli.* These greatly increase the surface area of the cell for absorption of dissolved substances. There may be as many as 3000 microvilli on one epithelial cell.

Büchner, Eduard (1860–1917) German chemist and biochemist who discovered that alcoholic fermentation could be initiated with a cell-free juice obtained from brewers yeast. He suggested that the active principle is a protein (zymase). He was awarded the Nobel Prize for chemistry in 1907 for his discovery of cell-free fermentation.

buffer A solution in which the pH remains reasonably constant when acids or alkalis are added to it; i.e. it acts as a buffer against (small) changes in pH. Buffer solutions generally contain a weak acid and one of its salts derived from a strong base; e.g. a solution of ethanoic acid and sodium ethanoate. If an acid is added, the H^+ reacts with the ethanoate ion (from dissociated sodium ethanoate) to form undissociated ethanoic acid; if a base is added the OH^- reacts with the ethanoic acid to form water and the ethanoate ion. The effectiveness of the buffering action is determined by the concentrations of the acid–anion pair:

$$K = [H^+][CH_3COO^-]/[CH_3COOH]$$

where K is the dissociation constant.

In biochemistry, the main buffer systems are the phosphate ($H_2PO_4^-$/ HPO_4^{2-}) and the carbonate (H_2CO_3/ HCO_3^-) systems.

bundle sheath The ring of parenchymatous or sclerenchymatous tissue, usually one cell thick, that surrounds the vascular bundle in an angiosperm leaf. The individual cells are closely packed with no apparent intercellular spaces, and conduct water and solutes from the vascular bundle to the surrounding tissues. Chloroplasts may be present in the bundle sheath, and are thought to be connected with starch storage in the tropical grasses.

buret A piece of apparatus used for the addition of variable volumes of liquid in a controlled and measurable way. The buret is a long cylindrical graduated tube of uniform bore fitted with a stopcock and a small-bore exit jet, enabling a drop of liquid at a time to be added to a reaction vessel. Burets are widely used for titrations in volumetric analysis. Standard burets permit volume measurement to 0.005 cm^3 and have a total capacity of 50 cm^3; a variety of smaller *microburettes* is available. Similar devices are used to introduce measured volumes of gas at regulated pressure in the investigation of gas reactions.

Burnet, (Sir) Frank Macfarlane (1899–1985) Australian physician, virologist, and immunologist who formulated the clonal selection theory of acquired immunity. He was awarded the Nobel Prize for physiology or medicine in 1960 jointly with P. B. Medawar for discovery of acquired immunological tolerance.

butanedioic acid (**succinic acid**) A crystalline carboxylic acid, $HOOC(CH_2)_2COOH$, that occurs in amber and certain plants. It forms during the fermentation of sugar (sucrose).

butanoic acid (**butyric acid**) A colorless liquid carboxylic acid. Esters of butanoic acid are present in butter.

Butenandt, Adolf Friedrich Johann (1903–95) German organic chemist and biochemist who made pioneering studies of steroid hormones and pheromones. He isolated and determined the structures of the first sex hormones (androsterone, estrone, and progesterone) and the first insect hormone (ecdysone), and also investigated the first pheromone (bombykol). He was awarded the Nobel Prize for chemistry in 1939 for his work on sex hormones. The prize was shared with L. Ružička.

butenedioic acid Either of two isomers. *Transbutenedioic acid* (fumaric acid) is a crystalline compound found in certain plants. *Cisbutenedioic acid* (maleic acid) is used in the manufacture of synthetic resins. It can be converted into the trans isomer by heating at 120°C.

butyric acid *See* butanoic acid.

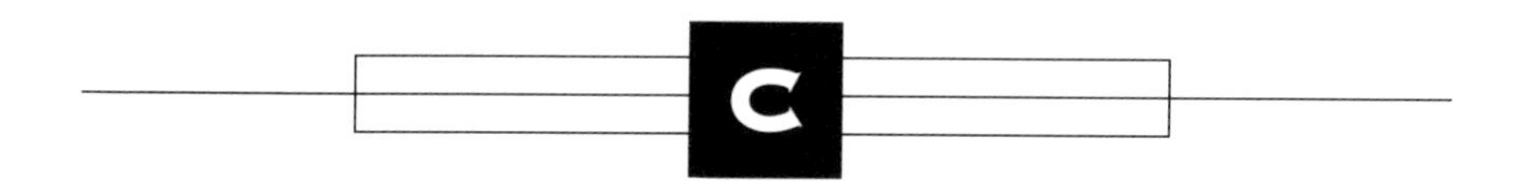

cadaverine An amine, $H_2N[CH_2]_5NH_2$, produced from lysine in decaying meat and fish.

cadherin Any of a group of CELL ADHESION MOLECULES that are important components of certain cell junctions, and also play a role in development by mediating interactions between cells. All are glycoproteins that extend from the cytoplasm, through the plasma membrane and into the intercellular space. Here they bind to similar cadherins extending from neighboring cells. Cadherins are found abundantly in junctions between epithelial cells, and require calcium ions for effective binding. During development, cells expressing one class of cadherin will bind to other cells expressing the same class of cadherin, thereby ensuring correct cell–cell interactions.

caffeine An alkaloid found in certain plants, especially tea and coffee. It is a stimulant of the central nervous system and a diuretic. The systematic name is *1,3,7-trimethylxanthine*.

calciferol *See* vitamin D.

calcitonin A polypeptide hormone, produced by the C-cells of the thyroid gland, that lowers the concentration of calcium in the blood. It acts by preventing calcium and phosphorus from leaving bone but not from entering it, thus antagonizing the effect of parathyroid hormone (*see* parathyroid glands).

calcium An essential mineral salt for animal and plant growth. It is present between plant cell walls as pectate, and is found in the bones and teeth of animals. Calcium ions, Ca^{2+}, are important in triggering muscle contraction where their rapid release from the cisternae of the sarcoplasmic reticulum triggers the reaction between ATP and the myofilaments. Calcium is important in resting muscles in maintaining the relative impermeability of the cell membranes. If the calcium concentration falls, the potential difference across the membrane also falls so that muscles may spontaneously contract without activation by acetylcholine, giving twitching and spasms. The concentration of calcium ions is also important in influencing the breakdown of glycogen in muscles. Calcium is important in the clotting of blood in the conversion of prothrombin to thrombin. In mammalian stomachs it is also important in precipitating casein from milk. The activity of some enzymes is regulated by calcium and it also acts as a secondary messenger.

calcium ion pump Any transport protein that actively carries calcium ions (Ca^{2+}) across a cell membrane against an electrochemical gradient. Such proteins are ATPases, in that they derive energy by the hydrolysis of ATP. *Plasma membrane calcium pumps* are found in animal and yeast cells, and export Ca^{2+} from the cell to maintain low intracellular Ca^{2+} concentrations. The rate of Ca^{2+} export is regulated by the Ca^{2+}-binding protein CALMODULIN. Another type of calcium ion pump occurs in the membrane that bounds the sarcoplasmic reticulum (SR) in skeletal muscle. This *muscle calcium pump* transports Ca^{2+} from the sarcoplasm into the lumen of the SR following muscle contraction. Two Ca^{2+} are transported for every ATP molecule hydrolyzed.

callose An insoluble carbohydrate that is laid down around the perforations in sieve plates. As the sieve tube ages the callose layers become thicker, eventually blocking the sieve element. Such blocking may be seasonal or permanent.

calmodulin A small calcium-binding protein found in the cytosol of cells that regulates many cellular activities. Each calmodulin molecule binds four calcium ions (Ca^{2+}) and in so doing undergoes a change in its three-dimensional shape, enabling it to bind to and activate various enzymes.

calorimetry The measurement of thermal changes involved in chemical or physical reactions, in either an *in vitro* system or an intact organism. For example, a *bomb calorimeter* is used to determine the calorific value of foods. This is a steel chamber in which the foodstuff is placed and ignited, in the presence of oxygen. The increase in temperature of water in a jacket surrounding the calorimeter indicates the heat produced by the oxidized food.

Calvin, Melvin (1911–97) American biochemist who pioneered the use of radioactive isotopes, especially carbon-14, as tracers in metabolism. He was awarded the Nobel Prize for chemistry in 1961 for his research on carbon dioxide assimilation in plants.

Calvin cycle (**reductive pentose-phosphate cycle**) *See* photosynthesis.

CAM *See* cell adhesion molecule.

camphor A naturally-occurring white organic compound, $C_{10}H_{16}O$, with a characteristic penetrating odor. It is a cyclic compound and a ketone, formerly obtained from the wood of the camphor tree but now made synthetically. Camphor is used as a platicizer for celluloid and as an insecticide against clothes moths.

Canada balsam A clear resin similar to glass in its optical properties, used for mounting microscope specimens.

cancer A malignant tumor, or disease caused by it. Malignant tumors are distinguished from benign ones in that they are not encapsulated, their cells show uncontrolled reproduction and lack differentiation of structure, and they are capable of producing secondary growths (*metastases*) in a part of the body distant from the original tumor. Cancer is caused by a failure to regulate cell growth and proliferation. This failure arises from mutations in genes that express regulatory proteins. Mutations can cause a loss of function or cause a protein to function when it should not. Multiple mutations are usually required to cause a cell to become a cancer cell. Most cancer-causing mutations are not hereditary. Cancers are classified into two main groups according to the tissue in which they arise: *carcinomas* arise in epithelial tissue; *sarcomas* in connective tissue.

cane sugar *See* sucrose.

caproic acid *See* hexanoic acid.

caprylic acid *See* octanoic acid.

capsid The protein coat of a virus, surrounding the nucleic acid. A capsid is present only in the inert extracellular stage of the life cycle. Capsids are composed of subunits called *capsomeres*. *See also* virus.

capsomere *See* capsid.

carbanion An intermediate in an organic reaction in which one carbon atom is electron-rich and carries a negative charge. Carbanions are usually formed by abstracting a hydrogen ion from a C–H bond using a base, e.g. from ethanal to form $^-CH_2CHO$, or from organometallic compounds in which the carbon atom is bonded to an electropositive metal.

carbene A transient species of the form RR′C:, with two valence electrons that do not form bonds. The simplest example is *methylene*, H_2C:. Carbenes are intermediates in some organic reactions.

carbohydrates A class of compounds occurring widely in nature and having the general formula type $C_x(H_2O)_y$. (Note that although the name suggests a hydrate of carbon these compounds are in no way hydrates and have no similarities to classes of hydrates.) Carbohydrates are generally divided into two main classes: SUGARS and POLYSACCHARIDES.

Carbohydrates are both stores of energy and structural elements in living systems; plants having typically 15% carbohydrate and animals about 1% carbohydrate. The body is able to build up polysaccharides from simple units (anabolism) or break the larger units down to more simple units for releasing energy (catabolism). Carbohydrates require neutral or slightly alkaline conditions for the operation of enzymes such as maltase and amylase. Thus, carbohydrate digestion is an intestinal rather than a stomach process.

carbon An essential element in plant and animal nutrition that occurs in all organic compounds and forms the basis of all living matter. It enters plants as carbon dioxide and is assimilated into carbohydrates, proteins, and fats. The element carbon is particularly suited to such a role as it can form stable covalent bonds with other carbon atoms, and with hydrogen, oxygen, nitrogen, and sulfur atoms. It is also capable of forming double and triple bonds as well as single bonds and is thus a particularly versatile building block. Carbon, like hydrogen and nitrogen, is far more abundant in living materials than in the Earth's crust, indicating that it must be particularly suitable to fulfill the requirements of living processes. *See also* carbon cycle.

carbon cycle The circulation of carbon between living organisms and the environment. The carbon dioxide in the atmosphere is taken up by autotrophic organisms (mainly green plants) and incorporated into carbohydrates during PHOTOSYNTHESIS by the reductive pentose-phosphate cycle. The carbohydrates so produced are the food source of the heterotrophs (mainly animals). All organisms return carbon dioxide to the air as a product of respiration and of decay. The burning of fossil fuels also releases CO_2. In water, carbon, combined as carbonates and bicarbonates, is the source for photosynthesis.

carbonium ion An intermediate in an organic reaction in which one carbon atom is electron-deficient and carries a positive charge. Carbonium ions are formed by breaking a covalent bond between a carbon atom and an electronegative atom:

$$C_2H_5Br \rightarrow C_2H_5^+ + Br^-$$

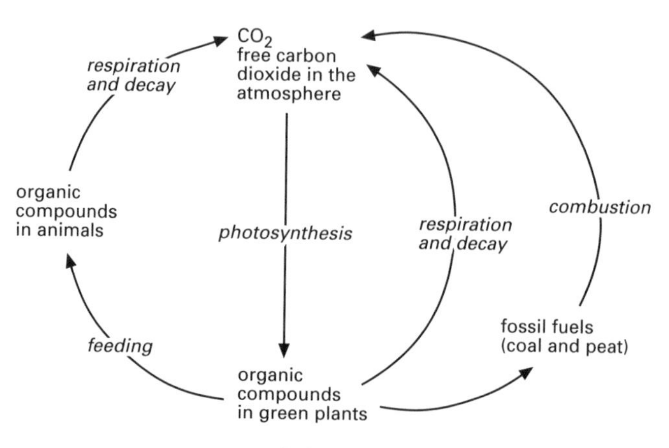

Carbon cycle

They are high-energy intermediates and can only be stabilized and isolated under special conditions.

carboxylase An enzyme that catalyzes the decarboxylation of ketonic acids. Carboxylases are found in yeasts, bacteria, plants, and animal tissues. Pyruvic carboxylase brings about the decarboxylation of pyruvic acid whilst oxaloacetic carboxylase helps the breakdown of oxaloacetic acid into pyruvic acid and carbon dioxide. Carboxylases are thus involved in the transfer of carbon dioxide in respiration.

carboxyl group The organic group –CO.OH, present in carboxylic acids.

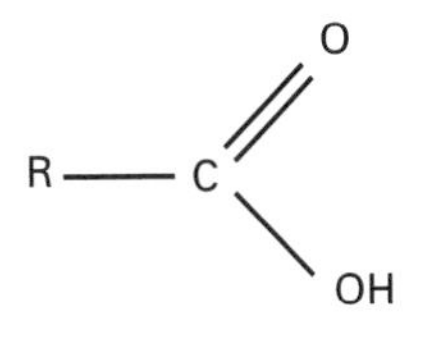

Carboxylic acid

carboxylic acid An organic compound of general formula RCOOH, where R is an organic group and –COOH is the carboxylate group. Many carboxylic acids are of biochemical importance. Those of particular significance are: **1.** The lower carboxylic acids (such as citric, succinic, fumaric, and malic acids), which participate in the KREBS CYCLE.
2. The higher acids, which are bound in lipids. These are also called *fatty acids*, although the term 'fatty acid' is often used to describe any carboxylic acid of moderate-to-long chain length. The fatty acids contain long hydrocarbon chains, which may be saturated (no double bonds) or unsaturated (C=C double bonds). Animal fatty acids are usually saturated, the most common being stearic acid and palmitic acid. Plant fatty acids are often unsaturated: oleic acid is the commonest example. Fatty acids have three major roles: components of membranes, intracellular messengers, and as a source of fuel. Fatty acids are degraded by the sequential removal of two-carbon units from the chain, yielding one molecule each of acetyl CoA, NADH, and $FADH_2$. NADH and $FADH_2$ are oxidized by the ELECTRON-TRANSPORT CHAIN and acetyl CoA enters the Krebs cycle. *See also* lipid; triglyceride.

carcinogen Any substance that causes living tissues to become cancerous. Chemical carcinogens include many organic compounds, e.g. hydrocarbons in tobacco smoke, as well as inorganic ones, e.g. asbestos. Carcinogenic physical agents include ultraviolet light, x-rays, and radioactive materials. Some viruses (e.g. hepatitis B) are also carcinogens. Many carcinogens are mutagenic, i.e. they cause changes in the DNA; dimethylnitrosamine, for example, methylates the bases in DNA. A potential carcinogen may therefore be identified by determining whether it causes mutations.

Carlsson, Arvid (1923–) Swedish biochemist noted for his work on signal transduction in the nervous system. He shared the 2000 Nobel Prize for physiology of medicine with P. Greengard and E. R. Kandel.

carnitine A zwitterionic derivative of lysine that shuttles long-chain acyl groups, derived from fatty acids, across the inner mitochondrial membrane. The acyl group is transferred from coenzyme A to carnitine by carnitine acyltransferase to form acylcarnitine and free coenzyme A on the cystolic side of the membrane. A translocase enzyme moves acylcarnitine into the mitochondrial matrix and at the same time moves a molecule of carnitine back to the cystolic side of the inner mitochondrial membrane. In the matrix, the acyl group is removed from carnitine and rejoined to coenzyme A. The free carnitine is moved back across the membrane as described above and acyl CoA is oxidised to acetyl CoA, NADH, and $FADH_2$ with the generation of ATP. Acetyl groups can also be transported across the inner mitochondrial membrane as acetylcarnitine using the same mechanism. See also fatty acid.

carotene A carotenoid pigment, examples being lycopene and α- and β-carotene. The latter compounds are important in animal diets as a precursor of vitamin A. *See* carotenoids; photosynthetic pigments.

carotenoids A group of yellow, orange, or red pigments comprising the CAROTENES and XANTHOPHYLLS. They are found in all photosynthetic organisms, where they function mainly as accessory pigments in photosynthesis, and in some animal structures, e.g. feathers. They contribute, with anthocyanins, to the autumn colors of leaves since the green pigment chlorophyll, which normally masks the carotenoids, breaks down first. They are also found in some flowers and fruits, e.g. tomato. Carotenoids have three absorption peaks in the blue-violet region of the spectrum.

Carotenes are hydrocarbons. The most widespread is β-carotene. This is the orange pigment of carrots whose molecule is split into two identical portions to yield vitamin A during digestion in vertebrates. *Xanthophylls* resemble carotenes but contain oxygen. *See also* photosynthetic pigments.

carotid body A vascular structure at the base of the external carotid artery that contains chemoreceptors, which monitor carbon dioxide and oxygen concentrations and pH of the blood. It responds to a change in any of these factors by firing nervous impulses that bring about reflex changes in respiratory and heart rates.

carotid sinus A swollen portion of the internal carotid artery, near its origin in the neck, containing sensory receptors that monitor changes in blood pressure. *See* baroreceptor.

cartilage A firm but flexible skeletal material (gristle) that makes up the entire skeleton of the cartilaginous fishes (sharks, etc.). In more advanced vertebrates the skeleton is first formed as cartilage in the embryo and then changed into bone; in adults cartilage persists in a few places, such as the end of the nose, the pinna of the ear, the disks between vertebrae, over the ends of bones, and in joints. Cartilage is a connective tissue containing cells (*chondroblasts*) embedded in a matrix of solid protein (*chondrin*), which may have elastic or tough white fibers in it. When the cells divide they cannot move apart, so they remain in groups of two or four.

cascade A sequence of reactions involving enzymes or hormones, in which the product of one reaction activates the next reaction in the sequence. In this way, a weak initial signal can be amplified. Enzyme cascades control blood clotting and glycogen metabolism, and also occur in signal transduction.

casein A phosphoprotein that is present in milk (as calcium caseinate). It belongs to a group of proteins whose main function it is to store amino acids as both nutrients and as building blocks for growing animals.

catabolism Metabolic reactions involved in the breakdown of complex molecules to simpler compounds. The main function of catabolic reactions is to provide energy, which is used in the synthesis of new structures, for work (e.g. contraction of muscles), for transmission of nerve impulses, and for the maintenance of functional efficiency. *See also* metabolism.

catalase An enzyme present in both plant and animal tissues that catalyzes the breakdown of hydrogen peroxide, a toxic compound produced during metabolism, into oxygen and water.

catalyst A substance that alters the rate of a chemical reaction without itself being changed chemically in the reaction. The catalyst can, however, undergo physical change; for example, large lumps of catalyst can, without loss in mass, be converted into a powder. Small amounts of catalyst are often sufficient to increase the rate of reaction considerably. A *positive catalyst* increases the rate of a reaction and a *negative catalyst* reduces it. *Homogeneous catalysts* are those that act in the same phase as the reactants (i.e. in gaseous and liquid systems). *Heterogeneous catalysts* act in a different phase from the reactants. For example, finely divided nickel (a solid) will catalyze the hydrogenation of oil (liquid).

The function of a catalyst is to provide a new pathway for which the rate-determining step has a lower activation energy than in the uncatalyzed reaction. A catalyst does not change the products in an equilibrium reaction and their concentration is identical to that in the uncatalyzed reaction; i.e. the position of the equilibrium remains unchanged. The catalyst simply increases the rate at which equilibrium is attained.

In *autocatalysis*, one of the products of the reaction itself acts as a catalyst. In this type of reaction the reaction rate increases with time to a maximum and finally slows down. For example, in the hydrolysis of ethyl ethanoate, the ethanoic acid produced catalyzes the reaction.

ENZYMES are highly efficient and specific biochemical catalysts.

catecholamines A group of chemicals (amine derivatives of catechol) that occur in animals, especially vertebrates, and act as neurohormones or neurotransmitters. Examples are norepinephrine, epinephrine, and dopamine.

catenation The formation of chains of atoms in molecules.

cathepsin One of a group of enzymes that break down proteins to amino acids within the various mammalian tissues. Several cathepsins have been isolated and are differentiated by their different activities.

cation A positively charged ion, formed by removal of electrons from atoms or molecules. In electrolysis, cations are attracted to the negatively charged electrode (the cathode). *Compare* anion.

cationic resin An ion-exchange material that can exchange cations, such as H^+ and Na^+, for ions in the surrounding medium. Such resins are used for a wide range of purification and analytical purposes.

cdc gene *See* cell division cycle gene.

CD marker Cluster of differentiation marker: any of a group of antigenic molecules occurring on the surface of white blood cells and other cells that are used to distinguish subsets of very similar cell populations, e.g. in immunology. For example, T-helper cells possess CD4 antigens, whereas CD8 antigens occur on cytotoxic T-cells. They are identified using specific monclonal antibodies. More than 80 individual markers have been identified. *See also* T-cell.

cDNA *See* complementary DNA.

Cech, Thomas Robert (1947–) American biochemist and molecular biologist. He was awarded the Nobel Prize for chemistry in 1989 jointly with S. Altman for their discovery of catalytic properties of RNA.

cell The basic unit of structure of all living organisms, excluding viruses. Cells were discovered by Robert Hooke in 1665, but Schleiden and Schwann in 1839 were the first to put forward a clear *cell theory*, stating that all living things were cellular. Prokaryotic cells (typical diameter 1 μm) are significantly smaller than eukaryotic cells (typical diameter 20 μm). The largest cells are egg cells (e.g. ostrich, 5 cm diameter); the smallest are mycoplasmas (about 0.1 μm diameter). All cells contain genetic material in the form of DNA, which controls the cell's activities; in eukaryotes this is enclosed in the nucleus. All contain cytoplasm, containing various organelles (see diagrams and relevant headwords), and are surrounded by a plasma membrane, which controls entry and exit of substances. Plant cells and most prokaryotic cells are surrounded by rigid cell walls. Differences between animal and plant cells can be seen in the diagrams; differences between prokaryotic and eukaryotic cells can be seen in the table. In multicellular organisms cells become specialized for different functions (differentiation); this is called division of labor. Within the cell, further division of labor occurs between the organelles.

cell adhesion molecule (CAM) Any of a class of proteins located in the plasma membrane that form various cell junctions in animal tissues, or mediate cell–cell interactions. CAMs effectively 'glue' cells to each other or to the extracellular matrix (ECM). Part of each CAM resides in the plasma membrane and part protrudes from the membrane to contact another molecule by means of a binding site. There are four main families of CAMs, the largest of which are the CADHERINS. These hold similar cells to each other, for example in epithelial cell layers such as those lining the small intestine and kidney tubules, and also occur in desmosomes. *Integrins* bind cells to components of the ECM, such as collagens and laminins, and can form cell–matrix junctions, such as hemidesmosomes and focal adhesions. *Nerve-cell adhesion molecules* (N-CAMs) are especially important in nervous tissue, and in differentiation of muscle and nerve cells during development. *Selectins* participate in interactions between endothelial cells lining blood vessels and leukocytes (white blood cells).

cell body The part of a nerve cell (neurone) that contains the nucleus and most of its organelles. It has a swollen appearance and contains Nissl granules. The cell body is a center of synthesis, supplying materials to the rest of the neurone. *See* neurone.

cell cycle The ordered sequence of phases through which a cell passes leading up to and including cell division (mitosis). It is divided into four phases G_1, S, and G_2 (collectively representing interphase), and M-phase, during which mitosis takes place. Synthesis of messenger RNA, transfer RNA, and ribosomes occurs in G_1, and replication of DNA occurs during the S phase. The materials required for spindle formation are formed in G_2. The time taken to complete the cell cycle varies in different tissues. For example, epithelial cells of the intestine wall may divide every 8–10 hours. Differentiated cells usually exist in a resting phase known as G_0 and do not divide. There are specific points (known as checkpoints) in the cell cycle where it can be halted.

cell division The process by which a cell divides into daughter cells. In unicellular organisms it is a method of reproduction. Multicellular organisms grow by cell division and expansion, and division may be very rapid in young tissues. Mature tissues may also divide rapidly when continuous replacement of cells is necessary, as in the epithelial layer of the intestine. In plants certain growth regulators (e.g. cytokinins) stimulate renewed cell division. *See* meiosis; mitosis.

cell division cycle gene (cdc gene) A gene that codes for proteins essential for cell division. Some are directly involved in controlling cell division.

cell fractionation The separation of the different constituents of the cell into homogenous fractions. This is achieved by breaking up the cells in a mincer or grinder and then centrifuging the resultant liquid. The various components settle out at different rates in a centrifuge and are thus separated by appropriately altering the speed and/or time of centrifugation.

cell membrane *See* plasma membrane.

cell theory The theory that all organisms are composed of cells and cell products and that growth and development results from the division and differentiation of cells. This idea resulted from numerous investigations that started at the beginning of the 19th century, and it was finally given form by Schleiden and Schwann in 1839.

cellulase An enzyme that hydrolyzes the β-1,4-glycosidic linkages in cellulose, yielding cellobiose (a disaccharide) and glucose. It is important in the degradation of plant cell walls in living plants (e.g. in leaf abscission). The cellulase enzymes of gut bacteria are essential for digestion in animals, such as ruminants, that consume plant material.

cellulose A POLYSACCHARIDE forming the framework of the rigid cell walls of all plants, many algae, and some fungi. Cellulose molecules are unbranched chains that together form a rigid structure of high tensile strength. Bundles of molecules form microfibrils, which may be aligned in the primary cell wall either transversely or longitudinally. Cellulose forms an important source of carbohydrate in the diets of herbivores, and is a major constituent of dietary fiber in human diets. The individual units are β-1,4 linked D-glucose molecules. *See* cell wall.

cell wall A rigid wall surrounding the cells of plants, fungi, bacteria, and algae. Plant cell walls are made of cellulose fibers in a cementing matrix of other polysaccharides. Fungi differ, with their walls usually containing chitin. The walls of some algae also differ, e.g. the silica boxes enclosing diatoms. Bacterial walls are more complex, containing peptidoglycans – complex polymers of amino acids and polysaccharides. Cell walls are freely permeable to gases, water, and solutes. They have a mechanical function, allowing the cell to become turgid by osmosis, but preventing bursting. This contributes to the support of herbaceous plants. Plant cell walls can be strengthened for extra support by addition of lignin (as in xylem and sclerenchyma) or extra cellulose (as in collenchyma). Plant cell walls are an important route for movement of water and mineral salts. Other modifications include the uneven thickening of guard cells, the sieve plates in phloem, and the waterproof coverings of epidermal and cork cells.

At cell division in plants the *primary wall* is laid down on the middle lamella of the cell plate as a loose mesh of cellulose fibers. This gives an elastic structure that allows cell expansion during growth. Later the *secondary wall* grows and acquires greater rigidity and tensile strength. New cellulose fibers are laid down in layers, parallel within each layer, but orientated differently in different layers. The Golgi apparatus provides polysaccharide-filled vesicles that deposit wall material by exocytosis, guided by microtubules.

centi- Symbol: c A prefix denoting 10^{-2}. For example, 1 centimeter (cm) = 10^{-2} meter (m).

central nervous system (CNS) The part of the nervous system that receives sensory information from all parts of the body and, through the many interconnections that are possible, causes the appropriate messages to be sent out to muscles and other organs. In vertebrates the CNS consists of the brain and spinal cord. The CNS of invertebrates consists of a connected pair of ganglia in each body segment and a pair of ventral nerve cords running the length of the body.

The development of a CNS is associated with the increasing sensory awareness and complex actions that are involved in locomotion, feeding, reproduction, etc., and the need for central integration of all sensory input and motor output. This compares with the simple localized integration found in the nerve net of coelenterates. *See also* autonomic nervous system; peripheral nervous system.

centrifuge An apparatus in which suspensions are rotated at very high speeds in order to separate the component solids by centrifugal force. If different components have different sedimentation coefficients then they may be separated by removing pellets of sediment at given intervals. *See also* ultracentrifuge.

centriole A CELL organelle consisting of two short tubular structures orientated at right angles to each other. It lies outside the nucleus of animal and protoctist cells, but is absent in cells of most higher plants. Prior to cell division it replicates, and the sister centrioles move to opposite ends of the cell to lie within the spindle-organizing structure, the centrosome. However, the centriole is not essential for spindle formation, although an analogous structure, the basal body, is responsible for organizing the microtubules of undulipodia (cilia and flagella). Under the electron microscope, each 'barrel' of the centriole is seen to consist of a cylinder of nine triplets of microtubules surrounding two central ones.

centromere (**kinetochore; kinomere; spindle attachment**) The region of the chromosome that becomes attached to the nuclear spindle during mitosis and meiosis. Following the replication of chromosomes, resultant chromatids remain attached at the centromere. The centromere is a specific genetic locus at which no crossing over occurs and remains relatively uncoiled during prophase, appearing as a primary constriction. It does not stain with basic dyes.

centrosome A structure found in all eukaryotic cells, except fungi, that forms the SPINDLE during cell division. It lies close to the nucleus in nondividing cells, but at the commencement of cell division it divides, and the sister centrosomes move to opposite ends of the cell, trailing the microtubules of the spindle behind them. In animal and protoctist cells the centrosome contains two short barrel-shaped structures, the CENTRIOLE, but this is not directly involved in spindle formation. In fungi, the function of the centrosome is served instead by the spindle pole body. *See also* mitosis.

cerebral cortex The surface layer of the cerebrum of the brain. It contains billions of nerve cell bodies, collectively called gray matter, and is responsible for the senses of vision, hearing, smell, and touch, for stimulating the contraction of voluntary muscles, and for higher brain activities, such as language and memory. Many of these activities occur in specific regions of the cortex. *See also* brain; neopallium.

cerebroside *See* glycolipid.

cerebrospinal fluid A clear watery fluid containing glucose, salts, and a few white blood cells, that is found in the internal cavities and between the surrounding membranes of the central nervous system. It is filtered from blood by the choroid plexuses in the brain and eventually returns via lymph vessels or in venous blood. It cushions and protects nerve tissues.

chain When two or more atoms form bonds with each other in a molecule, a chain of atoms results. This chain may be a *straight chain*, in which each atom is added to the end of the chain, or it may be a *branched chain*, in which the main chain of atoms has one or more smaller *side chains* branching off it.

Chain, (Sir) Ernst Boris (1906–79) German-born British biochemist noted for his part in the development of penicillin (with H. W. FLOREY) and his work on variants of penicillin as antibacterial agents. He was awarded the Nobel Prize for physiology or medicine in 1945 jointly with A. Fleming and H. W. Florey for the discovery of penicillin.

chair conformation *See* conformation.

chaperone A protein that ensures other proteins are folded correctly and protects them from denaturation.

chelate A metal coordination complex in which one ligand coordinates at two or more points to the same metal ion. The resulting complex contains rings of atoms that include the metal atom. An example of a *chelating agent* is 1,2-diaminoethane ($H_2NCH_2CH_2NH_2$), which can coordinate both its amine groups to the same atom. It is an example of a *bidentate ligand* (having two 'teeth'). Edta, which can form up to six bonds, is another example of a chelating agent. The word chelate comes from the Greek word meaning 'claw'. *See also* sequestration.

chemical bond A link between atoms that leads to an aggregate of sufficient stability to be regarded as an independent molecular species. Chemical bonds include covalent bonds, electrovalent (ionic) bonds, coordinate bonds, and metallic bonds. Hydrogen bonds and van der Waals forces are not usually regarded as true chemical bonds.

chemical dating A method of using chemical analysis to find the age of an archaeological specimen in which compositional changes have taken place over time. For example, the determination of the amount of fluorine in bone that has been buried gives an indication of its age because phosphate in the bone has gradually been replaced by fluoride ions from groundwater. Another dating technique depends on the fact that, in living organisms, amino acids are optically active. After death a slow racemization reaction occurs and a mixture of L- and D-isomers forms. The age of bones can be accurately determined by measuring the relative amounts of L- and D-amino acids present.

chemical fossils Particularly resistant organic chemicals (e.g. alkanes and porphyrins) present in geological strata that are thought to indicate the existence of life in the period when the rocks were formed. Chemical fossils are often the only evidence for life in rocks of Precambrian age.

chemiosmotic theory *See* electron-transport chain.

chemoautotrophism (**chemosynthesis**) *See* autotrophism; chemotrophism.

chemoheterotrophism *See* chemotrophism; heterotrophism.

chemoreceptor A receptor that responds to or binds to chemical compounds, e.g. the taste buds.

chemosynthesis *See* autotrophism; chemotrophism.

chemotrophism A type of nutrition in which the source of energy for the synthesis of organic requirements is chemical. Most chemotrophic organisms are heterotrophic (i.e. *chemoheterotrophic*) and their energy source is always an organic compound; animals, fungi, and most bacteria are chemoheterotrophs. If autotrophic (i.e. *chemoautotrophic* or *chemosynthetic*) the energy is obtained by oxidation of an inorganic compound; for example, by oxidation of ammonia to nitrite or a nitrite to nitrate (by nitrifying bacteria), or oxidation of hydrogen sulfide to sulfur (by colorless sulfur bacteria). Only a few specialized bacteria are chemoautotrophic. *Compare* phototrophism. *See* autotrophism; heterotrophism.

chemotropism (**chemotropic movement**) A tropism in which the stimulus is chemical. The hyphae of certain fungi (e.g. *Mucor*) are positively chemotropic, growing toward a particular source of food. Pollen tube growth is chemotropic. *See also* tropism.

chiasma A connection between homologous chromosomes seen during the prophase stage of meiosis. Chiasmata represent a mutual exchange of material between homologous, nonsister chromatids (crossing over) and provide one mechanism by which recombination occurs, through the splitting of linkage groups. *See also* recombination.

chiral Having the property of chirality. For example, lactic acid is a chiral compound because it has two possible structures that cannot be superposed. *See* optical activity.

chirality The property of existing in left- and right-handed forms; i.e. forms that are not superposable. In chemistry the term is applied to the existence of optical isomers. *See* isomerism; optical activity.

chirality element A part of a molecule that causes it to display chirality. The most common type of element is a *chirality center*, which is an atom attached to four different atoms or groups. This is also referred to as an *asymmetric atom*. Less commonly a molecule may have a *chirality axis*, as in the case of certain substituted allenes of the type $R_1R_2C=C=CR_3R_4$. In this form of compound the R_1 and R_2 groups do not lie in the same plane as the R_3 and R_4 groups because of the nature of the double bonds. The chirality axis lies along the C=C=C chain. It is also possible to have molecules that contain a *chirality plane*. *See also* optical activity.

chitin A nitrogen-containing heteropolysaccharide found in some animals and the cell walls of most fungi. It is a polymer of *N*-acetylglucosamine. It consists of many glucose units, in each of which one of the hydroxyl groups has been replaced by an acetylamine group (CH_3CONH). The outer covering of arthropods, the cuticle, is impregnated in its outer layers with chitin, which makes the exoskeleton more rigid. It is associated with protein to give a uniquely tough yet flexible and light skeleton, which also has the advantage of being waterproof. The chitinous plates are thinner for bending and flexibility or thicker for stiffness as required. The plates cannot grow once laid down and are broken down at each molt. Chitin is also found in the hard parts of several other groups of animals.

chlorine An element found in trace amounts in plants and one of the essential nutrients in animal diets. Common table salt, is made up of crystals of sodium chloride. The chloride ion (Cl^-) is important in buffering body fluids, and, because it can pass easily through cell membranes, it is also important in the absorption and excretion of various cations. The hydrogen chloride secreted in gastric juices is important in lowering the pH of the stomach so that the enzyme pepsin is able to act.

chlorophylls A group of photosynthetic pigments. They absorb blue-violet and red light and hence reflect green light, imparting the green color to green plants. The molecule consists of a hydrophilic (water-loving) head, containing magnesium at the center of a PORPHYRIN ring, and a long hydrophobic (water-hating) hydrocarbon tail (the phytol chain), which anchors the molecule in the lipid of the membrane. Different chlorophylls have different chemical groups attached to the head, which accounts for their differing absorption spectra. *See* absorption spectrum; photosynthetic pigments. *See also* bacteriochlorophyll.

chloroplast A photosynthetic plastid containing chlorophyll and other PHOTOSYNTHETIC PIGMENTS. It is found in all photosynthetic cells of plants and protoctists but not in photosynthetic prokaryotes. It has a membrane system containing the photosynthetic pigments on which the light reactions of photosynthesis occur. The surrounding gel-like ground substance, or *stroma*, is where the dark reactions occur. The typical higher plant chloroplast is lens-shaped and about 5 μm in length. Various other forms exist in the algae, e.g. spiral in *Spirogyra*, stellate in *Zygnema*, and cup-shaped in *Chlamydomonas*. The number per cell varies, e.g. one in *Chlorella* and *Chlamydomonas*, two in *Zygnema*, and about one hundred in palisade mesophyll cells of leaves.

Chloroplast membranes form elongated flattened fluid-filled sacs called *thylakoids*. The sheetlike layers of the thylakoids are called *lamellae*. In all plants except algae, the thylakoids overlap at intervals to form stacks, like piles of coins, called *grana*, which apparently improves the efficiency of the light reactions.

The stroma may contain storage products of PHOTOSYNTHESIS, e.g. starch grains. The chloroplasts of most algae contain one or more *pyrenoids*. These are dense protein bodies associated with polysaccharide storage. In green algae, for example, starch is deposited in layers around pyrenoids during development.

The stroma also typically contains *plastoglobuli*, spherical droplets of lipid staining intensely black with osmium tetroxide. They become larger and more numerous as the chloroplast senesces, when carotenoid pigments accumulate in them. Apart from enzymes of the dark reactions, the stroma also contains typical prokaryotic protein-synthesizing machinery including circular DNA and smaller ribosomes. There is now strong evidence that chloroplasts and other cell organelles, such as mitochondria, represent prokaryotic organisms that invaded heterotrophic eukaryotic cells early in evolution and are now part of an indispensable symbiotic union (*see* endosymbiont theory). Chloroplast DNA codes for some chloroplast proteins but there is dependence on nuclear DNA for others.

In C_4 plants there are two types of chloroplast. *See* photosynthesis; plastid.

chlorosis The loss of chlorophyll from plants resulting in yellow (*chlorotic*) leaves. It may be the result of the normal process of senescence, lack of key minerals for chlorophyll synthesis (particularly iron and magnesium), or disease.

cholecalciferol *See* vitamin D.

cholecystokinin (**pancreozymin**) A peptide hormone secreted by cells in the intestinal mucosa. It stimulates contraction of the gall bladder and hence discharge of bile into the duodenum. It also triggers the release of digestive enzyme precursors in pancreatic tissue. *See* duodenum.

cholesterol A neutral sterol (fat derivative) found in animal cells that can be obtained from the diet or synthesized in the liver or intestines. It occurs in bile, blood cells, cell membranes (where it regulates membrane fluidity), blood plasma, and egg yolk. Cholesterol (and other lipids) is transported in the blood by lipoproteins. The major carrier of cholesterol is *low-density lipoprotein* (LDL), which transports cholesterol to peripheral tissues. *High-density lipoprotein* (HDL) picks up cholesterol released into the blood plasma from dying cells and membrane turnover. An elevated level of LDL-cholesterol is a major contributory cause of hardening and narrowing of the arteries (atherosclerosis), which is a major cause of death in Westernized societies. Familial hypercholesterolamia is caused by deleterious mutations to the LDL-receptor gene. Cholesterol can also accumulate in the gall bladder as gallstones. *See* lipoprotein.

choline An amino alcohol often classified as a member of the vitamin B complex. It can be synthesized in humans from lecithin by putrefaction in the bowel, but is required as an essential nutrient for some animals and microorganisms. It acts to disperse fat from the liver or prevent its excess accumulation. Its ester acetylcholine functions in the transmission of nerve impulses.

cholinergic Designating the type of nerve fiber that releases acetylcholine from its ending when stimulated by a nerve impulse. In vertebrates, motor fibers to striated muscle, parasympathetic fibers to smooth muscle, and preganglionic sympathetic fibers are cholinergic. *Compare* adrenergic.

chorion **1.** One of the three embryonic membranes of reptiles, birds, and mammals. It arises from the outer layer of the amniotic hood and encloses the amnion (with embryo or fetus inside), yolk sac, and allantois. The trophoblast of mammals is part of the chorion. The allantois is usually fused with it, forming the chorio-allantoic membrane; this forms a 'lung' attached to the inside of the egg shell in birds and the embryonic part of the placenta in mammals.

2. The tough outer membrane of some eggs, notably those of insects. There is usually a pore, the *micropyle*, to admit spermatozoa.

chorionic gonadotropin (**choriogonadotropin**) A glycoprotein hormone secreted in higher mammals by the chorionic villi of the placenta (fingerlike projections of the chorion into the uterus). It prevents the regression of the corpus luteum in the earlier stages of pregnancy. The detection of human chorionic gonadotropin (HCG) in the urine is often used as a pregnancy test.

chromatid One of a pair of replicated chromosomes found during the prophase and metaphase stages of mitosis and meiosis. During mitosis, sister chromatids remain joined by their centromere until anaphase. In meiosis it is not until anaphase II that the centromere divides, the chromatids being termed daughter chromosomes after separation.

chromatin The loose network of threads seen in nondividing nuclei that represents the chromosomal material, consisting of DNA and protein (mainly histone). It has a regular repeating structure with about 200 bp of DNA wrapped around the outside of a core of histone proteins to form a nucleosome. Nucleosomes are joined together by *linker DNA* to make a flexible chain of nucleosomes. The nucleosome chains are then coiled to form hollow fibers 30 nm in diameter, which can be further coiled into 240 nm diameter fibers. It is this that is visible as the typical looped threadlike material of the nondividing chromosome. Chromatin is classified as *euchromatin* or *heterochromatin* on the basis of its staining properties, the latter staining much more intensely with basic stains because it is more coiled and compact. Euchromatin mainly consists of the 30 nm fibers and is thought to be actively involved in transcription and therefore protein synthesis, while heterochromatin is inactive. Euchromatin stains more intensively than heterochromatin during nuclear division. *See also* chromosome.

chromatography A technique used to separate or analyze complex mixtures. A number of related techniques exist; all depend on two phases: a *mobile phase*, which may be a liquid or a gas, and a *stationary phase*, which is either a solid or a liquid held by a solid. The sample to be separated or analyzed is carried by the mobile phase through the stationary phase. Different components of the mixture are absorbed or dissolved to different extents by the stationary phase, and consequently move along at different rates. In this way the components are separated. There are many different forms of chromatography depending on the phases used and the nature of the partition process between mobile and stationary phases. The main classification is into *column chromatography* and *planar chromatography*.

A simple example of column chromatography is in the separation of liquid mixtures. A vertical column is packed with an absorbent material, such as alumina (aluminum oxide) or silica gel. The sample is introduced into the top of the column and washed down it using a solvent. This process is known as *elution*; the solvent used is the *eluent* and the sample being separated is the *eluate*. If the components are colored, visible bands appear down the column as the sample separates out. The components are separated as they emerge from the bottom of the column. In this particular example of chromatography the partition process is adsorption on the particles of alumina or silica gel. Column chromatography can also be applied to mixtures of gases. *See* gas chromatography. In the other main type of chromatography, planar chromatography, the stationary phase is a flat sheet of absorbent material. *See* paper chromatography, thin-layer chromatography.

Components of the mixture are held back by the stationary phase either by adsorption (e.g. on the surface of alumina) or because they dissolve in it (e.g. in the moisture within chromatography paper).

chromomere A region of a chromosome where the chromosomal material is relatively condensed, and consequently stains darker. Clusters of chromomeres produce distinct bands, the pattern of which is characteristic for a particular chromosome and is used to distinguish the chromosomes of a particular organism.

chromophore A group of atoms in a molecule that is responsible for the color of the compound.

chromoplast (**chromatophore**) A colored plastid, i.e. one containing pigment. They include chloroplasts, which contain the green pigment chlorophyll and are therefore photosynthetic, and nonphotosynthetic chromoplasts. The term is sometimes confined to the latter, which are best known in flower petals, fruits (e.g. tomato) and carrot roots. They are yellow, orange, or red owing to the presence of carotenoid pigments.

chromosome One of a group of threadlike structures of different lengths and shapes in nuclei of eukaryotic cells. They consist of a single large DNA molecule and protein (mostly histones) and carry the genes. Some chromosomes also have an RNA component. (The name chromosome is also given to the genetic material of bacteria and viruses.) During nuclear division the chromosomes are tightly coiled and are easily visible through the light microscope. After division, they uncoil and become difficult to see individually, existing as less densely packaged chromatin.

The number of chromosomes per nucleus is characteristic of the species, for example, humans have 46. Normally one set (haploid) or two sets (diploid) of chromosomes are present in the nucleus. In early prophase of mitosis and later prophase of meiosis, the chromosomes split lengthwise into two identical chromatids held together by the centromere. In diploid cells, there is a pair of sex chromosomes; the remainder are termed autosomes. *See* cell division; centromere; chromatin; chromomere; chromosome map; gene.

chromosome map (**genetic map**) A diagram showing the order of genes along a chromosome. Such maps have traditionally been constructed from information gained by linkage studies (to give a *linkage map*) or by observations made on the polytene (giant) salivary-gland chromosomes of certain insects, e.g. *Drosophila*, to give a *cytological map*. The techniques employed differ according to the type of organism being studied. For example, many plants and animals can be crossed experimentally to study inheritance patterns of particular genes, but this is not possible in humans, where family pedigrees were, until recently, often the only available evidence. However, the advent of new molecular techniques has dramatically changed the nature of chromosome mapping in all organisms, including humans. The Human Genome Project is an international project which was set up to map all the 50 000 or so genetic loci present on human chromosomes. The first draft of the map was published in 2001, covering 97% of the human genome.

Mapping such a huge genome, distributed over 23 chromosomes, involves several steps. The first is to assign each gene to a particular chromosome. This can be achieved by, for example, somatic-cell hybridization or using a gene probe. The next step is to determine the relative positions of the genes on a particular chromosome. This involves comparing restriction fragment length polymorphisms between individuals and constructing a linkage map of all restriction sites, i.e. sites that are cleaved by restriction enzymes. These restriction sites can then be used as markers for closely neighboring genes. The last step is to construct a physical map of the base sequence of the chromosomal DNA. One approach uses cloned DNA segments obtained from a gene library of the chromosome. These clones can be fitted together to form a series of overlapping segments (contig) that corresponds to a particular region of the chromosome. The base sequence of the contig is then determined (*contig mapping*), and hence the base sequence of the chromosomal DNA. This is a painstaking, time consuming, and labor-intensive process. Shotgun cloning combined with the use of massive computer power to fit the cloned sequences together gives a (draft) base sequence much more quickly but the results need to be checked and gaps filled in by contig mapping. *See* gene library; gene probe; restriction fragment length polymorphism; restriction map.

chromosome mutation A change in the number or arrangement of genes in a chromosome. If chromosome segments break away during nuclear division they may rejoin the chromosome the wrong way round, giving an *inversion*. Alternatively, they may rejoin a different part of the same chromosome, or another chromosome, giving a *translocation*. If the segment becomes lost, this is termed a *deficiency* or *deletion*; it is often fatal. A part of a chromosome may be duplicated and occur either twice on the same chromosome or on two different nonhomologous chromosomes: this is a *duplication*. Chromosome mutations can occur naturally but their frequency is increased by the effect of x-rays and chemical mutagens. Many human tumors are associated with specific chromosome mutations.

chylomicron *See* lipoprotein.

chymosin (**rennin**) An enzyme found in gastric juices and responsible for the coagulation of milk. It acts by hydrolyzing peptide links. At 37°C, rennin can coagulate 10^7 times its own weight of milk in ten minutes. It has been crystallized out but is manufactured from the stomach of animals and sold under the name *rennet*. It is used in the manufacture of cheese and junkets.

chymotrypsin An enzyme used to carry out the partial hydrolysis of peptide chains. It is found in pancreatic tissue and in pancreatic juices as an inactive form, chymotrypsinogen. Chymotrypsin will catalyze the hydrolysis of the peptide bonds formed by the carbonyl groups from only certain amino acids, notably phenylalanine, tryosine, and tryptophan residues.

cilium A whiplike extension of certain eukaryotic cells that beats rapidly, thereby causing locomotion or movement of fluid over the cell. Cilia and flagella represent the two types of eukaryotic undulipodium. Cilia are identical in structure to flagella, though shorter, typically 2–10 μm long and 0.5 μm in diameter and usually arranged in groups. Each cilium has a basal body at its base. In ciliated protoctists, sperm, and some marine larvae they allow locomotion. In multicellular animals they may function in respiration and nutrition, wafting water containing respiratory gases and food over cell surfaces, e.g. filter-feeding mollusks. In mammals the respiratory tract is lined with ciliated cells, which waft mucus, containing trapped dust, bacteria, etc., towards the throat.

Cilia and flagella have a '9 + 2' structure (the axoneme), consisting of 9 outer pairs of microtubules with 2 single central microtubules enclosed in an extension of the plasma membrane. The beat of each cilium comprises an effective downward stroke followed by a gradual straightening (limp recovery). Cilia beat in such a way that each is slightly out of phase with its neighbor (*metachronal rhythm*), thus producing a constant rather than a jerky flow of fluid. The basal bodies of cilia are connected by threadlike strands (neuronemes), which coordinate the beating of neighboring cilia. *Compare* flagellum. *See* undulipodium.

cinnamic acid *See* 3-phenylpropenoic acid.

circadian rhythm (diurnal rhythm) A daily rhythm of various metabolic activities in animals and plants. Such rhythms persist even when the organism is not exposed to 24-hour cycles of light and dark, and are thought to be controlled by an endogenous biological clock. Circadian rhythms are found in the most primitive and the most advanced of organisms. Thus *Euglena* shows a diurnal rhythm in the speed at which it moves to a light source, while humans are believed to have at least 40 daily rhythms. *See also* biorhythm.

cis- Designating an isomer with groups that are adjacent. *See* isomerism.

cisterna A flattened membrane-bounded sac of the endoplasmic reticulum or the Golgi apparatus, being the basic structural unit of these organelles.

***cis-trans* isomerism** *See* isomerism.

cistron A unit of function, i.e. a segment of DNA that codes for a single polypeptide chain or a protein molecule. Its extent may be defined by the *cis-trans* test. *See* gene.

citric acid A carboxylic acid, occurring in the juice of citrus fruits, particularly lemons, and present in many other fruits. Citric acid is biologically important because it participates in the KREBS CYCLE. The systematic name is 2-hydroxypropane-1,2,3-tricarboxylic acid. The formula is:

$$HOOCCH_2C(OH)(COOH)CH_2COOH$$

citric acid cycle *See* Krebs cycle.

citrulline *See* amino acids.

Claude, Albert (1899–1983) Belgian-born American cell biologist who developed methods of separating cellular components by centrifugation. He also identified mitochondria as the primary cellular site for biological oxidation. He was awarded the Nobel Prize for physiology or medicine in 1974 jointly with C. R. M. J.

de Duve and G. E. Palade for their work concerning the structural and functional organization of the cell.

clone A group of organisms or cells that are genetically identical. In nature, clones are derived from a single parental organism or cell by asexual reproduction or parthenogenesis. Clones of sexually reproducing higher animals have been produced experimentally by new embryo-splitting techniques, and even from single adult body cells. This was first accomplished in 1997 when DNA from a sheep's udder cell was transferred to a fertilized egg cell from which the DNA had been removed. The egg cell was then implanted in the womb of a ewe, where it developed to produce a normal lamb, named Dolly. This is a complex technique and took many hundreds of attempts to achieve success. Other animals have since been produced by similar techniques. The effects on lifespan and quality of life of producing animals by cloning mature cells is not yet known. The formation of a mature animal from a fully differentiated cell shows that all cells retain their full genetic potential but they must be 'reprogrammed' by instructions in the cytoplasm of the egg cell in order to exhibit it. In genetic engineering, multiple identical copies of a gene are produced in *cloning vectors* (or *vehicles*), such as plasmids and phages (*see* gene cloning).

clotting factors A group of substances that are activated when blood leaves the circulatory system, usually by injury, and cause blood clotting. They include 14 or so proteins, calcium ions, and platelets. *See* blood clotting.

CNS *See* central nervous system.

CoA *See* coenzyme A.

cobalt *See* trace element.

cocaine An alkaloid obtained from the dried leaves of a South American shrub *Erythroxylon Coca*. It is a stimulant and narcotic. Its use is restricted in many countries.

codeine A derivative of morphine, methylmorphine. It is less potent than morphine and is used as an analgesic. It is a controlled substance in the US.

coding strand The DNA strand of a double helix that has the same sequence of bases (with T substituted for U) as the mRNA transcribed from that double helix. The coding strand is therefore not transcribed.

co-dominance The situation in which two different alleles are equally dominant. If they occur together the resulting phenotype is intermediate between the two respective homozygotes. For example, if white antirrhinums (AA) are crossed with red antirrhinums (A′A′) the progeny (AA′) will be pink. Sometimes one allele may be slightly more dominant than the other (*partial* or *incomplete dominance*) in which case the offspring, though still intermediate, will resemble one parent more than the other.

codon A group of three nucleotide bases (i.e. a nucleotide triplet) in a messenger RNA (mRNA) molecule that codes for a specific amino acid or signals the beginning or end of the message (start and stop codons). Since four different bases are found in nucleic acids there are 64 ($4 \times 4 \times 4$) possible triplet combinations. The arrangement of codons along the mRNA molecule constitutes the *genetic code*.

coenzyme A nonprotein group without which certain enzymes are inactive or incomplete. The protein part of an enzyme is known as the apoenzyme and when united with a coenzyme, either permanently or temporarily, the two form an active enzyme known as a holoenzyme.

coenzyme A (CoA) A complex nucleotide containing an active –SH group that is readily acetylated to CoAS–$COCH_3$ (acetyl CoA). Acetyl CoA is the source of the two-carbon units that feed into the KREBS CYCLE. It is produced from glycolysis and the breakdown of fatty acids and some amino acids. It is also a key intermediate in the biosynthesis of lipids and other anabolic reactions.

coenzyme Q *See* ubiquinone.

cofactor A nonprotein substance that helps an enzyme to carry out its activity. Cofactors may be cations or organic molecules, known as coenzymes. Unlike enzymes they are, in general, stable to heat. When a catalytically active enzyme forms a complex with a cofactor a *holoenzyme* is produced. An enzyme without its cofactor is termed an *apoenzyme*.

Cohen, Stanley (1922–) American biochemist. He was awarded the Nobel Prize for physiology or medicine in 1986 jointly with R. Levi-Montalcini for their discoveries of growth factors.

colchicine A drug obtained from the autumn crocus *Colchicum autumnale* that is used to prevent spindle formation in mitosis or meiosis. It has the effect of halting cell division at metaphase, the stage at which the chromosomes have duplicated to give four homologs for each chromosome. If a resting nucleus forms after colchicine treatment it is thus likely to be tetraploid. Colchicine is also used to double the chromosome number of haploid plants derived from cultured pollen grains.

collagen The protein of fibrous connective tissues, present in bone, skin, and cartilage. It is the most abundant of all the proteins in the higher vertebrates. Collagen contains about 35% glycine, 11% alanine, 12% proline and small percentages of other amino acids. The amino acid sequence is remarkably regular with almost every third residue being glycine. Collagen is chemically inert (and insoluble) which suggests that its reactive side groups are immobilized by ionic bonding. Collagen fibrils are highly complex and have a variety of orientations depending on the biological function of the particular type of connective tissue. The secondary structure of collagen is that of a triple helix of peptide chains. Its tertiary structure is one of three alpha helices in a 'super helix', which is responsible for its high tensile strength and therefore its role in support tissues.

Collip, James Bertram (1892–1965) Canadian biochemist and endocrinologist. With J. J. R. Macleod he developed a method for the fractionation of pancreatic extracts containing insulin. These were pure enough to permit the first clinical trials of the hormone to take place. Macleod divided his share of the Nobel Prize with Collip.

colloid A heterogeneous system in which the interfaces between phases, though not visibly apparent, are important factors in determining the system properties. The three important attributes of colloids are:
1. They contain particles, commonly made up of large numbers of molecules, forming the distinctive unit or *disperse phase*.
2. The particles are distributed in a continuous medium (the *continuous phase*).
3. There is a stabilizing agent, which has an affinity for both the particle and the medium; in many cases the stabilizer is a polar group.

Particles in the disperse phase typically have diameters in the range 10^{-6}–10^{-4} mm. Milk, rubber, and emulsion paints are typical examples of colloids.

colony-stimulating factor (CSF) Any of a group of cytokines that control the growth and differentiation of blood cells. Various ones are responsible for different types of cells. For example, *erythropoietin*, produced in the kidney and liver, promotes the formation of red blood cells (erythrocytes); interleukin-3 controls the production of certain white blood cells, namely granulocytes and monocytes/macrophages. *See* cytokine; interleukin.

color blindness Imperfect perception of color thought to be caused by a malfunction or absence of one of the three pigments in the light-sensitive cells (cones) of the retina of the eye. Although it can occasionally be acquired by disease or injury, the defect is usually inherited as a sex-linked recessive character on the X chromosome and is therefore more common in men (about 8% of the population) than in women (about 0.5%). However, women can be carriers of the gene. Complete color blindness is extremely rare, the most common form (*Daltonism*) being the inability to distinguish between reds and greens.

colostrum Liquid secreted by the mammary glands immediately and for the first few days after parturition, preceding the secretion of milk. It is rich in nutrients and antibodies and contains enzymes to clear mucus from the digestive tract of the newborn.

column chromatography *See* chromatography.

compensation point The light intensity at which the rate of photosynthesis is exactly balanced by the combined rates of respiration and photorespiration, so that net exchange of oxygen and carbon dioxide is zero. At normal daylight intensities the rate of photosynthesis exceeds respiration. Shade plants tend to reach their compensation points faster than sun plants but are unable to utilize high light intensities to the same extent. The point at which photosynthesis does not increase with increased light intensity is termed the *light saturation point.* This point occurs at much higher light intensities in C_4 plants than C_3 plants. *See* photorespiration.

competent Describing embryonic tissue that is able to respond to natural (induction) or experimental (evocation) stimuli by becoming or making a specialized tissue. For example, the ectoderm over the optic cup of vertebrate embryos is competent to produce lens tissue.

complement A complex system of proteins (mainly β and γ globulins) normally found in vertebrate blood that react in an ordered sequence when an antigen-antibody complex has formed. The reaction causes lysis of the foreign cells or bacteria, attracts phagocytic cells to the reaction site, and promotes ingestion of antigen-bearing cells by phagocytes. Complement plays a role in inflammation, and is also involved in the tissue damage associated with certain autoimmune disorders.

complementary DNA (**cDNA**) A form of DNA synthesized by genetic engineering techniques from a messenger RNA template using a reverse transcriptase, which therefore has a complementary base sequence to the mRNA. It is used in cloning to obtain gene sequences from mRNA isolated from the tissue to be cloned. It differs from the original DNA sequence in that it lacks intron and promoter sequences. Labeled single-stranded cDNA is used as a gene probe to identify common gene sequences in different tissues and species. A cDNA library contains cDNAs prepared from all the mRNAs of a certain species or organ. *See* gene cloning.

complement fixation The combination of complement with antibody–antigen complexes. It is a property used to test for the presence of a specific antigen or antibody.

complex (**coordination compound**) A type of compound in which molecules or ions form coordinate bonds with a metal atom or ion. The coordinating species (called *ligands*) have lone pairs of electrons, which they can donate to the metal atom or ion. They are molecules such as ammonia or water, or negative ions such as Cl^- or CN^-. The resulting complex may be neutral or it may be a *complex ion.*

The formation of such coordination complexes is typical of transition metals. Often the complexes contain unpaired electrons and are paramagnetic and colored.

compound A chemical combination of atoms of different elements to form a substance in which the ratio of combining atoms remains fixed and is specific to that substance. The constituent atoms cannot be separated by physical means; a chemical reaction is required for the compounds to be formed or to be changed.

condensation A type of chemical reaction in which two molecules join together to form a larger molecule, with the associated production of a small molecule such as water (H_2O).

cone One of the two types of light-sensitive cells in the RETINA of the vertebrate eye. They are concerned with vision in bright light and color vision and have a high visual acuity. There are three classes of cones, containing photopigment (iodopsin) sensitive to red, green, or blue light. Iodopsins each consist of a glycoprotein (opsin) combined with retinal, derived from vitamin A.

conformation A particular shape of molecule that arises through the normal rotation of its atoms or groups about single bonds. Any of the possible conformations that may be produced is called a *conformer*, and there will be an infinite number of these possibilities, differing in the angle between certain atoms or groups on adjacent carbon atoms. Sometimes, the term 'conformer' is applied more strictly to possible conformations that have minimum energies.

conjugated Describing compounds that have alternating double and single bonds in their structure. In such compounds there is delocalization of the electrons in the double bonds.

conjugated protein A protein that on hydrolysis yields not only amino acids but also other organic and inorganic substances. They are simple proteins combined with nonprotein groups (prosthetic groups). *See also* glycoprotein; lipoprotein; phosphoprotein.

connective tissue A type of tissue in which the cells are isolated from each other by a large amount of extracellular matrix. It supports, binds, connects, and holds in position the organs of the body and arises from the mesoderm germ layer of the embryo. Connective tissue usually contains varying amounts of branching yellow elas-

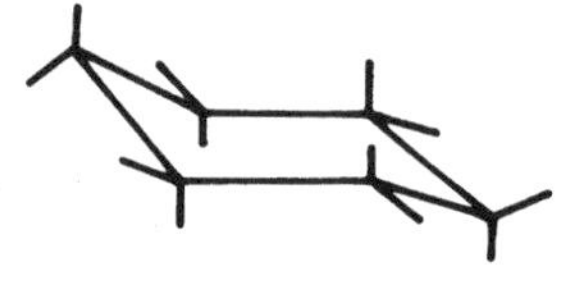

chair conformation

boat conformation

Conformations of cyclohexane

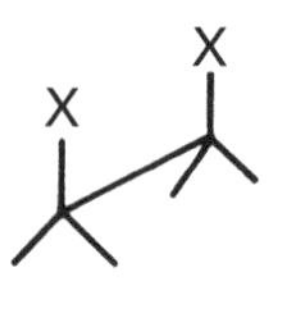

eclipsed

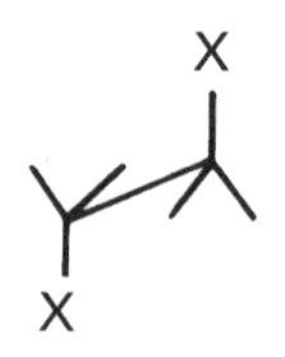

staggered

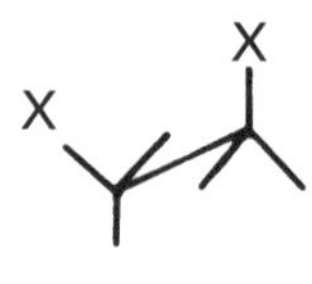

gauche

Conformations for rotation about a single bond

Conformation

tic and tough white collagen fibers. The cells that secrete the matrix are called mast cells; those that produce the fibers are fibroblasts. Macrophages and lymphocytes are also present. *See also* macrophage; mast cell.

conserved (of sequences) A measure of the similarity of corresponding sequences of two or more proteins (and other macromolecules) taken from different sources. Proteins showing a high degree of similarity are said to be highly conserved.

constitutive enzyme One of a group of enzymes that are always present in nearly constant amounts in a given cell. These enzymes are formed at constant rates and in constant amounts regardless of the metabolic state of the organism.

constitutive genes Genes that are essential for cell activity and are expressed in almost all cells. Expression does not require any specific activation factors.

contig mapping *See* chromosome map.

convoluted tubule *See* kidney tubule.

coordinate bond (dative bond) A covalent bond in which the bonding pair is visualized as arising from the donation of a lone pair from one species to another species, which behaves as an electron acceptor. The definition includes such examples as the 'donation' of the lone pair of the ammonia molecule to H^+ (an acceptor) to form NH_4^+ or to Cu^{2+} to form $[Cu(NH_3)_4]^{2+}$.

The donor groups are known as Lewis bases and the acceptors are either hydrogen ions or Lewis acids. Simple combinations, such as $H_3N \rightarrow BF_3$, are known as *adducts*.

coordination compound *See* complex.

copper *See* trace element.

corepressor *See* repressor molecule.

Cori, Carl Ferdinand (1896–1984) and **Gerty Theresa Cori** (née Radnitz, 1896–1957) Czech-born American biochemists distinguished for their studies of glucose and glycogen metabolism, largely carried out in collaboration. They were jointly awarded the Nobel Prize for physiology or medicine (1947) for their discovery of the course of the catalytic conversion of glycogen. The prize was shared with B. A. Houssay.

Cornforth, (Sir) John Warcup (1917–) Australian-born British organic chemist and biochemist. Cornforth elucidated the steps involved in the biosynthesis of cholesterol from acetate. He was awarded the Nobel Prize for chemistry in 1975 for his work on the stereochemistry of enzyme-catalyzed reactions. The prize was shared with V. Prelog.

cornification *See* keratinization.

corn rule *See* optical activity.

corpus luteum A yellow mass of glandular tissue formed temporarily within the Graafian follicle after ovulation. It secretes progesterone.

cortex 1. (*Botany*) A primary tissue in roots and stems of vascular plants derived from the corpus meristem, that extends inwards from the epidermis to the phloem. It usually consists of parenchyma cells but other tissues (e.g. collenchyma) may be present. Some algae, fungi, mosses, and lichens have a well defined region that is termed the cortex although this is different in origin and composition from the cortex of vascular plants.
2. (*Zoology*) The outermost layer of an organ or part. For example, the outer region of the kidney is called the *renal cortex* and the surface layer of gray matter in the cerebral hemispheres of the brain is the *cerebral cortex*. *Compare* medulla.

cortical granules Membrane-bound vesicles in the cortex of many animal eggs, whose contents are extruded at fertilization. The contents turn the vitelline membrane into the fertilization membrane and effect the zona reaction in mammals, preventing further spermatozoa penetrating the egg.

corticosteroid Any steroid hormone produced (from cholesterol) by the adrenal cortex. The release of corticosteroids is controlled by corticotropin. They are classified into two groups: mineralocorticoids and glucocorticoids. Natural and synthetic corticosteroids have widespread use in the treatment of adrenal insufficiency, allergy, and skin and inflammatory diseases. *See* glucocorticoid; mineralocorticoid.

corticosterone A steroid hormone produced by the adrenal cortex and having glucocorticoid activity. *See* glucocorticoid.

corticotropin (**ACTH**; **adrenocorticotropic hormone**) A polypeptide hormone secreted by the anterior pituitary gland. It acts on the adrenal cortex, stimulating the secretion of corticosteroid hormones. Its release is controlled by the hypothalamus and by circulating corticosteroids, whose production it stimulates. Stress also stimulates its secretion. It is used in the diagnosis of disorders of the anterior pituitary gland and adrenal cortex and may be used therapeutically, for example to stimulate corticosteroid production in children.

cortisol *See* hydrocortisone.

cortisone A steroid hormone, produced by the adrenal cortex, that is mainly inactive until converted into hydrocortisone.

cotransport The concerted movement of two chemical species across a plasma membrane. A symport moves the two species in the same direction and an antiport moves them in opposite directions.

coupling A chemical reaction in which two groups or molecules join together. An example is the formation of azo compounds.

covalent bond A bond formed by the sharing of an electron pair between two atoms. The covalent bond is conventionally represented as a line, thus H–Cl indicates that between the hydrogen atom and the chlorine atom there is an electron pair formed by electrons of opposite spin implying that the binding forces are strongly localized between the two atoms. Molecules are combinations of atoms bound together by covalent bonds; covalent bonding energies are of the order 10^3 kJ mol^{-1}.

C_3 plant A plant in which carbon dioxide is fixed by ribulose biphosphate carboxylase (RUBP carboxylase) to form a 3-carbon acid, phosphoglyceric acid. Most temperate and many other plants are C_3 plants. They are characterized by high carbon dioxide compensation points owing to photorespiration, and are not as efficient photosynthetically as C_4 plants. *Compare* C_4 plant.

C_4 plant A plant in which carbon dioxide is initially fixed by phosphoenolpyruvate carboxylase (*PEP carboxylase*) to form a 4-carbon dicarboxylic acid. C_4 plants have evolved from C_3 plants by a modification in carbon dioxide fixation, leading to more efficient photosynthesis. The modified pathway is called the *Hatch–Slack pathway* or the *C_4 dicarboxylic acid pathway*. C_4 plants are mainly tropical or subtropical, including many tropical grasses (e.g. maize and sorghum). PEP carboxylase has a higher affinity for carbon dioxide than *RUBP carboxylase*. The product of carbon dioxide fixation is oxaloacetate, which is rapidly converted to the C_4 acids malate and aspartate. The decarboxylation of C_4 acids releases CO_2, which is then refixed as in C_3 plants.

C_4 plants are more efficient than C_3 plants: they are capable of utilizing much higher light intensities and temperatures; have up to double the maximum rate of photosynthesis; and lose less water by transpiration because, with more efficient CO_2 uptake, smaller stomatal apertures are needed to obtain sufficient CO_2 for photosynthesis. *Compare* C_3 plant.

cranial nerves The paired nerves that originate directly from the brain of vertebrates. Most supply the sense organs and muscles of the head although some, such as the vagus nerve, supply other parts of the body also. In humans and other mammals there are 12 cranial nerves: olfactory (I), optic (II), oculomotor (III), trochlear (IV), trigeminal (V), abducens (VI), facial (VII), acoustic (VIII), glossopharyngeal (IX), vagus (X), accessory (XI), and hypoglossal (XII).

creatine A constituent of vertebrate muscle, averaging 0.3 to 0.4 per cent of the tissue. The greatest concentration is found in voluntary muscle and the least in involuntary muscle. Vertebrate animals, especially carnivores, obtain some creatine in the diet, but all, and particularly herbivores, can synthesize it. In the resting muscle, creatine is combined with phosphoric acid to form *phosphocreatine* (creatine phosphate). When the muscle contracts, phosphocreatine is split back into creatine and phosphoric acid, with the release of energy. This is energy responsible for the actual muscle contraction through the medium of ATP. The energy necessary for synthesis of the creatine phosphate is provided by carbohydrate breakdown (glycolysis).

creatinine A characteristic constituent of the urine of all mammals. It is a waste product produced by catabolism of creatine.

Crick, Francis Harry Compton (1916–) British physicist and molecular biologist who, with J. D. Watson in 1953, put forward the double-helix structure of DNA, formulated the adaptor hypothesis of protein synthesis and the central dogma of molecular biology, and defined the codon. He was awarded the Nobel Prize for physiology or medicine in 1962 jointly with J. D. Watson and M. H. F. Wilkins.

crista The structure formed by folding of the inner mitochondrial membrane. The extent and nature of the folding varies, active cells having complex and closely packed cristae, less active cells having fewer and less complex cristae. The surface of cristae is covered with stalked particles (respiratory granules) that contain the oxidative enzymes (e.g. ATPase and the cytochromes).

crossing over The exchange of material between homologous chromatids by the formation of chiasmata. *See also* chiasma.

cross linkage An atom or short chain joining two longer chains in a polymer.

crystallography The study of the geometric structure and internal arrangement of crystals. It is often used in the identification of macromolecules, as each type of crystal has a characteristic refractive index, i.e. a light ray passing through the crystal will change direction at a constant angle. *See also* x-ray crystallography.

CSF *See* cerebrospinal fluid.

culture A population of microorganisms or tissue cells grown on or within a solid or liquid medium for experimental purposes. This is done by inoculation and incubation of the nutrient medium. *See also* tissue culture.

culture medium A mixture of nutrients used, in liquid form or solidfied with agar, to cultivate microorganisms, such as bacteria or fungi, or to support tissue cultures.

cuticle A protective layer secreted by an epidermis.

In plants, the cuticle is a waterproof layer of waxy cutin covering the epidermis, mainly of aerial parts and some seeds. It may be covered by a layer of wax, e.g. apple, and sometimes resin, e.g. horse chestnut buds. Its thickness varies with the species and environment.

Cuticles are found in a variety of animals, e.g. the thick cuticles of endoparasites such as tapeworms and flukes; the thin collagen cuticle of earthworms; the chitin-containing cuticle of arthropods, which is calcium-impregnated in crustaceans for extra hardness; and the calcium-impregnated shells of mollusks. The arthropod cuticle contributes greatly to the success of the group. Apart from its protective role, it acts as an exoskeleton, serving as an attachment for muscles and being flexible at the joints. The insect cuticle is extremely waterproof and is covered by a thin waxy layer. Arthropods must, however, periodically molt their cuticles to allow growth during which time vulnerability to predators is high. *See* chitin.

cuticularization The formation of cuticle by the secretion of fluid materials, which subsequently harden.

Cyanobacteria A phylum of bacteria containing the blue-green bacteria (formerly called blue-green algae) and the green bacteria (chloroxybacteria). Both groups convert carbon dioxide into organic compounds using photosynthesis, generally using water as a hydrogen donor to yield oxygen, like green plants. However, under certain circumstances they use hydrogen sulfide instead of water, yielding sulfur. Cyanobacteria are an ancient group, and their fossils (stromatolites) have been dated at up to 2500 million years old. Today, most species are found in soil and freshwater. They are spherical (coccoid) or form long microscopic filaments of individual cells. Many species, e.g. *Nostoc* and *Oscillatoria*, are nitrogen fixers. They reproduce asexually by binary fission, or by releasing sporelike propagules or filament fragments. It is thought that certain ancient cyanobacteria became permanent symbionts of ancestral algae and green plants, taking up residence in their cells as photosynthetic organelles (plastids). This theory would account for the striking similarities between cyanobacteria and plastids. *See* stromatolite.

cyanocobalamin (**vitamin B_{12}**) One of the water-soluble B-group of vitamins. It has a complex organic ring structure at the center of which is a single cobalt atom. Foods of animal origin are the only important dietary source. A deficiency in humans leads to the development of pernicious anemia since the vitamin is required for the development of red blood cells. *See also* vitamin B complex.

cyclic AMP (**cAMP; adenosine-3′,5′-monophosphate**) A form of adenosine monophosphate (*see* AMP) formed from ATP in a reaction catalyzed by the enzyme adenylate cyclase. It has many functions, acting as an enzyme activator, genetic regulator, chemical attractant, secondary messenger, and as a mediator in the activity of many hormones, including epinephrine, norepinephrine, vasopressin, ACTH, and the prostaglandins.

cyclic compound A compound containing a ring of atoms. If the atoms forming the ring are all the same the compound is *homocyclic*; if different atoms are involved it is *heterocyclic*.

cyclins A group of proteins involved with the regulation of the cell cycle whose concentrations vary with the phases of the cell cycle. Each cyclin activates a cyclin-dependent kinase (cdk).

cyclization A reaction in which a straight-chain compound is converted into a cyclic compound.

cyclohexane (C_6H_{12}) A colorless liquid alkane that is commonly used as a solvent and in the production of hexanedioic acid (adipic acid) for the manufacture of nylon. It exists as a 'puckered' six-membered ring, having all bonds between carbon atoms at 109.9° (the tetrahedral angle). The molecule undergoes rapid interconversion between two 'chair-like' conformations, which are energetically equivalent, passing through a 'boat-like' structure of higher energy. It is commonly represented by a hexagon.

cysteine *See* amino acids.

cystine A compound formed by the joining of two cysteine amino acids through a –S–S– linkage (a *cystine link*). Bonds of this type are important in forming and maintaining the tertiary structure of proteins.

cytidine (**cytosine nucleoside**) A nucleoside formed when cytosine is linked to D-ribose via a β-glycosidic bond.

cytochrome P-450 Any of a family of CYTOCHROME enzymes, found in many types of organism, that catalyze the oxidation of various lipid-soluble substances, such as steroids, fatty acids, or certain toxins. The iron atom in the heme prosthetic group has an associated sulfur atom. For instance, in animal cells oxidation of toxins, such as hydrocarbons, insecticides, or drugs, by this enzyme starts the process of converting these often highly insoluble compounds into a soluble form that can be excreted by the body.

cytochromes Conjugated proteins containing heme, that act as intermediates in the electron-transport chain. There are four main classes, designated *a*, *b*, *c*, and *d*.

cytogenetics The area of study that links the structure and behavior of chromosomes with inheritance.

cytokeratin *See* keratin.

cytokine Any of a large group of proteins released by mammalian cells that act as highly potent chemical messengers for other cells. They may cause a wide variety of responses in their target cells, for example, triggering differentiation or stimulating secretion. Cytokines released by lymphocytes are called *lymphokines*; these regulate and coordinate the activities of the different types of lymphocytes that participate in immune responses. Other cytokines include the growth factors, transforming growth factors, interferons, and interleukins. *See* colony-stimulating factor; insulin-like growth factor; interferon; interleukin; lymphokine.

cytokinesis The division of the cytoplasm after nuclear division (mitosis or meiosis). In animal cells cytokinesis involves constriction of cytoplasm between daughter nuclei; in plant cells it involves formation of a new plant cell wall.

cytokinin One of a class of plant hormones concerned with the stimulation of cell division, nucleic acid metabolism, and root-shoot interactions. Cytokinins are often purine derivatives: e.g. *kinetin* (6-furfuryl aminopurine), an artificial cytokinin commonly used in experiments; and zeatin, found in maize cobs.

Cytokinins are produced in roots, where they stimulate cell division. They are also transported from roots to shoots in the transpiration stream, where they are essential for healthy leaf growth. Subsequent movement from the leaves to younger leaves, buds, and other parts may occur in the phloem and be important in sequential leaf senescence up the stem. Senescence of detached leaves can be delayed by adding kinins, which mobilize food from other leaf parts and preserve green tissue in their vicinity.

Cytokinins promote bud growth, working antagonistically to auxins in causing bud regeneration in tobacco callus tissue and in releasing lateral buds from apical dominance. They work synergistically with auxins and gibberellins in stimulating cambial activity. The richest sources of cytokinins have been fruit and endosperm tissues notably coconut milk.

cytology The study of cells; cell biology.

cytoplasm The living contents of a cell, excluding the nucleus and large vacuoles, in which many metabolic activities occur. It is contained within the plasma membrane and comprises a colorless substance (*hyaloplasm*) containing organelles and various inclusions (e.g. crystals and insoluble food reserves). The cytoplasm is about 90% water. It is a true solution of ions (e.g. potassium, sodium, and chloride) and small molecules (e.g. sugars, amino acids, and ATP); and a colloidal solution of large molecules (e.g. proteins, lipids, and nucleic acids). It can be gel-like, usually in its outer regions, or sol-like. *See* organelle; protoplasm.

cytoplasmic inheritance The determination of certain characters by genetic material contained in plasmids or organelles other than the nucleus, e.g. mitochondria and chloroplasts. Characters controlled by the DNA of extranuclear organelles are not inherited according to Mendelian laws and are transmitted only through the female line, since only the female gametes have an appreciable amount of cytoplasm. Cytoplasmic inheritance is known in a wide variety of animals, plants, and unicellular organisms, e.g. *Paramecium*.

cytosine A nitrogenous base found in DNA and RNA. Cytosine has the pyrimidine ring structure.

cytoskeleton A network of fibers within the cytoplasm of a cell that maintains its shape, enables movement of the cell, and provides anchorage and movement of its organelles, is actively engaged in cell division, and also plays a role in endocytosis. It comprises various elements, including microtubules, microfilaments, and intermediate filaments. *See* microfilament; microtubule; spindle.

cytosol The soluble fraction of cytoplasm remaining after all particles have been removed by centrifugation.

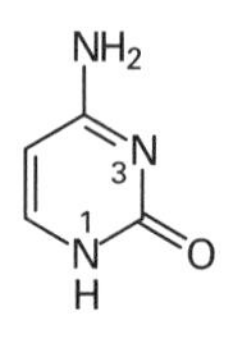

Cytosine

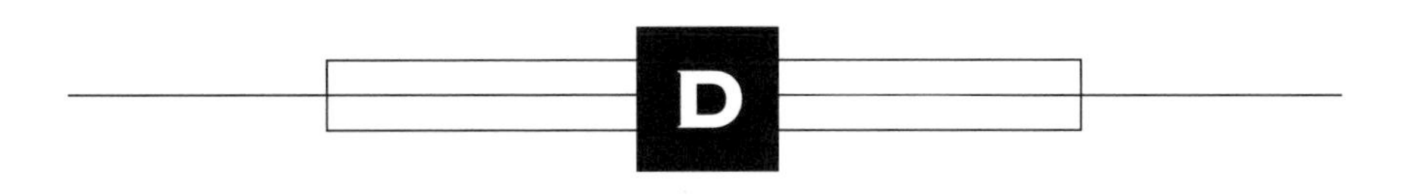

2,4-D (**2,4-dichlorophenoxyacetic acid**) A synthetic auxin used as a potent selective weedkiller. Monocotyledenous species with narrow erect leaves (e.g. cereals and grasses) are generally resistant to 2,4-D while dicotyledenous plants are often very susceptible. The compound is thus used for controlling weeds in cereal crops and lawns. *See* auxin.

Dale, (Sir) Henry Hallett (1875–1968) British physiologist and pharmacologist who discovered acetylcholine, made fundamental physiological studies of the actions of acetylcholine, epinephrine, and histamine. He was awarded the Nobel Prize for physiology or medicine in 1936 jointly with O. Loewi for their discoveries concerning chemical transmission of nerve impulses.

Daltonism *See* color blindness.

Dam, Henrik Carl Peter (1895–1976) Danish biochemist noted for his discovery in green plants of a principle that he named vitamin K (later called vitamin K_1 or phylloquinone). He was awarded the Nobel Prize for physiology or medicine in 1943. The prize was shared with E. A. Doisy.

dark reactions A group of reactions that follow the light reaction in photosynthesis and form glucose and other reduced products from carbon dioxide. They are not dependent on light, although they can and, in growing plants, do take place in the light. *See* photosynthesis.

Darwinism Darwin's explanation of the mechanism of evolutionary change, namely, that in any varied population of organisms only the best adapted to that environment will tend to survive and reproduce. Individuals that are less well adapted will tend to perish without reproducing. Hence the unfavorable characteristics, possessed by the less well-adapted individuals, will tend to disappear from a species, and the favorable characteristics will become more common. Over time the characteristics of a species will therefore change, eventually resulting in the formation of new species. The main weakness of Darwin's theory was that he could not explain how the variation, which natural selection acts upon, is generated, since at the time it was believed that the characteristics of the parents become blended in the offspring. This weakness was overcome with the discovery of Mendel's work and its description of particulate inheritance. *See also* neo-Darwinism.

dative bond *See* coordinate bond.

Dausset, Jean Baptiste Gabriel Joachim (1916–) French physician who first demonstrated the existence in humans of the major histocompatability complex. He was awarded the Nobel Prize for physiology or medicine in 1980 jointly with B. Benacerraf and G. D. Snell.

DDT (**dichlorodiphenyltrichloroethane**) An organochlorine insecticide introduced in the late 1930s and subsequently widely used to control insect carriers of diseases, such as malaria, typhus, and yellow fever. However, its persistence in the environment led to widespread poisoning of certain wild predators, especially hawks and other birds of prey, which accumulated high concentrations of DDT in their body tissues. It has now been banned in many countries.

deaminase An enzyme that catalyzes deamination.

deamination A type of chemical reaction in which an amino group (NH_2) is removed. It occurs in animals when excess amino acids are to be excreted, by the action of deaminating enzymes, known as deaminases (in the liver and kidneys of mammals). Depending on the type of organism, ammonia produced by the reaction may be excreted directly or first converted to urea or to uric acid. *Compare* transamination.

deca- Symbol: da A prefix denoting 10. For example, 1 decameter (dam) = 10 meters (m).

decarboxylase An enzyme that catalyzes the decarboxylation of carboxylic acids, including the conversion of amino acids to amines.

decarboxylation The process in which a carboxyl group, –COOH, is removed from a molecule, with production of carbon dioxide (CO_2). In biochemical processes it is catalyzed by a decarboxylase enzyme.

decomposition A chemical reaction in which a compound is broken down into compounds with simpler molecules.

de Duve, Christian René Marie Joseph (1917–) Belgian biochemist and cell biologist noted for his discoveries of the lysosome and the peroxisome. He was awarded the Nobel Prize for physiology or medicine in 1974 jointly with A. Claude and G. E. Palade.

deficiency disease A disease caused by deficiency of a particular essential nutrient (such as a vitamin or a trace element), usually with a characteristic set of symptoms. Green plants, being autotrophic, are only likely to suffer mineral deficiency diseases, whereas animals, which are heterotrophic, are susceptible to a wider range of such diseases due to dietary deficiencies. *See* chlorosis; micronutrient; vitamins.

degradation A chemical reaction involving the decomposition of a molecule into simpler molecules, usually in stages.

dehydration **1.** Removal of water from a substance.
2. Removal of the elements of water (i.e. hydrogen and oxygen in a 2:1 ratio) to form a new compound. An example is the dehydration of propanol to propene over hot pumice:

$$C_3H_7OH \rightarrow CH_3CH{:}CH_2 + H_2O$$

dehydrogenase An enzyme that catalyzes the removal of certain hydrogen atoms from specific substances in biological systems. Many require the coenzyme NAD or NADP to act as a hydrogen acceptor. They are usually called after the name of their substrate, e.g. lactate dehydrogenase. Some dehydrogenases are highly specific, both with respect to their substrate and coenzyme, whilst others catalyze the oxidation of a wide range of substrates.

Deisenhofer, Johann (1943–) German-born American biochemist who worked on photosynthesis. He was awarded the Nobel Prize for chemistry in 1988 jointly with R. Huber and H. Michel.

Delbrück, Max (1906–81) German-born American physicist and molecular geneticist. He developed the plaque technique for studying viruses and worked on the interaction of bacteriophages with bacteria. Delbrück was awarded the Nobel Prize for physiology or medicine in 1969 jointly with A. D. Hershey and S. E. Luria for their work the replication mechanism and the genetic structure of viruses.

deletion *See* chromosome mutation.

denaturation A process that causes unfolding of the peptide chain of proteins or of the double helix of DNA. These changes may be brought about by a variety of physical factors: change in pH, temperature, violent shaking, and radiation. The primary structure remains intact. Denatured proteins and nucleic acids show changes in physical and biological properties; proteins, for example, are often insoluble in solvents in which they were originally soluble.

dendrite One of several slender branching projections that arise from or near the cell body of a nerve cell (neurone) to make contact with other nerve cells. The entire array of dendrites is sometimes called the *dendritic tree*. *See* neurone.

denitrification The chemical reduction of nitrate by soil bacteria. The process is important in terms of soil fertility since the products of denitrification (e.g. nitrites and ammonia) cannot be used by plants as a nitrogen source. *Compare* nitrification. *See* nitrogen cycle.

dentine A hard substance, closely resembling bone, that makes up the bulk of a tooth. It has a higher mineral content than bone.

deoxycorticosterone (**deoxycortone**) A steroid hormone, produced by the adrenal cortex, having MINERALOCORTICOID activity.

deoxyribonuclease *See* DNase.

deoxyribonucleic acid *See* DNA.

deoxy sugar A sugar in which oxygen has been lost by replacement of a hydroxyl group (OH) with hydrogen (H). The most important example is deoxyribose, the sugar component of DNA.

depolarization A reduction in the potential difference that exists across the membrane of a nerve or muscle cell; i.e. a reduction in the resting potential. It occurs during the passage of an impulse when the membrane becomes more permeable to ions, which previously have accumulated on one side and caused the difference. The ions diffuse across the membrane, tending to equalize their concentration on both sides.

derivative A compound obtained by reaction from another compound. The term is most often used in organic chemistry of compounds that have the same general structure as the parent compound.

deuterated compound A compound in which one or more ^1H atoms have been replaced by deuterium (^2H) atoms.

deuterium Symbol: D, ^2H A naturally occurring stable isotope of hydrogen in which the nucleus contains one proton and one neutron. The atomic mass is thus approximately twice that of ^1H; deuterium is known as 'heavy hydrogen' and deuterium oxide (D_2O) as heavy water. Chemically it behaves almost identically to hydrogen, forming analogous compounds, although reactions of deuterium compounds are often slower than those of the corresponding ^1H compounds. This is made use of in kinetic studies where the rate of a reaction may depend on transfer of a hydrogen atom (i.e. a kinetic isotope effect).

dextrin Any of a class of intermediates produced by the hydrolysis of starch. Further hydrolysis eventually produces the monosaccharide glucose.

dextrorotatory Describing compounds that rotate the plane of polarized light to the right (clockwise as viewed facing the oncoming light). Indicated by the symbol (+). *Compare* levorotatory. *See* optical activity.

dextrose (**grape sugar**) Naturally occurring glucose belongs to the stereochemical series D and is dextrorotatory. Thus the term 'dextrose' has been traditionally used to indicate D-(+)-glucose. As other stereochemical forms of glucose have no significance in biological systems the term 'glucose' is often used interchangeably with dextrose. *See also* glucose.

D-form *See* optical activity.

diabetes (**diabetes mellitus**) A condition caused by deficiency of the hormone insulin and characterized by large quantities of glucose in the blood and urine. The volume of urine also increases. Diabetes that starts early in life is usually more severe (*insulin-dependent diabetes*); such patients require regular injections of insulin. A mild form of diabetes (*noninsulin-dependent diabetes*) is also common in middle-aged to elderly overweight people. In such patients, insulin is not usually required and the condition may be treated by weight reduction, dietary control, and (sometimes) the administration of drugs to lower the blood-glucose level.

diakinesis The last stage of the prophase in the first division of meiosis. Chiasmata are seen during this stage, and by the end of diakinesis the nucleoli and nuclear membrane have disappeared.

dialysis A technique for separating compounds with small molecules from compounds with large molecules by selective diffusion through a semipermeable membrane. For example, a mixed solution of starch (large molecules) and glucose (small molecules) is placed in a bag or piece of tubing made of thin cellophane or other suitable material. If the container is put in water, the glucose molecules diffuse out, leaving the starch behind. Dialysis is performed naturally by the kidneys to extract wastes from the blood, or, in the case of lost or damaged kidneys, artificially by machine. *See also* osmosis.

diene An organic compound containing two carbon–carbon double bonds.

differentially permeable membrane *See* osmosis.

differentiation A process of change during which cells with generalized form become morphologically and functionally specialized to produce the different cell types that make up the various tissues and organs of the organism. Differentiation has been best studied in experimental organisms, such as the fruit fly *Drosophila*. Here, proteins called *morphogens*, encoded by maternal genes of follicle cells, diffuse into the developing early embryo where they lay the foundations of the general body plan. Gradients of concentration of the various morphogens cause genes in different zones of the embryo to be activated to different extents, creating a rudimentary pattern of body segments. This pattern is reinforced and refined by the embryo's own genes – the so-called *segment genes*. Within each segment the differentiation of limbs and other appendages is controlled by a class of master genes, called *homeotic genes*. Mutations of these in *Drosophila* can result in, for example, legs developing on the head instead of antennae. Homeotic genes show remarkable similarities in base sequence across a wide range of species, from plants to humans, and produce a protein that binds to DNA, acting as a switch for various other genes. *See also* totipotency.

diffusion pressure deficit (DPD) *See* osmosis.

digestion The breakdown of complex organic foodstuffs by enzymes into simpler soluble substances, which can be absorbed and assimilated by the tissues. In most animals (e.g. vertebrates and arthropods) it is extracellular, occurring in an alimentary canal or gut into which the enzymes are secreted. In simpler animals (e.g. protoctists, cnidarians, and some other invertebrates) it is intracellular, with solid particles being engulfed and digested by ameboid cells. *See also* endocytosis; phagocyte.

dimer A compound (or molecule) formed by combination or association of two molecules of a monomer.

2,4-dinitrophenylhydrazine (**Brady's reagent**) An orange solid commonly used in solution with methanol and sulfuric acid to produce crystalline derivatives by condensation with aldehydes and ketones. The derivatives, known as 2,4-dinitrophenylhydrazones, can easily be purified by recrystallization and have characteristic melting points, used to identify the original aldehyde or ketone.

dinucleotide A compound of two nucleotides linked by their phosphate groups. Important examples are the coenzymes NAD and FAD.

diol (**dihydric alcohol; glycol**) An alcohol that has two hydroxyl groups (–OH) per molecule of compound.

dioxins A related group of highly toxic chlorinated compounds. Particularly important is the compound 2,3,7,8-tetrachlorodibenzo-*p*-dioxin (TCDD), which is produced as a by-product in the manufacture of 2,4,5-T, and may consequently occur as an impurity in certain types of weedkiller. The defoliant known as agent orange used in Vietnam contained significant amounts of TCDD. Dioxins cause a skin disease (chloracne) and birth defects. Dioxins have been released into the atmosphere as a result of explosions at herbicide manufacturing plants, most notably at Seveso, Italy, in 1976.

diploid A cell or organism containing twice the haploid number of chromosomes (i.e. $2n$). In animals the diploid condition is generally found in all but the reproductive cells and the chromosomes exist as homologous pairs, which separate at meiosis, one of each pair going into each gamete. In plants exhibiting an alternation of generations the sporophyte is diploid, while higher plants are normally always diploid. Exceptions are those species in which polyploidy occurs.

diplotene In meiosis, the stage in late prophase I when the pairs of chromatids begin to separate from the tetrad formed by the association of homologous chromosomes. Chiasmata can often be seen at this stage.

disaccharide A sugar with molecules composed of two monosaccharide units. Sucrose and maltose are examples. These are linked by a –O– linkage (*glycosidic link*). *See also* sugar.

dispersion force A weak type of intermolecular force. *See* van der Waals force.

displacement reaction A chemical reaction in which an atom or group displaces another atom or group from a molecule.

dissociation constant The equilibrium constant of a dissociation reaction. For example, the dissociation constant of a reaction:

$$AB \rightleftharpoons A + B$$

is given by:

$$K = [A][B]/[AB]$$

where the brackets denote concentration (activity).

Often the degree of dissociation is used – the fraction (α) of the original compound that has dissociated at equilibrium. For an original amount of AB of n moles in a volume V, the dissociation constant is given by:

$$K = \alpha^2 n/(1 - \alpha)V$$

Note that this expression is for dissociation into two molecules.

Acid dissociation constants (or *acidity constants*, symbol: K_a) are dissociation constants for the dissociation into ions in solution:

$$HA + H_2O \rightleftharpoons H_3O^+ + A^-$$

The concentration of water can be taken as unity, and the acidity constant is given by:

$$K_a = [H_3O^+][A^-]/[HA]$$

The acidity constant is a measure of the strength of the acid. Base dissociation constants (K_b) are similarly defined. The expression:

$$K = \alpha^2 n/(1 - \alpha)V$$

applied to an acid is known as *Ostwald's dilution law*. In particular if α is small (a weak acid) then $K = \alpha^2 n/V$, or $\alpha = C\sqrt{V}$, where C is a constant. The degree of dissociation is then proportional to the square root of the dilution.

distillation The process of boiling a liquid and condensing the vapor. Distillation is used to purify liquids or to separate components of a liquid mixture. *See also* destructive distillation, fractional distillation, steam distillation, vacuum distillation.

distal Denoting the part of an organ, limb, etc., that is furthest from the origin or point of attachment. *Compare* proximal.

diurnal rhythm *See* circadian rhythm.

***dl*-form** *See* optical activity.

DNA (deoxyribonucleic acid) A nucleic acid, mainly found in the chromosomes, that contains the hereditary information of organisms. The molecule is made up of two antiparallel helical polynucleotide chains coiled around each other to give a *double helix*. It is also known as the *Watson-Crick model* after James Watson and Francis Crick who first proposed this model in 1953. Phosphate molecules alternate with deoxyribose sugar molecules along both chains and each sugar molecule is also joined to one of four nitrogenous bases – adenine (A), guanine (G), cytosine (C), or thymine (T). The two chains are joined to each other by bonding between bases. The two purine bases (adenine and guanine) always bond with the pyrimidine bases (thymine and cytosine), and the pairing is quite specific: adenine with thymine and guanine with cytosine. The two chains are therefore complementary. The sequence of bases along the chain makes up a code – the genetic code – that determines the precise sequence of amino acids in proteins (*see* messenger RNA; protein synthesis; transcription).

DNA is the hereditary material of all organisms with the exception of RNA viruses. Together with histones (and RNA in some instances) it makes up the chromosomes of eukaryotic cells. *See* chromosome; junk DNA; replication; selfish DNA. *See also* RNA.

DNA chip *See* DNA microarray.

DNA fingerprinting *See* genetic fingerprinting.

DNA hybridization *See* nucleic acid hybridization.

DNA methylation An inherited occurrence in animals that reduces gene expression. Cytosine residues in the dinucleotide CpG are methylated to 5-methylcytosine by sequence-specific DNA methylase enzymes. Regions of DNA that contain a high number of methylated CpG sequences are called CpG islands. They have a high mutational frequency because spontaneous deamination of 5-methylcytosine gives thymine, which is not recognized by DNA repair enzymes.

DNA microarray (DNA chip; genome chip) An ordered arrangement of sam-

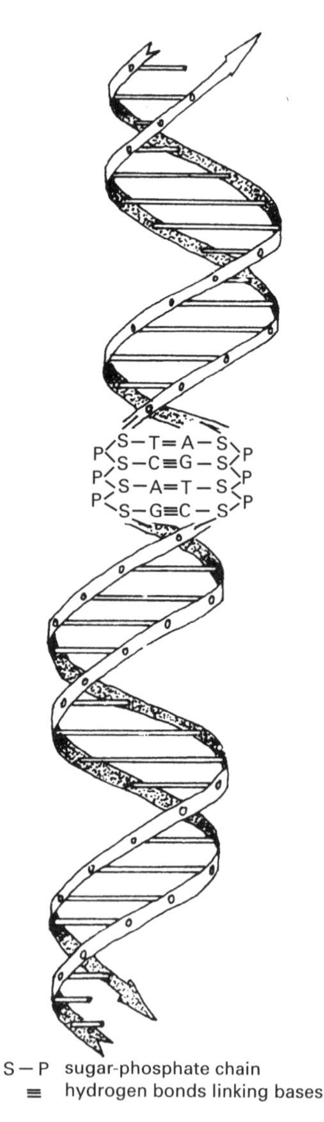

DNA

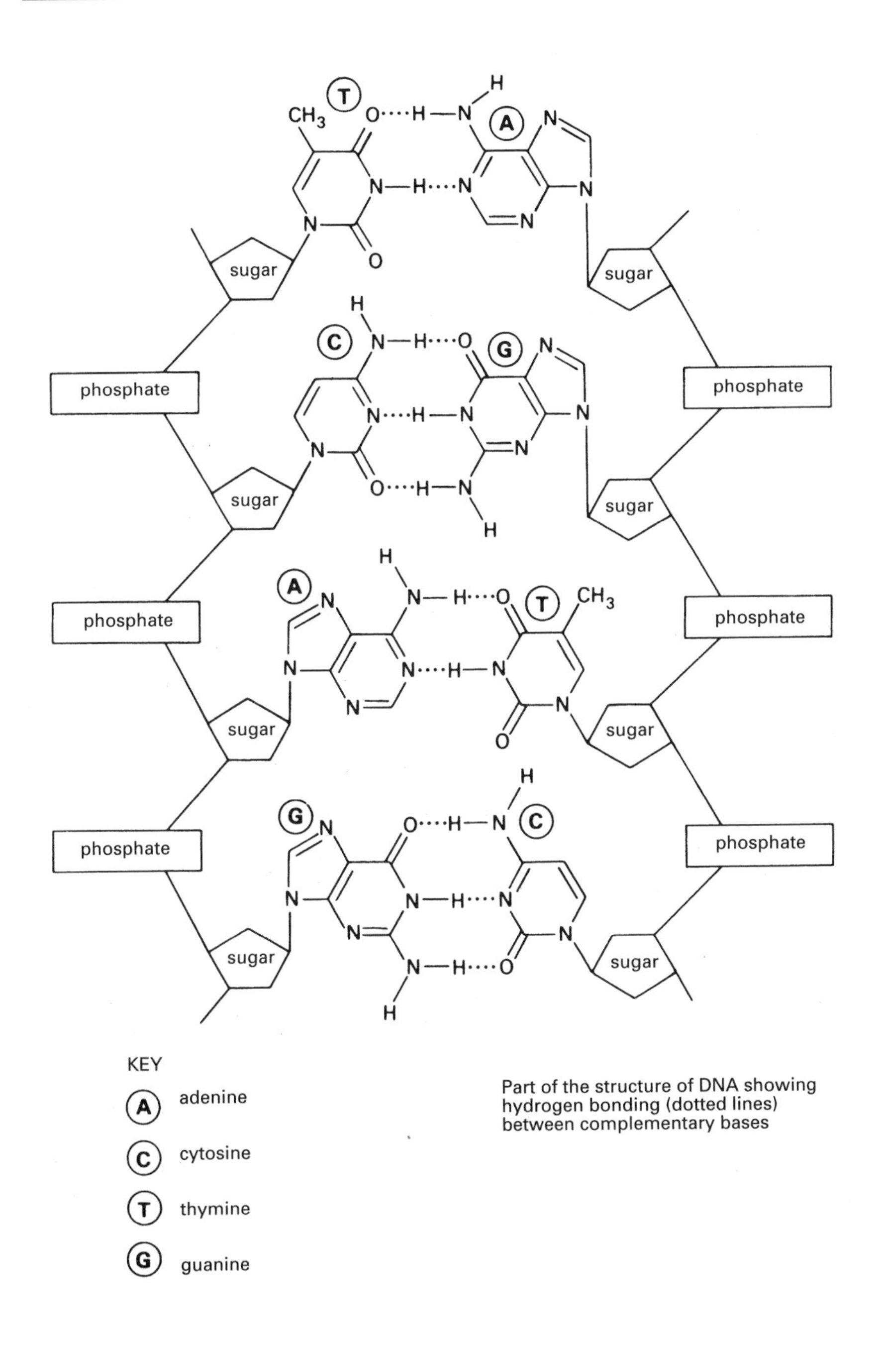

Part of the structure of DNA showing hydrogen bonding (dotted lines) between complementary bases

ples that provides the means for comparing known and unknown DNA samples using hybridization. The process is automated and a single microarray can provide information on thousands of genes simultaneously. DNA microarrays are used for the identification of DNA sequences and for investigating patterns of gene expression. The ability to investigate many genes at the same time is a revolution in molecular biology and microarray technology is used in gene discovery, disease diagnosis, drug discovery, and toxicology.

DNA polymerase *See* polymerase.

DNA probe (gene probe) A nucleic acid consisting of a single strand of nucleotides whose base sequence is complementary to that of a particular DNA fragment being sought, for example a gene on a chromosome or a restriction fragment in a DNA digest. The probe is labeled (e.g. with a radioisotope or a fluorescent compound) so that when it binds to the target sequence, both it and the target can be identified (by autoradiography or fluorescence microscopy).

DNase (deoxyribonuclease) Any enzyme that hydrolyzes the phosphodiester bonds of DNA. DNases are classified into two groups, according to their site of action in the DNA molecule (*see* endonuclease; exonuclease).

dodecanoic acid (lauric acid) A white crystalline carboxylic acid, $CH_3(CH_2)_{10}$-COOH, used as a plasticizer and for making detergents and soaps. Its glycerides occur naturally in coconut and palm oils.

Doherty, Peter C. (1940–) Australian biochemist who was jointly awarded the 1996 Nobel Prize for physiology or medicine with R. M. Zinkernagel for work on the specificity of the cell-mediated immune defense.

Doisy, Edward Adelbert (1893–1986) American biochemist and endocrinologist who isolated the female sex hormones estradiol, estriol, and estrone. Doisy also isolated a principle named vitamin K_2 (now termed menaquinone) and was able to characterize vitamins K_1 and K_2. He was awarded the Nobel Prize for physiology or medicine in 1943. The prize was shared with H. C. P. Dam.

dominant An allele that, in a heterozygote, prevents the expression of another (recessive) allele at the same locus. Organisms with one dominant and one recessive allele thus appear identical to those with two dominant alleles, the difference in their genotypes only becoming apparent on examination of their progenies. The dominant allele usually controls the normal form of the gene, while mutations are generally recessive.

donor The atom, ion, or molecule that provides the pair of electrons in forming a covalent bond.

dopamine A catecholamine precursor of epinephrine and norepinephrine. In mammals it is found in highest concentration in the corpus striatum of the brain, where it functions as an inhibitory neurotransmitter. High levels of dopamine are associated with Parkinson's disease in humans.

dormin (abscisin II) A former name for *abscisic acid.*

double bond A covalent bond between two atoms that includes two pairs of electrons, one pair being the single bond equivalent (the sigma pair) and the other forming an additional bond, the pi bond (π bond). It is conventionally represented by two lines, for example $H_2C=O$.

double helix *See* DNA.

DPD Diffusion pressure deficit. *See* osmosis.

Dulbecco, Renato (1914–) Italian-born American virologist who demonstrated how certain viruses are able to transform certain types of cell into a cancerous state. He was awarded the Nobel

Prize for physiology or medicine in 1975 jointly with D. Baltimore and H. M. Temin.

duodenum The first part of the small intestine into which the food passes when it leaves the stomach. It forms a loop (30 cm long in man) into which the duct from the pancreas and the bile duct from the liver open. The lining is covered with villi, between which are glands secreting intestinal juice (*succus entericus*) containing enzymes. When acid chyme from the stomach enters the duodenum, the lining cells secrete a hormone (cholecystokinin) that stimulates the pancreas to release pancreatic juice containing enzymes. Cholecystokinin also causes contraction of the gall bladder, resulting in the passage of bile into the duodenum. These alkaline secretions neutralize the acid from the stomach and continue the process of digestion.

duplex Double, or having two distinct parts. The term is particularly used to describe the double helix of the Watson–Crick DNA model. *See* DNA.

duplication The occurrence of extra genes or segments of a chromosome in the genome. *See* chromosome mutation.

du Vigneaud, Vincent (1901–78) American biochemist noted for the synthesis of the posterior pituitary hormones oxytocin and vasopressin. He also established the structure of biotin and worked on the synthesis of penicillin. He was awarded the Nobel Prize for chemistry in 1955 for his work on biochemically important sulfur compounds, especially for the first synthesis of a polypeptide hormone.

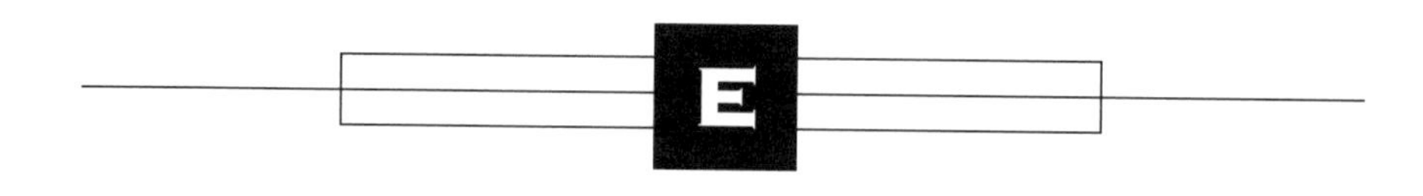

Eccles, (Sir) John Carew (1903–97) Australian neurophysiologist. He was awarded the Nobel Prize for physiology or medicine in 1963 jointly with A. L. Hodgkin and A. F. Huxley for their discoveries concerning the ionic mechanisms involved in excitation and inhibition in the peripheral and central portions of the nerve cell membrane.

ecdysis (molting) **1.** The periodic shedding of the rigid cuticle of arthropods, especially insects and crustaceans, to enable growth to occur. Some useful materials are reabsorbed from the old cuticle, which is then split along lines of weakness, revealing a soft new cuticle underneath. The animal then enlarges its body, by taking in air or water, so the new cuticle hardens a size larger than the old. Ecdysis is controlled by the hormone ecdysterone (β-ecdysone).
2. The periodic shedding of the outer epidermal layer of reptiles (except crocodiles). It is shed in a single piece by snakes, but in small patches by lizards.

A similar process occurs continuously in mammals, including humans, in which very small flakes of epidermis are shed.

ecdysterone (β-ecdysone) A steroid hormone, produced by arthropods (e.g. insects, spiders, and scorpions), that induces molting and metamorphosis. It acts on DNA to initiate the synthesis of new proteins and enzymes involved in the process of molting and cuticle formation. Ecdysterone is formed from its inactive precursor, ecdysone, which is secreted by the prothoracic glands.

eclipsed conformation *See* conformation.

ECM *See* extracellular matrix.

E. coli *See* Escherichia coli.

Edelman, Gerald Maurice (1929–) American biochemist and molecular biologist. He was awarded the Nobel Prize for physiology or medicine in 1972 jointly with R. R. Porter for their discoveries concerning the chemical structure of antibodies.

edta (ethylenediamine tetraacetic acid) A compound with the formula

$$(HOOCCH_2)_2N(CH_2)_2N(CH_2COOH)_2$$

It is used in forming chelates of transition metals. *See* chelate.

effector An organ or cell that responds in a particular way to a nervous impulse. Effectors include muscles and glands.

Ehrlich, Paul (1854–1915) German medical scientist noted for his early work in chemotherapy, especially his use of arsenic compounds. He also put forward the side-chain theory of immunity. He was awarded the Nobel Prize for physiology or medicine in 1908 jointly with I. I. Mechnikov in recognition of their work on immunity.

Eijkman, Christiaan (1858–1930) Dutch physician, bacteriologist, and nutritionist who first observed that unpolished rice grains contained an antiberiberi principle (later recognized as vitamin B_1 or thiamine). He was awarded the Nobel Prize for physiology or medicine in 1929. The prize was shared with F. G. Hopkins.

elastin A structural protein found in mammalian connective tissues, especially in elastic fibers. Glycine is the main component; proline, alanine, and valine are the other main residues.

electrolyte A liquid containing positive and negative ions that conducts electricity by the flow of those charges. Electrolytes can be solutions of acids or metal salts ('ionic compounds'), usually in water. Alternatively they may be molten ionic compounds – again the ions can move freely through the substance. Liquid metals (in which conduction is by free electrons rather than ions) are not classified as electrolytes.

electron micrograph *See* micrograph.

electron microscope *See* microscope.

electron spin resonance (**ESR**) A similar technique to nuclear magnetic resonance, but applied to unpaired electrons in a molecule (rather than to the nuclei). It is a powerful method of studying free radicals and transition-metal complexes.

electron-transport chain (**respiratory chain**) A chain of chemical reactions involving proteins and enzymes, resulting in the formation of ATP and the transfer of hydrogen atoms to oxygen to form water. The enzymes and other proteins are, in eukaryotic cells, located in the inner membrane of the mitochondria and are grouped into discrete complexes. The reduced coenzyme NADH gives up two electrons to the first component in the chain, NADH dehydrogenase, and two hydrogen ions (H^+) are discharged from the matrix of the mitochondrion into the intermembrane space. The electrons are transferred along the chain to a carrier molecule (ubiquinone). Ubiquinone passes them to the next complex, which contains cytochromes b and c_1. Another carrier (cytochrome c) transfers the electrons to the final complex in the chain. There they act with the enzyme cytochrome oxidase to reduce an oxygen atom, which combines with two H^+ ions to form water. During this electron transfer, a further two pairs of H^+ ions are pumped into the intermembrane space by the complexes, making a total of six per molecule of NADH. If $FADH_2$ is the electron donor, only four H^+ ions are pumped across, as it donates electrons directly to ubiquinone.

The function of electron transport in the mitochondrion is to provide the energy required to phosphorylate ADP to ATP. According to the *chemiosmotic theory*, the H^+ ions in the intermembrane space diffuse back to the matrix through the inner mitochondrial membrane down a concentration gradient. As they do so they pass through the protein channel (the F_0 unit) of the enzyme ATP synthase. The energy released allows the catalytic F_1 unit of ATP synthase to synthesize ATP from ADP and inorganic phosphate. Each pair of H^+ ions catalyzes the formation of one molecule of ATP, so for each NADH molecule, three molecules of ATP may be synthesized (two molecules of ATP per molecule of $FADH_2$). A similar mechanism is involved in ATP formation by components of the light reaction in photosynthesis. *See* chemiosmotic theory; photosynthesis; oxidative phosphorylation.

electrophile An electron-deficient ion or molecule that takes part in an organic reaction. The electrophile can be either a positive ion (H^+, NO_2^+) or a molecule that can accept an electron pair (SO_3, O_3). The electrophile attacks negatively charged areas of molecules, which usually arise from the presence in the molecule of an electronegative atom or group or of pi-bonds. *Compare* nucleophile.

electrophoresis The migration of electrically charged particles towards oppositely charged electrodes in solution under an electric field – the positive particles to the cathode and negative particles to the anode. The rate of migration varies with molecular size and shape. The technique can be used to separate or analyze mixtures (e.g. of proteins or nucleic acids). Wetted filter paper, starch gel, polyacrylamide gel, or a similar inert porous medium is used in the technique. It can be carried out in conjunction with paper chromatography. In *immunoelectrophoresis*, components are firstly separated by electrophoresis on a gel. Then a solution of antibodies is added to a trough in the gel. The antibodies diffuse through the gel, and where they encounter their specific antigen they form a precipitate. This allows specific antigenic components to be identified. Polyacrylamide gel electrophoresis (PAGE) is particularly useful for separating polymers.

electrovalent bond (**ionic bond**) A binding force between the ions in compounds in which the ions are formed by complete transfer of electrons from one element to another element or radical. For example, Na + Cl becomes $Na^+ + Cl^-$. The electrovalent bond arises from the excess of the net attractive force between the ions of opposite charge over the net repulsive force between ions of like charge. The magnitude of electrovalent interactions is of the order 10^2–10^3 kJ mol^{-1} and electrovalent compounds are generally solids with rigid lattices of closely packed ions.

elimination reaction A reaction involving the removal of a small molecule, e.g. water or hydrogen chloride, from an organic molecule to give an unsaturated compound. An example is the elimination of a water molecule from an alcohol to produce an alkene.

Elion, Gertrude Belle (1918–99) American biochemist and pharmacologist distinguished for her introduction of a range of widely used synthetic drugs designed as antimetabolites, including the antifolate bactericidal agent co-trimoxazole, the immunosuppressants mercaptopurine and azathioprine, and the antiviral compound acycloguanosine. She was awarded the Nobel Prize for physiology or medicine in 1988 jointly with J. W. Black and G. H. Hitchings.

ELISA (**enzyme-linked immunosorbent assay**) A sensitive and convenient type of immunoassay used for determining the concentration of proteins or other (potentially) antigenic substances in biological samples. Specific antibodies against the substance being tested are adsorbed onto an insoluble carrier surface, such as a PVC sheet. Then a known amount of the sample is added, so that molecules of the test substance are bound by the antibodies. The carrier is rinsed and a second antibody, specific to a second site on the test substance, is added. Molecules of this also carry an enzyme, which causes a color change in a fourth reagent. The intensity of the color change can then be measured photometrically and compared against known standard solutions of the test substance. ELISA is widely used in medical and veterinary diagnostics, and in research. *See* immunoassay.

elution The removal of an adsorbed substance in a chromatography column or ion-exchange column using a solvent (*eluent*), giving a solution called the *eluate*. The chromatography column can selectively adsorb one or more components from the mixture. to ensure efficient recovery of these components graded elution is used. The eluent is changed in a regular manner starting with a nonpolar solvent and gradually replacing it by a more polar one. This will wash the strongly polar components from the column.

Embden–Meyerhoff pathway *See* glycolysis.

embryo **1.** (*Zoology*) The organism formed after cleavage of the fertilized ovum and before hatching or birth. In mammals the embryo in its later well-differentiated stages is called a fetus.
2. (*Botany*) The organism that develops from the zygote of bryophytes, pteridophytes, and seed plants before germination.

embryology The study of the development of organisms, especially animals, usually restricted to the period from fertilization to hatching or birth.

empirical formula The formula of a compound showing the simplest ratio of the atoms present. The empirical formula is the formula obtained by experimental analysis of a compound and it can be related to a molecular formula only if the molecular weight is known. For example P_2O_5 is the empirical formula of phosphorus(V) oxide although its molecular formula is P_4O_{10}.

emulsion A colloid in which a liquid phase (small droplets with a diameter range 10^{-5}–10^{-7} cm) is dispersed or suspended in a liquid medium. Emulsions are classed as lyophobic (solvent-repelling and generally unstable) or lyophilic (solvent-attracting and generally stable).

enamel The hard white outer coating of the teeth of vertebrates. It is a protective layer, epidermal in origin. Enamel is constructed from hexagonal crystals of calcium phosphate, calcium carbonate, and calcium fluoride bound together by keratin fibers.

enantiomer (**enantiomorph**) A compound whose structure is not superimposable on its mirror image: one of any pair of optical isomers. *See also* isomerism; optical activity.

encephalin *See* endorphin.

endocrine gland (**ductless gland**) A gland that has no duct or opening to the exterior. It produces hormones, which pass directly into the bloodstream. The circulatory system then transmits them to other body tissues or organs, where activity is modified. *Compare* exocrine gland.

endocrinology The study of the endocrine glands and their secretions (hormones).

endocytosis The bulk transport of materials into cells across the plasma membrane. It is described as *pinocytosis* (cell drinking) or *phagocytosis* (cell eating) depending on whether the material is fluid, containing molecules in solution, or solid respectively. The process involves extension and invagination of the plasma membrane to form small vesicles in pinocytosis (*pinocytotic vesicles*) or vacuoles in phagocytosis (*food vacuoles*). The contents are often digested by enzymes from LYSOSOMES. Pinocytosis occurs in plant and animal cells. Sometimes it is used simply to transport molecules, e.g. proteins and hormones through the cells lining blood capillaries. Phagocytosis is carried out particularly by protozoan protoctists during feeding, e.g. *Amoeba*, and by certain white blood cells (hence called phagocytes) when engulfing bacteria. *Compare* exocytosis.

endoderm (**entoderm**) The innermost germ layer of most metazoans (including vertebrates) that develops into the gut lining and its derivatives (e.g. liver, pancreas). It also forms the yolk sac and allantois in birds and mammals.

endogenous Produced or originating within an organism. *Compare* exogenous.

endomitosis The duplication of chromosomes without division of the nucleus. Endomitosis may take two forms: the chromatids may separate causing endopolyploidy, e.g. in the macronucleus of ciliates, or the chromatids may remain joined leading to multistranded chromosomes or *polyteny*, e.g. during larval development of dipteran flies. Both processes lead to an increase in nuclear and cytoplasmic volume. *Compare* mitosis.

endonuclease An enzyme that catalyzes the hydrolysis of internal bonds of polynucleotides such as DNA and RNA, producing short segments of linked nucleotides (oligonucleotides). *See also* DNase; restriction endonuclease.

endoplasmic reticulum (**ER**) A system of membranes forming tubular channels and flattened sacs (cisternae), running through the cytoplasm of all eukaryotic cells and continuous with the nuclear envelope. Although often extensive, it was only discovered with the advent of electron microscopy. Its surface is often covered with ribosomes, forming *rough ER*. ER-bound ribosomes make lysosomal proteins, secretory proteins (including transmembrane proteins), and proteins that will reside in the ER cisternae. After synthesis these proteins are sorted and targeted to their final location by the Golgi apparatus. ER lacking ribosomes is called *smooth ER* and is involved with lipid synthesis, including phospholipids and steroids.

In muscle cells a specialized form of ER called sarcoplasmic reticulum is present. *See* Golgi apparatus; sarcoplasmic reticulum. *See* cell.

endoribonuclease *See* ribonuclease.

endorphin (**encephalin; enkephalin**) One of a group of peptides produced in the brain and other tissues that are released after injury and have pain-relieving effects similar to those of opiate alkaloids, such as morphine. They include the *enkephalins*, which consist of just five amino acids. Other larger endorphins occur in the pituitary, while some are polypeptides, found mainly in pancreas, adrenal gland, and other tissues. Pain relief from acupuncture may be due to stimulated production of endorphins.

endosymbiont theory The theory that eukaryotic organisms evolved from symbiotic associations between bacteria. It proposes that integration of photosynthetic bacteria, for example purple bacteria and cyanobacteria, into larger bacterial cells led to their permanent incorporation as forerunners of the mitochondria and plastids (e.g. chloroplasts) seen in modern eukaryotes. There is compelling supporting evidence for the theory, particularly from studies of mitochondrial and plastid DNA and ribosomes, which demonstrate remarkable similarities with those of bacteria.

endothelium The tissue lining the blood vessels and heart. It consists of a single layer of thin flat cells fitting very close together with only a little cement substance between them. In capillaries it is the only layer, providing the barrier between the blood and the fluid bathing the cells. Water and all dissolved substances with small molecules pass through the cells. White blood cells pass between the endothelial cells by an ameboid movement known as *diapedesis*.

endotoxins Toxic substances formed inside the cells of Gram-negative bacteria (e.g. *Salmonella*) and released on disintegration of the cell. They are heat-stable polysaccharide-protein complexes causing nonspecific effects in their hosts, e.g. fevers. *Compare* exotoxins. *See* toxin.

enkephalin *See* endorphin.

enol An organic compound containing the C:CH(OH) group; i.e. one in which a hydroxyl group is attached to the carbon of a double bond. *See* keto–enol tautomerism.

enterokinase *See* enteropeptidase.

enteropeptidase (**enterokinase**) A peptidase enzyme that converts trypsinogen (inactive form) to trypsin (active form).

enthalpy Symbol: H A thermodynamic property of a system defined as $U + pV$, where U is the internal energy, p the pressure, and V the volume. Changes in enthalpy are important for chemical reactions, for which the heat absorbed (or evolved) is equal to the change of internal energy (ΔU) plus the external work done during the change (ΔpV). In many biochemical systems no work is done and the enthalpy change is equal to the change in internal energy.

entropy Symbol: S In any system that undergoes a reversible change, the change of entropy is defined as the heat absorbed (δQ) divided by the thermodynamic temperature:

$$\delta S = \delta Q/T$$

A given system is said to have a certain entropy, although absolute entropies are seldom used: it is change in entropy that is important. The entropy of a system measures the availability of energy to do work.

In any real (irreversible) change in a closed system the entropy always increases. Although the total energy of the system has not changed (first law of thermodynamics) the available energy is less – a consequence of the second law of thermodynamics.

The concept of entropy has been widened to take in the general idea of disorder – the higher the entropy, the more disordered the system. For instance, a chemical reaction involving polymerization may well have a decrease in entropy because there is a change to a more ordered system. The 'thermal' definition of entropy is a special case of this idea of disorder – here the entropy measures how the energy transferred is distributed among the particles of matter.

enzyme A macromolecule that catalyzes biochemical reactions. Enzymes act with a given compound (the substrate) to produce a complex, which then forms the products of the reaction. The enzyme itself is unchanged in the reaction; its presence allows the reaction to take place. The names of most enzymes end in -ase, added to the substrate (e.g. lactase) or the reaction (e.g. hydrogenase).

Enzymes are extremely efficient catalysts for chemical reactions, and very specific to particular reactions. Most enzymes are proteins that They may have a nonprotein part (cofactor), which may be an inorganic ion or an organic constituent (coenzyme). The mechanism of action of most enzymes appears to be by *active sites* on the enzyme molecule. The substrate acting with the enzyme changes shape to fit the active site, and the reaction proceeds. Enzymes are very sensitive to their environment – e.g. temperature, pH, and the presence of other substances. Catalytic activity has also been found in some RNA molecules (RIBOZYMES) and monoclonal antibodies (abzymes). *See* inhibition; Michaelis constant; Michaelis kinetics.

enzyme-linked immunosorbent assay *See* ELISA.

enzyme technology (**enzyme engineering**) A branch of biotechnology that utilizes enzymes for industrial purposes. For example rennet (impure rennin) is manufactured on a large scale to make cheese and junkets. Enzymes are also used to determine the concentration of reactants or products in specific reactions catalyzed by them.

eosin *See* staining.

eosinophil A white blood cell (leukocyte) with a lobed nucleus and cytoplasmic granules that stain with acidic dyes. Eosinophils comprise 1.5% of all LEUKOCYTES, but the number increases in allergic conditions, such as asthma and hay fever, as they have antihistamine properties. They can also destroy bacteria by phagocytosis and by releasing hydrogen peroxide.

epidemiology The study of diseases affecting large numbers within a population. These include both epidemics of infectious diseases and also diseases associated with environmental factors and dietary habits (e.g. lung cancer, some forms of heart disease, etc.).

epidermis 1. (*Botany*) The outer protective layer of cells in plants. In aerial parts of the plant the outer wall of the epidermis is usually covered by a waxy cuticle that prevents desiccation, protects the underlying cells from mechanical damage, and increases protection against fungi, bacteria, etc. The cells are typically platelike and closely packed together except where they are modified for a particular function, as are guard cells. The epidermis arises from the tunica meristem. When it is damaged it is replaced by a secondary layer, the periderm. The specialized epidermal area of the roots from which the root hairs arise is termed the *piliferous layer*.
2. (*Zoology*) The outer layer of cells or outer tissue of an animal that generally protects the tissues beneath and insures that the body is waterproof. In vertebrates, the epidermis consists of several layers of cells and forms the outer layer of skin. As it wears away at the surface it is renewed continuously by growth of new cells in the Malpighian layer, which is immediately beneath. The harder cornified cells of the stratum corneum are the chief protective cells. Products of the epidermis of vertebrates include hair, claws, nails, hooves, horns of cattle and sheep, feathers and beaks of birds, and the scales on the legs of birds and on the shells of tortoises. The epidermis of invertebrates is a single layer of

cells, often secreting a protective cuticle. In arthropods, this cuticle forms the exoskeleton.

epimerism A form of isomerism exhibited by carbohydrates in which the isomers (*epimers*) differ in the positions of –OH groups. The α- and β- forms of glucose are epimers. *See* sugar.

epinephrine (**adrenaline**) A hormone produced by the adrenal glands. The middle part of these glands, the adrenal medulla, secretes the hormone, which is chemically almost identical to the transmitter substance norepinephrine produced at the ends of sympathetic nerves. Epinephrine secretion into the bloodstream in stress causes acceleration of the heart, constriction of arterioles, and dilation of the pupils. In addition, epinephrine produces a marked increase in metabolic rate thus preparing the body for emergency.

episome A genetic element that exists inside a cell, especially a bacterium, and can replicate either as part of the host cell's chromosome or independently. Homology with the bacterial chromosome is required for integration, therefore a plasmid may behave as an episome in one cell but not in another. Examples of episomes are temperate phages. *See* plasmid.

epistasis The action of one gene (the *epistatic gene*) in preventing the expression of another, nonallelic, gene (the *hypostatic gene*). Epistatic and hypostatic genes are analogous to dominant and recessive alleles.

epithelium A tissue consisting of a sheet (or sheets) of cells that covers a surface or lines a cavity. The cells are close together, with very little cement substance between them, and they rest on a basement membrane. Epithelia may be *cubical*, *columnar*, *ciliated*, or *squamous*, depending on the shape of the cells, and in some cases there are several layers, as in the epidermis of the skin.

epitope The region of an antigen molecule that is unique to the antigen and therefore responsible for its specificity in an antigen–antibody reaction. The epitope combines with the complementary region on the antibody molecule.

equatorial plate The equator of the nuclear spindle upon which the centromeres of the chromosomes become aligned during metaphase of mitosis and meiosis.

equilibrium In a reversible chemical reaction:

$$A + B \rightleftharpoons C + D$$

The reactants are forming the products:

$$A + B \rightarrow C + D$$

which also react to give the original reactants:

$$C + D \rightarrow A + B$$

The concentrations of A, B, C, and D change with time until a state is reached at which both reactions are taking place at the same rate. The concentrations (or pressures) of the components are then constant – the system is said to be in a state of chemical equilibrium. Note that the equilibrium is a dynamic one; the reactions still take place but at equal rates. The relative proportions of the components determine the 'position' of the equilibrium, which may be *displaced* by changing the conditions (e.g. temperature or pressure).

equilibrium constant In a chemical equilibrium of the type

$$xA + yB \rightleftharpoons zC + wD$$

The expression:

$$[A]^x[B]^y/[C]^z[D]^w$$

where the square brackets indicate concentrations, is a constant (K_c) when the system is at equilibrium. K_c is the equilibrium constant of the given reaction; its units depend on the stoichiometry of the reaction. For gas reactions, pressures are often used instead of concentration. The equilibrium constant is then K_p, where $K_p = K_c^n$. Here n is the number of moles of product minus the number of moles of reactant; for instance, in

$$3H_2 + N_2 \rightleftharpoons 2NH_3$$

n is $2 - (1 + 3) = -2$.

ER *See* endoplasmic reticulum.

ergosterol A sterol present in plants. It is converted, in animals, to vitamin D_2 by ultraviolet radiation, and is the most important of vitamin D's provitamins.

Ernst, Richard Robert (1933–) Swiss physical chemist who used pulsed radiofrequency radiation in nuclear magnetic resonance spectroscopy and extended the technique to biological macromolecules. He was awarded the Nobel Prize for chemistry in 1991.

erythroblast One of the cells in the red bone marrow from which erythrocytes develop. At first they are colorless, but by the time they are released into the blood the cytoplasm is full of hemoglobin and, in mammals, the nucleus has disappeared. In humans, 200 000 million new erythrocytes are made each day to replace those that are worn out. *See also* erythrocyte.

erythrocyte (**red blood cell**) A type of blood cell that contains hemoglobin and is responsible for the transport of oxygen in the blood. Mammalian red cells are circular biconcave disks without nuclei (the red cells of other vertebrates are oval and nucleated). Human blood contains 5 million red cells per cubic millimeter; each cell lives for about 120 days, after which it is destroyed in the liver and replaced by a new cell from the red bone marrow. The number of red cells increases in regions of oxygen shortage, such as high altitudes. In addition to hemoglobin, erythrocytes also contain an enzyme, carbonic anhydrase, and therefore have an important role in transporting carbon dioxide and maintaining a constant pH. *See also* blood; hemoglobin.

Escherichia coli A bacterium widely used in genetic research, occurring naturally in the intestinal tract of animals and in soil and water. It is Gram-negative and the cells are typically straight round-ended rods, usually occurring singly or in pairs. Some strains are pathogenic, causing diarrhea or more serious gastrointestinal infections. The strains can be distinguished serologically on the basis of their antigens. *E. coli* is killed by pasteurization and many common disinfectants.

ESR *See* electron spin resonance.

essential amino acid *See* amino acids.

essential element An element that is indispensable for the normal growth, development and maintenance of a living organism. Some, the *major elements*, are required in relatively large quantities and may be involved in several different metabolic reactions (*see* carbon; hydrogen; oxygen; nitrogen; sulfur; phosphorus; potassium; magnesium, calcium). Others are required in only small or minute amounts, such as iron, manganese, molybdenum, boron, zinc, copper, cobalt, iodine, and selenium (*see* trace element).

essential fatty acids Polyunsaturated fatty acids (*see* carboxylic acid) required for growth and health that cannot be synthesized by the body and therefore must be included in the diet. Linoleic acid and (9,12,15)-linolenic acid are the only essential fatty acids in humans, being required for cell membrane synthesis and fat metabolism. Arachidonic acid is essential in some animals, such as the cat, but in humans it is synthesized from linoleic acid. Essential fatty acids occur mainly in vegetable-seed oils, e.g. safflower-seed and linseed oils.

essential oil Any pleasant-smelling volatile oil obtained from various plants, widely used in making flavorings and perfumes. Most consist of terpenes and they are obtained by steam distillation or solvent extraction.

ester A compound formed by reaction of a carboxylic acid with an alcohol:

$$RCOOH + HOR_1 \rightarrow RCOOR_1 + H_2O$$

Glycerides are esters of long-chain fatty acids and glycerol.

estradiol The most active ESTROGEN produced by the body. It promotes proliferation of the endometrium of the womb during the first half of the estrous cycle to prepare the womb for ovulation, and is also important in female development. It is metabolized to estrone and then estriol.

estriol A metabolite of estradiol and estrone, produced mainly by the placenta during pregnancy, when it occurs in the urine in high concentrations. It has only weak estrogenic activity. *See* estrogen.

estrogen Any of various female sex hormones (steroids) involved in the development and maintenance of accessory sex organs and secondary sex characteristics (e.g. growth of the breasts). During the menstrual cycle estrogens act on the female sex organs to produce an environment suitable for fertilization of the egg cell and implantation and growth of the embryo. They are used therapeutically to correct estrogen deficiency, for example at the menopause, and to treat some forms of breast cancer. In oral contraceptives, synthetic estrogens and progesterone act to prevent ovulation and the release of gonadotropins, e.g. follicle-stimulating hormone. Excessive blood estrogen levels lead to sickness, as in pregnancy morning sickness. The ovary produces mainly two estrogens: estradiol, and estrone. These hormones are produced in smaller amounts by the adrenal cortex, testis, and placenta. *See also* progesterone.

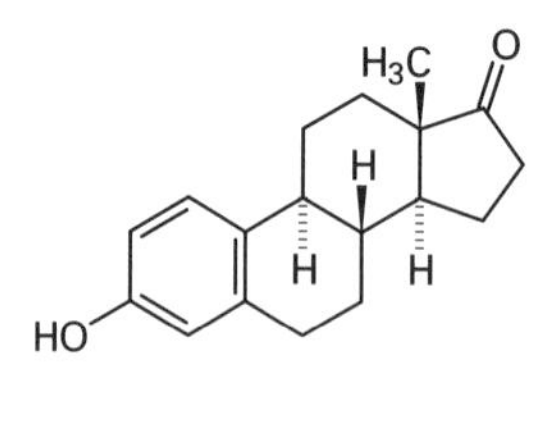

Estrone

estrone An estrogen hormone produced by the ovary and by peripheral tissues, with actions similar to estradiol. *See* estradiol; estrogen.

ethanal (**acetaldehyde**) A water-soluble liquid aldehyde, CH_3CHO, used as a starting material in the manufacture of several other organic compounds.

ethanedioic acid (**oxalic acid**) A white crystalline organic acid, $(COOH)_2$, that occurs naturally in rhubarb, sorrel, and other plants of the genus *Oxalis*. It is slightly soluble in water, highly toxic, and used in dyeing and as a chemical reagent.

ethanoic acid *See* acetic acid.

ethanol (**ethyl alcohol; alcohol**) A colorless volatile liquid alcohol, C_2H_5OH. Ethanol occurs in intoxicating drinks, in which it is produced by fermentation of a sugar:

$$C_6H_{12}O_6 \rightarrow 2C_2H_5OH + 2CO_2$$

Yeast is used to cause the reaction. At about 15% alcohol concentration (by volume) the reaction stops because the yeast is killed. Higher concentrations of alcohol are produced by distillation. Apart from its use in drinks, alcohol is used as a solvent and to form ethanal.

ethene *See* ethylene.

ether A type of organic compound containing the group –O–. Simple ethers have the formula R–O–R′, where R and R′ are alkyl or aryl groups, which may or may not be the same. They are either gases or very volatile liquids and are very flammable. The commonest example is ethoxyethane (diethylether, $C_2H_5OC_2H_5$) used formerly as an anesthetic. Ethers now find application as solvents.

ethylene (**ethene**) A gaseous hydrocarbon (C_2H_4), produced in varying amounts by many plants, that functions as a plant hormone. Its production is usually stimulated by auxins and the amino acid methionine may be a precursor. It is involved in the control of germination, cell growth, fruit ripening, abscission, and senescence, and it inhibits longitudinal growth and promotes radial expansion.

eucaryote *See* eukaryote.

euchromatin *See* chromatin.

Eukarya One of the three cellular kingdoms (Archaea and Bacteria are the other two). Eukaryotic cells are distinguished by the presence of a membrane-bound nucleus that contains true chromosomes.

eukaryote An organism whose cells have their genetic material packaged in chromosomes within a membrane-bound nucleus. Eukaryotic cells possess mitochondria and (in plants and other photosynthetic eukaryotes) chloroplasts in the cell cytoplasm. *Compare* prokaryote.

Euler, Ulf Svante von *See* von Euler

Euler-Chelpin, Hans Karl August von (1873–1964) German-born Swedish chemist and biochemist, father of U. S. Euler, distinguished for his work on the enzymic mechanism of fermentation. He isolated and studied cozymase (now known as NAD). He was awarded the Nobel Prize for chemistry in 1929, the prize being shared with A. Harden.

euploidy The normal state in which an organism's chromosome number is an exact multiple of the haploid number characteristic of the species. For example, if the haploid number is 7, the euploid number would be 7, 14, 21, 28, etc., and there would be equal numbers of each different chromosome. Aneuploidy is an abnormal state where the chromosome number is not an exact multiple of the haploid number and results from the nondisjunction of homologous chromosomes of meiosis.

euryhaline Describing organisms that are able to tolerate wide variations of salt concentrations (and hence osmotic pressure) in the environment, for example the eel can live in both fresh and salt water. *Compare* stenohaline.

eutrophic Describing lakes or ponds that are rich in nutrients and consequently are able to support a dense population of plankton and littoral vegetation. *Eutrophication* is the process that results when an excess of nutrients enters a lake, for example as sewage or from water draining off land treated with fertilizers. The nutrients stimulate the growth of the algal population giving a great concentration or 'bloom' of such plants. When these die they are decomposed by bacteria, which use up the oxygen dissolved in the water, so that aquatic animals such as fish are deprived of oxygen and die from suffocation. *Compare* oligotrophic.

evolution The gradual process of change that occurs in populations of organisms over a long period of time. It manifests itself as new characteristics in a species, and the formation of new species. *See* Darwinism; natural selection.

excitation–contraction coupling A process by which excitation of the muscle fiber membrane at a neuromuscular junction results in contraction of a muscle. The resulting depolarization spreads along infoldings of the membrane, the transverse tubule (or T) system, which activates the sarcoplasmic reticulum to release calcium ions into the sarcoplasm. The calcium ions act by removing the effect of an inhibitory protein so that the filaments which make up a myofibril become linked by crossbridges. Each crossbridge changes its configuration in rapid succession to result in the filaments sliding over one another (contraction). The energy for this process is derived from the breakdown of ATP. On cessation of stimulation (when the T system is no longer depolarized) calcium ions are resorbed into the sarcoplasmic reticulum and relaxation occurs.

excretion The process by which excess, waste, or harmful materials, resulting from the chemical reactions that occur within the cells of living organisms, are eliminated from the body. The main excretory products in animals are water, carbon dioxide, salts, and nitrogenous compounds: in unicellular or simple multicellular animals these substances are excreted by diffusion through the cell or body surface, but in more complex animals excretion occurs largely from special organs. In humans and other vertebrates the main excretory organs are the kidneys: they eliminate excess water, salts, and nitrogenous compounds as urine. In addition, the lungs excrete carbon dioxide and water from respiration; the liver excretes bile pigments derived from the breakdown of hemoglobin, and small amounts of water, sodium chloride, and urea are lost from the skin in sweat.

exocrine gland A gland that produces a secretion that passes along a duct to an epithelial surface. The ducts may pass to the body surface (e.g. sweat, lacrimal, and mammary glands), or they may be internal (e.g. in the mouth, stomach, and intestines). *Compare* endocrine gland.

exocytosis The bulk transport of materials out of the cell across the plasma membrane. It involves fusion of vesicles or vacuoles with the plasma membrane in a reversal of endocytosis. The materials thus lost may be secretory, excretory (e.g. from autophagic vacuoles), or may be the undigested remains of materials in food vacuoles. Typical secretions are enzymes and hormones from gland cells, often brought to the plasma membrane by Golgi vesicles. *See also* lysosome.

exogenous Produced or originating outside an organism. *Compare* endogenous.

exon A segment of DNA that is both transcribed and translated and hence carries part of the code for the gene product. Most eukaryotic genes consist of exons interrupted by noncoding sequences (*see* intron). Both exons and introns are transcribed to heterogeneous nuclear RNA (hnRNA), an intermediary form of messenger RNA (mRNA); the introns are then removed leaving mRNA, which has only the essential sequences and is translated into the protein. DNA recombination can create novel combinations of exons – a process known as exon *shuffling*. *See* intron.

exonuclease An enzyme that catalyzes the hydrolysis of the terminal linkages of polynucleotides such as DNA and RNA, thereby removing terminal nucleotides. *See also* DNase.

exoribonuclease *See* ribonuclease.

exoskeleton The hard outer covering of the body of certain animals, such as the thick cuticle of arthropods (e.g. insects and crustaceans). It forms a rigid skeleton, which protects and supports the body and its internal organs and provides attachment for muscles. Growth of the body may only occur in stages, by a series of molts (ecdyses) of the cuticle. The term is also applied to other hard external protective structures, including a mollusk shell and the shell of a tortoise.

exotoxins A group of heat-labile proteinaceous toxic substances that are produced by bacteria and secreted into the tissues of their host. Exotoxins may act by interfering with a vital biochemical pathway or the molecular structure of the host cell. Others are neurotoxins. Diseases caused by exotoxins include tetanus, diphtheria, and botulism. *Compare* endotoxins. *See* toxin.

explantation The culture of isolated tissues of adults or embryos in an artificial environment, usually *in vitro*, for maintenance, growth, and/or differentiation. *Compare* implantation; transplantation.

exponential growth A type of growth in which the rate of increase in numbers at a given time is proportional to the number of individuals present. Thus, when the population is small multiplication is slow, but as the population gets larger, the rate of multiplication also increases. An exponential growth curve starts off slowly and increases faster and faster as time goes by. However, at some point factors such as lack of nutrients, accumulated wastes, etc., limit further increase, when the curve of number against time begins to level off. The total curve is thus sigmoid (S-shaped).

expression vector *See* gene cloning.

extracellular Occurring or situated outside a cell.

extracellular matrix (ECM) A complex network of secreted polysaccharides and proteins that forms a sheet beneath cells such as endothelial and epithelial cells.

extrachromosomal DNA In eukaryotes, DNA found outside the nucleus of the cell and replicating independently of the chromosomal DNA. It is contained within self-perpetuating organelles in the cytoplasm, for example in mitochondria, chloroplasts, and plastids. Extrachromosomal DNA is responsible for cytoplasmic inheritance

facilitated diffusion A passive transport of molecules across a cell membrane along a concentration gradient, mediated by carrier molecules or complexes. No energy is expended in this process, but it enables the passage through the membrane of molecules that otherwise could not pass through.

facilitation The phenomenon in which passage of an impulse across a synapse renders the synapse more sensitive to successive impulses so increasing the postsynaptic response. Eventually one stimulus will evoke a response large enough to trigger an impulse. *Compare* summation.

FAD (flavin adenine dinucleotide) A derivative of riboflavin that is a coenzyme in electron-transfer reactions. Its reduced form is written as $FADH_2$. *See also* flavoprotein.

fascia A sheet of connective tissue. For example, the layer of adipose tissue under the human dermis and the sheets of tough connective tissue around muscles are types of fasciae.

fast green *See* staining.

fat Triglycerides of long-chain carboxylic acids (fatty acids) that are solid below 20°C. They commonly serve as energy storage material in higher animals and some plants. *See also* adipose tissue; lipid.

fat body A mass of fatty (adipose) tissue forming a definite structure within the body cavity of some animals. In amphibians and reptiles, a pair of solid fat bodies are attached to the kidneys or near the rectum and act as a food store for use during hibernation and breeding. In insects the fat forms a more diffuse tissue around the gut and reproductive organs and stores protein and glycogen as well as fat.

fatty acid *See* carboxylic acid.

feces Solid or semisolid material, consisting of undigested food, bacteria, mucus, bile, and other secretions, that is expelled from the alimentary canal through the anus.

feedback inhibition The inhibition of the activity of an enzyme (often the first) in a reaction sequence by the product of that sequence. When the product accumulates beyond an optimal amount it binds to a site (allosteric site) on the enzyme, changing the shape so that it can no longer react with its substrate. However, once the product is utilized and its concentration drops again, the enzyme is no longer inhibited and further formation of product results. The mechanism is used to regulate the concentration of certain substances (e.g. amino acids) within a cell.

Fehling's solution A freshly mixed solution used for testing for the presence of reducing sugars (e.g. glucose) and aldehydes in solution. When boiled with equal amounts of Fehling's A (copper(II) sulfate solution) and Fehling's B (sodium potassium tartrate and sodium hydroxide solution) reducing sugars and aldehydes produce a brick red precipitate of copper(I) oxide.

femto- Symbol: f A prefix denoting 10^{-15}. For example, 1 femtometer (fm) = 10^{-15} meter (m).

fermentation The breakdown of organic substances, particularly carbohydrates, under anaerobic conditions. It is a form of anaerobic respiration and is seen in certain bacteria and in yeasts. The incompletely oxidized products of alcoholic fermentation – ethanol and carbon dioxide – are important in the brewing and baking industries. *See also* glycolysis; lactic acid bacteria.

ferredoxins A group of red-brown proteins found in green plants, many bacteria and certain animal tissues. They contain nonheme iron in association with sulfur at the active site. They are strong reducing agents (very negative redox potentials) and function as electron carriers, for example in photosynthesis and nitrogen fixation. They have also been isolated from mitochondria.

fertilization **(syngamy)** The fusion of a male gamete with a female gamete to form a zygote; the essential process of sexual reproduction. In animals, a fertilization membrane forms around the egg after the penetration of the sperm, preventing the entry of additional sperm. *External fertilization* occurs when gametes are expelled from the parental bodies before fusion; it is typical of aquatic animals and lower plants. *Internal fertilization* takes place within the body of the female and complex mechanisms exist to place the male gametes into position. Internal fertilization is usually an adaptation to life in a terrestrial environment, although it is retained in secondarily aquatic organisms, such as pondweeds or sea turtles.

Internal fertilization is necessary for terrestrial animals because the male gametes are typically very small and require external water for swimming towards the female gametes. In addition, the propagules produced on land require waterproof integuments, which would be impenetrable to male gametes, so they must be fertilized before being discharged from the female's body. Internal fertilization also allows a considerable degree of nutrition and protection of the early embryo, which is seen in both mammals and seed plants. As plants are relatively immotile, they are dependent on other agents such as wind or insects to carry the male gamete to the female plant.

fertilizer A substance added to soil to increase its fertility. Artificial fertilizers generally consist of a mixture of chemicals designed to supply nitrogen (N), as well as some phosphorus (P) and potassium (K), known as PKN fertilizers. Typical chemicals include ammonium nitrate, ammonium phosphate, ammonium sulfate, and potassium carbonate.

Feulgen's stain *See* staining.

fiber 1. (*Botany*) A form of sclerenchyma cell that is often found associated with vascular tissue. Fibres are long narrow cells, with thickened walls and finely tapered ends. Their function is more as supporting tissue than conducting tissue. Where they occur interspersed with the xylem they may be distinguished from tracheids by their narrower lumen.
2. (*Zoology*) A narrow thread of material, usually flexible and having high tensile strength. Examples include the fibers in such tissues as skin, cartilage, and tendons, which are strengthened by the protein collagen; the silk of the web of a spider; the fibroin fibers of the horny sponges; and the fibrin fibers formed from fibrinogen at the site of a wound. The elongated cells of muscles and the axons of neurones are also called fibers.
3. (*Nutrition*) The indigestible fraction of the diet, consisting of various plant cell-wall materials, that passes through the body largely unchanged. Adequate dietary fiber (over 30 g per day in humans) is considered important in the prevention of certain disorders of the digestive system common in Western societies, e.g. diverticulosis and bowel cancer. Foods high in fiber include cereals, fruit, and vegetables.

fibrin An insoluble protein material that aids blood clotting. It is not present as such in any quantity in blood but is formed from fibrinogen, which is normally present in blood plasma. The conversion from soluble fibrinogen to insoluble fibrin is brought about by the enzyme thrombin. If fresh blood is rapidly whipped a stringy mass of fibrin is obtained.

fibrinogen (**Factor I**) A protein present in blood; the precursor of fibrin, the structural element of blood clots.

fibrinolysis The destruction of blood clots as a result of dissolution of fibrin by the enzyme plasmin (fibrinolysin).

fibroblast A cell that produces fibers in connective tissue. Usually they are long flat cells found alongside the fibers. *See also* connective tissue.

Fischer, Edmond Henri (1920–) Swiss-born biochemist. He was awarded the Nobel Prize for physiology or medicine in 1992 jointly with E. G. Krebs for their discoveries concerning reversible protein phosphorylation as a biological regulatory mechanism.

Fischer, Emil Hermann (1852–1919) German organic chemist, father of H. O. L. Fischer, noted for his work on a wide range of natural products (including sugars, purines, amino acids, and polypeptides). Fischer determined the configurations of all the aldohexoses and aldopentoses. He also recognized the stereochemical specificity of enzyme action (lock-and-key model). One of the greatest chemists of the nineteenth century, Fischer was awarded the Nobel Prize for chemistry in 1902.

Fischer, Hans (1881–1945) German organic chemist noted who worked on tetrapyrroles. He was awarded the Nobel Prize for chemistry in 1930 for his research on hemin and chlorophyll, especially his synthesis of hemin.

Fischer, Hermann Otto Laurenz (1888–1960) German organic chemist and biochemist, son of Emil H. Fischer, noted for his synthesis of glyceraldehyde 3-phosphate and of glycerone phosphate.

FISH Fluorescence in situ hybridization. *See* fluorescence.

fission A type of asexual reproduction in which a parent cell divides into two (binary fission) or more (multiple fission) similar daughter cells. Binary fission occurs in many unicellular organisms (protoctists, bacteria); multiple fission occurs in apicomplexans, such as the malaria parasite *Plasmodium*. Fission begins with division of the nucleus by mitosis, followed by cytoplasmic division and sometimes sporulation.

flagellum A whiplike extension of prokaryote cells with a basal body at its base, whose beat causes locomotion of the cell.

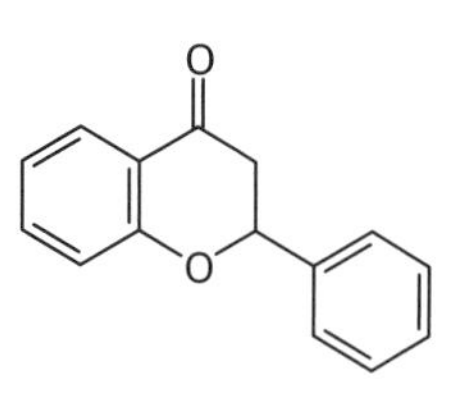

Flavanone

flavanone A type of flavonoid. Flavanone glycosides are found in flowering plants.

flavin A derivative of riboflavin occurring in the flavoproteins; i.e. FAD or FMN.

flavin adenine dinucleotide *See* FAD.

flavin mononucleotide *See* FMN.

flavone *See* flavonoid.

flavonoid One of a common group of plant compounds having the C_6–C_3–C_6 chemical skeleton in which C_6 is a benzene ring. They are an important source of nonphotosynthetic pigments in plants. They are classified according to the C_3 portion and include the yellow chalcones and aurones; the pale yellow and ivory flavones and flavonols and their glycosides; the red, blue, and purple anthocyanins and antho-

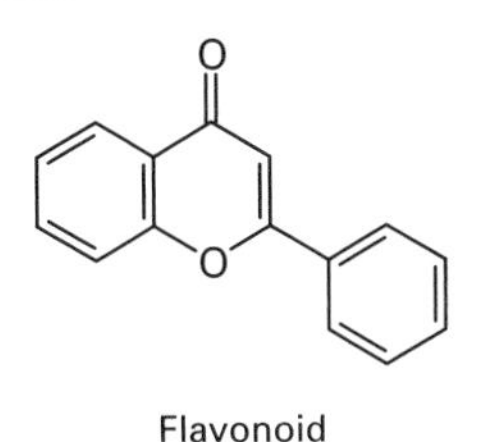

Flavonoid

cyanidins; and the colorless isoflavones, catechins, and leukoanthocyanidins. They are water soluble and usually located in the cell vacuole. *See* anthocyanin.

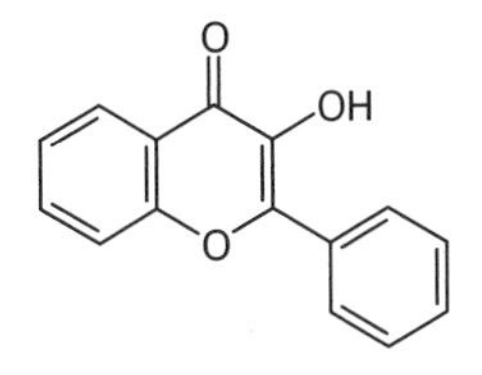

Flavonol

flavonol A plant pigment that modifies the effects of certain growth substances. *See* flavonoid.

flavoprotein A conjugated protein in which a flavin (FAD or FMN) is joined to a protein component. Flavoproteins are enzymes in the electron-transport chain.

Fleming, (Sir) Alexander (1881–1955) British bacteriologist famous for his discovery of the bacteriolytic agents lysozyme and penicillin. He was awarded the Nobel Prize for physiology or medicine in 1945 jointly with E. B. Chain and H. W. Florey.

flocculation (**coagulation**) The combining of the particles of a finely divided precipitate, such as a colloid, into larger particles or clumps that sink and are easier to filter off.

flocculent Describing a precipitate that has aggregated in wooly masses.

Florey, (Sir) Howard Walter, Lord Florey (1898–1968) Australian physiologist noted for his role in the development of penicillin (with E. B. CHAIN) and cephalosporin C. He was awarded the Nobel Prize for physiology or medicine in 1945 jointly with E. B. Chain and A. Fleming for the discovery of penicillin.

fluorescein A fluorescent dye used as an absorption indicator.

fluorescence The emission of electromagnetic radiation (usually visible light) following the absorption of energy. It is a type of luminescence. Emission of energy is immediate and the excited state only short-lived. Fluorescent probes are useful for studying macromolecules. For example, chromosomes are studied by *fluorescence in situ hybridization* (*FISH*), which uses fluorescent gene probes. FISH can also be used to monitor when and where a particular gene is expressed. *See* gene probe; nucleic acid hybridization.

fluoridation The introduction of small quantities of fluoride compounds into the water supply as a public-health measure to reduce the incidence of tooth decay in children.

fluorine *See* trace element.

FMN (**flavin mononucleotide**) A derivative of riboflavin that is a coenzyme in electron-transfer reactions. *See also* flavoprotein.

folic acid (**pteroylglutamic acid**) One of the water-soluble B-group of vitamins. The principal dietary sources of folic acid are leafy vegetables, liver, and kidney. Deficiency of the vitamin exhibits itself in anemia in a similar manner to vitamin B_{12} deficiency, while deficiency during pregnancy increases the risk of birth defects in children.

Folic acid is important in metabolism in various coenzyme forms, all of which are specifically concerned with the transfer and utilization of the single carbon (C_1) group. Before functioning in this manner folic acid must be reduced to either dihydrofolic acid (FH_2) or tetrahydrofolic acid

(FH_4). It is important in the growth and reproduction of cells, participating in the synthesis of purines and thymine. *See also* vitamin B complex.

follicle 1. (*Botany*) A dry dehiscent fruit formed from one carpel that splits along one edge to release its seed, for example columbine fruit.
2. (*Zoology*) A small cavity or sac within an organ or tissue. Follicles within the ovary, for example, contain developing ova.

follicle-stimulating hormone (FSH) A gonadotropin, also called *follitropin*, produced by the anterior pituitary gland. It acts on the ovary to stimulate the growth and maturation of the tissues forming follicles and ova, which, under the action of luteinizing hormone, mature and are released from the ovary. It also stimulates spermatogenesis in males. It has been used in the treatment of female sterility.

food chain The chain of organisms existing in any natural community, through which food energy is transferred. Each link in the food chain obtains energy by eating the one preceding it and is in turn eaten by the organisms in the link following it. At each transfer a large proportion (80–90%) of the potential energy is lost as heat, therefore the number of links in a sequence is limited, usually to 4 or 5. The shorter the food chain, the greater the available energy. The food chains in a community are interconnected with one another, because most organisms consume more than one type of food, and the interlocking pattern is referred to as a *food web* or *food cycle*. *See* trophic level.

food web *See* food chain.

formaldehyde *See* methanal.

formalin A mixture of about 40% formaldehyde, 8% methyl alcohol, and 52% water (the methyl alcohol is present to prevent polymerization of the formaldehyde). It is a powerful reducing agent and is used as a disinfectant, germicide, and fungicide and also as a general preserving solution. In contact with the skin formalin may cause irritant dermatitis and ingestion can cause severe abdominal pain.

formate *See* methanoate.

formic acid *See* methanoic acid.

formula A representation of a chemical compound using symbols for the atoms and subscript numbers to show the numbers of atoms present. *See* empirical formula; general formula; molecular formula; structural formula.

formyl group The group HCO–.

fossil fuel A mineral fuel that forms underground from the remains of living organisms. Fossil fuels include coal, natural gas, peat, and petroleum.

fraction A mixture of liquids with similar boiling points collected by fractional distillation.

fractional crystallization Crystallization of one component from a mixture in solution. When two or more substances are present in a liquid (or in solution), on cooling to a lower temperature one substance will preferentially form crystals, leaving the other substance in the liquid (or dissolved) state. Fractional crystallization can thus be used to purify or separate substances if the correct conditions are known.

fractional distillation (fractionation) A distillation carried out with partial reflux, using a long vertical column (fractionating column). It utilizes the fact that the vapor phase above a liquid mixture is generally richer in the more volatile component. If the region in which refluxing occurs is sufficiently long, fractionation permits the complete separation of two or more volatile liquids.

Unlike normal reflux, the fractionating column may be insulated to reduce heat loss, and special designs are used to maximize the liquid-vapor interface.

fractionation *See* fractional distillation.

Franklin, Rosalind Elsie (1920–58) British biophysicist who worked on the structure of DNA using x-ray crystallography. Her results contributed to the ideas of J. D. Watson and F. H. C. Crick concerning their double-helix structure for DNA.

free energy A measure of the ability of a system to do useful work. *See* Gibbs function.

free radical A molecule with an unpaired electron. The need to form an electron pair makes them very reactive and therefore highly toxic. They are powerful mutagens. Molecular oxygen can accept an electron to form a superoxide radical and all eukaryotic cells have an enzyme (superoxide dismutase) whose function is to convert superoxide radicals into hydrogen peroxide and water.

freeze fracturing A method of preparation of material for electron microscopy, particularly useful for studying membranes. Material is frozen rapidly (e.g. by immersion in liquid nitrogen) thus preserving it in lifelike form. It is then fractured, usually with a sharp knife. The fracture plane tends to follow lines of weakness, such as between the two lipid layers of membranes, revealing their internal surfaces. Replicas made of the surfaces are shadowed for examination in the electron microscope. In *freeze etching* the fractured surface is etched, i.e. some ice is allowed to sublime away, before shadowing. This exposes further structure, such as the outer surface of the membrane. *See* shadowing.

fructan A polysaccharide made entirely of fructose residues. They are used as food stores in many plants.

fructose A sugar ($C_6H_{12}O_6$) found in fruit juices, honey, and cane sugar. It is a ketohexose, existing in a pyranose form when free. In combination (e.g. in sucrose) it exists in the furanose form.

FSH *See* follicle-stimulating hormone.

fucoxanthin A xanthophyll pigment of diatoms, brown algae, and golden brown algae. The light absorbed is used with high efficiency in photosynthesis, the energy first being transferred to chlorophyll *a*. It has three absorption peaks covering the blue and green parts of the spectrum.

fumaric acid An unsaturated dicarboxylic acid, which occurs in many plants. The fumarate ion participates in several important metabolic pathways, e.g. the Krebs cycle, purine pathways, and the urea cycle. Fumaric acid is isomeric with maleic acid (both are butenedioic acids). *See also* butenedioic acid.

functional group A group of atoms in a compound that is responsible for the characteristic reactions of the type of compound. Examples are:

alcohol	–OH
aldehyde	–CHO
amine	$–NH_2$
ketone	=CO.
carboxylic acid	–CO.OH
acyl halide	–CO.X (X = halogen)
nitro compound	$–NO_2$
sulfonic acid	$–SO_2.OH$
nitrile	–CN
diazonium salt	$–N_2^+$
diazo compound	–N=N–

Fungi A kingdom of nonphotosynthetic mainly terrestrial organisms that are now regarded as quite distinct from plants or other living kingdoms. They are characterized by having cell walls made chiefly of chitin, not the cellulose of plant cell walls, and they all develop directly from spores without an embryo stage. Moreover, undulipodia (cilia or flagella) are never found in any stage of their life cycles. Fungi are generally saprophytic or parasitic, and may be unicellular or composed of filaments (termed hyphae) that together comprise the fungal body or mycelium. Hyphae may grow loosely or form a compacted mass of pseudoparenchyma giving well-defined structures, as in toadstools.

furan (**furfuran**) A heterocyclic liquid organic compound. Its five-membered ring contains four carbon atoms and one oxygen atom. The ring structure of furan is characteristic of some monosaccharide sugars (furanoses).

furanose A sugar that has a five-membered ring (four carbon atoms and one oxygen atom). *See also* sugar.

Furchgott, Robert F. (1916–) American biochemist who was awarded the 1998 Nobel Prize for physiology or medicine jointly with L. J. Ignarro and F. Murad for work on nitric oxide acting as a signaling molecule in the cardiovascular system.

furfuran *See* furan.

fused ring *See* ring.

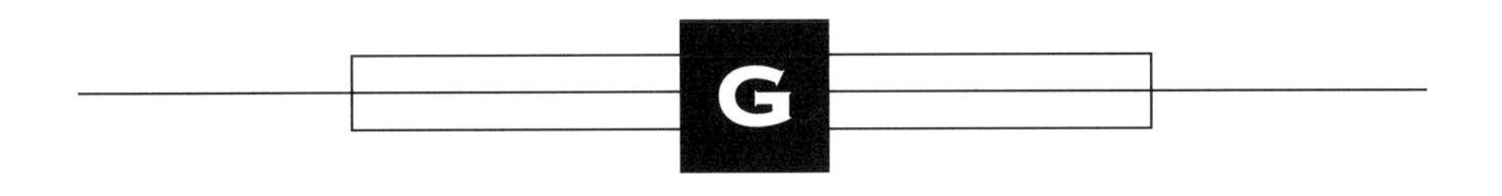

GAG *See* glycosaminoglycan.

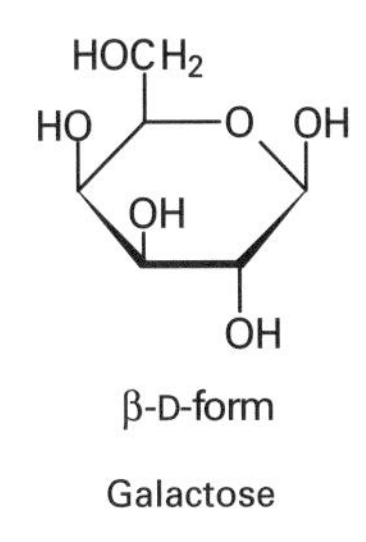

β-D-form

Galactose

galactose A SUGAR found in lactose and many polysaccharides. It is an aldohexose, isomeric with glucose.

gamete A cell capable of fusing with another cell to produce a zygote, from which a new individual organism can develop. Gametes may have similar structure and behavior (*isogametes*), as in many simple organisms, but are usually dissimilar in appearance and behavior (*anisogametes*). The typical male gamete is small, motile, and produced in large numbers. The typical female gamete is large because of the food reserves it contains, immotile, and is produced in small numbers. Fusion of gametes results in the nucleus of the zygote having exactly twice the number of chromosomes present in the nucleus of each gamete. *See also* ovum; spermatozoon.

gametogenesis The formation of sex cells or gametes, i.e. ova or spermatozoa.

gamma globulin *See* globulin.

ganglioside *See* glycolipid.

gas chromatography A technique widely used for the separation and analysis of mixtures. Gas chromatography employs a column packed with either a solid stationary phase (*gas-solid chromatography* or *GSC*) or a solid coated with a nonvolatile liquid (*gas-liquid chromatography* or *GLC*). The whole column is placed in a thermostatically controlled heating jacket. A volatile sample is introduced into the column using a syringe, and an unreactive carrier gas, such as nitrogen, passed through it. The components of the sample will be carried along in this mobile phase. However, some of the components will cling more readily to the stationary phase than others, either because they become attached to the solid surface or because they dissolve in the liquid. The time taken for different components to pass through the column is characteristic and can be used to identify them. The emergent sample is passed through a detector, which registers the presence of the different components in the carrier gas.

gas–liquid chromatography *See* gas chromatography.

gas–solid chromatography *See* gas chromatography.

gastric juice An agent of digestion in the stomach secreted by gastric glands situated in the thick stomach wall. It contains two main enzymes: pepsin, which breaks proteins down into short polypeptide chains, and, in milk-feeding young mammals, chymosin (rennin), which coagulates caseinogen to form casein. Gastric juice also contains mucus (to lubricate movement of food) and has an acid pH. The mechanical and chemical stimulation of the stomach lining by food itself causes secretion of gastric juice and of a hormone (gastrin). This hormone circulates in the blood and causes the gastric glands to secrete hydrochloric acid, thus generating the acid pH of the stomach.

gastrin A polypeptide hormone, secreted by the stomach, that stimulates secretions of gastric acid and pepsin in the stomach. It is released in response to the presence of components of the meal in the stomach.

gel A lyophilic colloid that is normally stable but may be induced to coagulate partially under certain conditions (e.g. lowering the temperature). This produces a pseudo-solid or easily deformable jelly-like mass, called a gel, in which intertwining particles enclose the whole dispersing medium. Gels may be further subdivided into elastic gels (e.g. gelatin) and rigid gels (e.g. silica gel).

gelatin (**gelatine**) A pale yellow protein obtained from the bones, hides, and skins of animals, which forms a colloidal jelly when dissolved in hot water. It is used in jellies and other foods, to make capsules for various medicinal drugs, as an adhesive and sizing medium, and in photographic emulsions.

gel electrophoresis *See* electrophoresis.

gel filtration (**gel-permeation chromatography**) A chromatographic method using a column packed with porous gel particles. It is a standard technique used for separating and identifying macromolecules of various sizes, e.g. proteins or nucleic acids. A solution of the mixture of macromolecules is added to the top of the column and allowed to flow through by gravity. The smaller molecules are hindered in their passage down the column because they are better able to penetrate the hydrated pores within the particles of the gel. Molecules too large to penetrate the pores are excluded, and thus flow more rapidly through the column. By analyzing the liquid that drips from the bottom of the column (the eluate) at set intervals and comparing it with a standard (obtained by running a known macromolecule through the column) information about the sizes and molecular weights of the components of the mixture is gathered. The most frequently used commercial gel is Sephadex.

gem positions Positions in a molecule on the same atom. For example, 1,1-dichloroethane (CH_3CHCl_2) is a gem dihalide.

gene In classical genetics, a unit of hereditary material located on a chromosome that, by itself or with other genes, determines a phenotypic characteristic in an organism. It corresponds to a segment of the genetic material, usually DNA (although the genes of some viruses consist of RNA). Genes may exist in a number of forms, termed *alleles*. For example, a gene controlling the characteristic 'height' in peas may have two alleles, one for 'tall' and another for 'short'. In a normal diploid cell only two alleles can be present together, one on each of a pair of homologous chromosomes: the alleles may both be of the same type, or they may be different. The segregation of alleles at meiosis and their dominance relationships are responsible for the particulate nature of inheritance. However, many phenotypic characteristics are now known to be caused by multiple genes and also environmental factors.

Although the DNA molecules of the chromosomes account for the great majority of genes, genes are also found as plasmagenes in certain DNA-containing cytoplasmic bodies (e.g. mitochondria, plastids).

A gene has also been defined as the smallest hereditary unit capable either of recombination or of mutation or of controlling a specific function. These three definitions do not necessarily describe the same thing and a unit of function, a cistron, may be much larger than a unit of recombination or mutation. Research with bacteria has shown that the smallest unit of recombination or mutation is one base pair, while a unit of function can be determined by the *cis-trans* test.

Molecular genetics now defines three types of gene: (i) *structural genes*, which code for polypeptides of enzymes and other proteins; (ii) *RNA genes*, which code for ribosomal RNA and transfer RNA molecules used in polypeptide assembly; and (iii) *regulator genes*, which regulate the expression of the other two types. Many genes of higher eukaryotes e.g. mammals, have a mosaic structure composed of coding (exons) and noncoding regions (introns).

gene cloning (**DNA cloning; molecular cloning**) A technique of genetic engineering whereby a gene sequence is replicated, giving many identical copies. The gene sequence is isolated by using restriction endonucleases, or by making a complementary DNA from a messenger RNA template using a reverse transcriptase. It is then inserted into the circular chromosome of a cloning vector, i.e. a plasmid or a bacteriophage. The hybrid is used to infect a bacterium, usually *Escherichia coli*, and is replicated within the bacterial cell. A culture of such cells produces many copies of the gene, which can subsequently be isolated and purified.

Genes can also be cloned into so-called *expression vectors*, which allow expression of that gene in a different cellular system (known as an expression system) from which it originated. Bacterial cells cannot always express eukaryotic proteins and eukaryotic cells e.g. yeast are required.

Entire genomes can be cloned at one go using *shotgun cloning*. The DNA is randomly broken into fragments, which are then cloned into vectors. This technique was used in sequencing the human genome. *See also* genetic engineering; restriction endonuclease.

gene expression The exhibition of genetic information in the phenotype. For genes that encode proteins, it involves transcription and translation. Gene expression is tightly regulated.

gene knockout A technique for selectively inactivating a certain gene within a living cell or organism, by replacing the normal alleles with mutant nonfunctional alleles. It is applied to experimental organisms such as yeasts and mice, in order to study how normal function is disrupted by loss (i.e. 'knockout') of the gene. A specific gene is cloned, and then inactivated by making precise changes in its sequence using a technique called *in vitro mutagenesis*. The inactivated gene is incorporated into a suitable targeting vector containing marker genes, and introduced to normal cells, for example, by electroporation or direct injection into the nucleus. In a small fraction of diploid cells, the gene becomes incorporated into the chromosome at its target site by a process called homologous recombination, producing a heterozygote with one normal allele and one mutant allele. The marker genes enable selection of cells containing the inactivated gene at the correct target site.

Gene knockout is relatively easy to perform in single-celled organisms, such as yeasts, and the effects of the knockout can be readily assessed by inducing the diploid heterozygotes to undergo meiosis. Half the haploid spores from such a cell will contain the knockout allele alone, and the viability of these spores reflects the functional importance of the inactivated gene. In multicellular organisms, like the mouse, gene knockout requires a more complex procedure, but nonetheless provides a powerful tool for studying the roles of specific genes in development, physiology, disease, etc. In the mouse, the knockout gene is incorporated into its target site in embryonic stem cells. Successfully targeted cells are introduced to a second embryo from a different mouse strain. This embryo then develops inside a surrogate mother, to produce a chimera – i.e. a newborn mouse comprising cells from both strains. When mature, the chimeric mouse is used to breed generations of progeny mice, some of which are homozygous for the knockout gene. Knockout mice are especially valuable as model systems for studying various human genetic diseases, such as cystic fibrosis.

gene library A collection of cloned DNA fragments derived from the entire genome of an organism. The genetic material of the organism is first broken up randomly into fragments using restriction enzymes, for example. Then each fragment is cloned using a vector (e.g. plasmid or bacteriophage) inside a suitable host, such as the bacterium *E. coli*. Particular genes or DNA sequences are identified by a suitable DNA probe.

gene pool All the genes carried by the gametes of a sexually reproducing population.

gene probe *See* DNA probe.

gene product Any RNA (transcription product) or protein (translation product) synthesized from genetic information.

general formula A representation of the chemical formula common to a group of compounds. For example, C_nH_{2n+2} is the general formula for an alkane, whose members form a homologous series.

gene sequencing Determination of the order of bases of a DNA molecule making up a gene. Sequencing requires multiple cloned copies of the gene; long DNA sequences are cut into more manageable lengths using restriction enzymes. Since these cleave DNA at specific points, it is possible to reconstitute the overall sequence once the constituent fragments have been analyzed individually by the methods outlined below.

There are two methods of sequencing DNA. One is the *chemical cleavage method*, or *Maxam–Gilbert method*. This involves firstly labeling one end of the DNA of the gene or DNA segment with radioactive ^{32}P. The segment is then subjected to a chemical reaction that cleaves the sequence at positions occupied by one of the four bases, say, adenine. Starting with numerous cloned DNA segments, the result is a set of radioactive fragments extending from the ^{32}P label to each successive position of adenine in the segment. This process is repeated for the other three bases, and the four sets of fragments are then separated according to the number of nucleotides they contain by gel electrophoresis, in adjacent lanes on the gel. The sequence can then be deduced directly from the autoradiograph of the gel. This method is used for sequences of more than 150 nucleotides.

The second method is the *chain-termination* or *dideoxy method* (also called the *Sanger method*). A single-stranded segment of DNA taken from the gene is used as a template to replicate a new DNA strand using the enzyme DNA polymerase. The enzyme is provided with the four normal nucleoside triphosphates (ATP, GTP, CTP, TTP), plus the dideoxy (dd) derivative of one of them, say ddATP. Incorporation of the dideoxy derivative causes replication of the new strand to cease at that point. The result is a set of new strands of varying length, terminating at all the different positions where adenine, say, normally occurs in the sequence. The process is repeated in turn for each of the three remaining bases, and the set of fragments from each incubation are separated according to size by gel electrophoresis. The sequence of bases in the newly synthesized strand can be deduced directly from the gel. This method is used for sequences of less than 200 nucleotides.

Automation of both procedures, for example by using laser scanning of fluorescent dye markers instead of autoradiography, has greatly increased the speed with which DNA can be sequenced, and made possible the sequencing of entire genomes, including the human genome.

gene splicing The joining of DNA fragments by the action of the enzyme DNA ligase.

genetic code The sequence of bases in either DNA or messenger RNA that conveys genetic instructions to the cell. The basic unit of the code consists of a group of three consecutive bases, the base triplet or codon, which specifies instructions for a particular amino acid in a polypeptide, or acts as a start or stop signal for translation of the message into polypeptide assembly. The genetic code is almost universal with the only differences being found in some mitochondrial DNA. For example, the

DNA triplet CAA (which is transcribed as GUU in mRNA) codes for the amino acid valine. There are 64 different triplet combinations but only 20 amino acids; thus many amino acids can be coded for by two or more triplets. The code is therefore said to be *degenerate*, and it appears that only the first two bases, and in certain cases only one base, are necessary to insure the coding of a specific amino acid. Three triplets (UAA, UAG, and UGA), termed 'nonsense triplets', do not code for any amino acid and mark the end of a polypeptide chain. AUG (or sometimes GUG) is the signal to start translation. *See also* codon.

genetic engineering (**recombinant DNA technology**) The direct introduction of foreign genes into an organism's genetic material by micromanipulation at the cell level. Genetic engineering techniques bypass crossbreeding barriers between species to enable gene transfer between widely differing organisms. Gene transfer can be achieved by various methods, many of which employ a replicating infective agent, such as a virus or plasmid, as a vector (*see* gene cloning). Other methods include microinjection of DNA into cell nuclei and direct uptake of DNA through the cell membrane. Recognizing whether or not transfer has occurred may be difficult unless the new gene confers an obvious visual or physiological characteristic. Consequently the desirable gene may be linked to a marker gene, e.g. a gene conferring resistance to an antibiotic in the growth medium. The transferred gene must also be linked to appropriate regulatory DNA sequences to insure that it works in its new environment and is regulated correctly and predictably.

Initial successes in DNA transfer were achieved with bacteria and yeast. Human genes coding for medically useful proteins have been transferred to bacteria. Human insulin, growth hormone, and interferon are now among a wide range of therapeutic substances produced commercially from genetically engineered bacteria. Genetically engineered vaccines have also been produced by transfer of antigen-coding genes to bacteria.

Modified microorganisms are grown in large culture vessels and the gene product harvested from the culture medium. However, the problems associated with scaling up laboratory systems are still limiting the exploitation of genetic engineering. Genetic manipulation of higher animals and plants has been achieved more recently. Transgenic mammals, including mice, sheep, and pigs, have been produced by microinjection of genes into the early embryo, and it is also now possible to clone certain mammals from adult body cells (*see* clone). Such technology may have considerable impact on livestock production, e.g. by injection of growth hormone genes. Dicotyledonous plants, including tobacco and potato, have been transfected using the natural plasmid vector of the soil bacterium *Agrobacterium tumefaciens* (*see* Agrobacterium). Genes have been introduced to crop plants for various reasons, for instance to reduce damage during harvest or to make them resistant to the herbicides used in controlling weeds. Genetically modified tomatoes and soya beans are now widely available. There is also hope that in future many genetic diseases will be treatable by manipulating the faulty genes responsible (gene therapy). However, genetic engineering raises many legal and ethical issues, and the introduction of genetically modified organisms into the environment requires strict controls and monitoring. *See also* recombinant DNA.

genetic fingerprinting (**DNA fingerprinting; genetic profiling**) A technique for identifying individuals by means of their DNA. The DNA being tested is extracted from cells (from blood, semen, tissue fragments, etc.) and broken into fragments of 600–700 bases each, using restriction enzymes. The human genome contains many loci (known as *minisatellites*) where short base sequences are repeated in tandem, with great variation between individuals in the number of such repeats. These so-called *variable number tandem repeats* (VNTR) can be identified using special DNA probes, thus providing a virtually unique set of markers for any given individual. This technique is used in veterinary and human medicine to establish the parentage of individuals, and in forensic science to identify individuals from traces of body tissue or fluids. Even minute amounts of DNA can now be amplified, using the POLYMERASE CHAIN REACTION, to provide sufficient material for genetic fingerprinting.

genetic imprinting Maternal and paternal genes are not always expressed to the same degree in an offspring. Genetic imprinting controls the extent to which maternal and paternal genes are expressed by differing amounts of DNA methylation. *See* X chromosome.

genetic map *See* chromosome map.

genetic marker A character or chromosomal site (locus) that can be used as a signpost to track closely linked genes or DNA sequences. Any marker must exist in at least one form (preferably more) that is distinct from the normal (wild-type) condition. Markers are used in establishing inheritance patterns or in chromosome mapping, for example, and there are several types. *Phenotypic markers* are variants of genetically determined traits, such as curly wing shape or white eye color in fruit flies (*Drosophila*), whose inheritance can be easily assessed. They serve as labels to flag the inheritance of the determining alleles. Since the early 1900s, numerous examples have been described in various experimental organisms, and used in recombinational analysis to yield linkage maps of chromosomes (*see* chromosome map). The advent of modern analytical techniques in the 1980s led to the discovery of a vast collection of *molecular markers*. These are variations in DNA sequence that have no apparent effect on the phenotype of the organism concerned, but can be detected by biochemical analysis. The main examples are RESTRICTION FRAGMENT LENGTH POLYMORPHISMS, which result from silent mutations at cleavage sites for restriction enzymes; and VARIABLE NUMBER TANDEM REPEATS – variations in the number of certain tandemly repeated DNA sequences. Molecular markers have filled in the large gaps between existing phenotypic markers, and provide the signposts needed for fine-structure mapping.

genetic profiling *See* genetic fingerprinting.

genetics The term coined by Bateson to describe the study of inheritance and variation and the factors controlling them. Today the subject has three main subdivisions – Mendelian genetics, population genetics, and molecular or biochemical genetics.

genome The complete genetic information of an organism. *See also* proteome.

genome chip *See* DNA microarray.

genomic library A library of cloned DNA fragments that cover the whole genome. The fragments are made solely from genomic DNA.

genomics The branch of molecular genetics concerned with the study of genomes. It has emerged since the 1980s as one of the fastest-growing fields in all biology, following the development of automated techniques for nucleic acid and protein sequencing, and computerized systems for handling and analyzing the resultant data (*see* bioinformatics). It can be divided into three main areas. *Structural genomics* deals with determining the DNA sequence of all the genetic material of a particular organism. Such a definitive physical map of the genome is the ultimate objective of all genome projects. *Functional genomics* focuses on characterizing all the messenger RNAs produced by transcription of an organism's genome, and the polypeptides that they encode. Complementary DNAs can be constructed from such transcripts, and compared with the genome sequence data to identify corresponding coding regions. *Comparative genomics* is concerned with comparing the genomic sequences of different species, to see how genomes change in the course of evolution, and which parts remain unchanged. The presence of such highly conserved sequences sheds light on their functional significance, and helps in making predictions about similar sequences in other organisms.

genotype The genetic composition of an organism. The actual appearance of an individual (the phenotype) depends on the dominance relationships between alleles in the genotype and the interaction between genotype and environment.

genus A collection of similar species. Genera may be subdivided into subgenera, and also, especially in plant taxonomy, into sections, subsections, series, and subseries. Similar genera are grouped into families.

geometrical isomerism *See* isomerism.

geotropism (**gravitropism; geotropic movement**) A directional growth movement of part of a plant in response to gravity. Primary roots (tap roots) grow vertically towards gravity (positive geotropism) whereas primary shoots grow vertically away from gravity (negative geotropism), though the direction of shoot growth may also be modified by light. Dicotyledon leaves and some stem structures (e.g. rhizomes and stolons) grow horizontally (*diageotropism*). Secondary (lateral) roots and stem branches may grow at an intermediate angle with respect to gravity (*plagiogeotropism*).

Geotropic responses are believed to involve hormones, maybe auxins or gibberellins. It is proposed that if a shoot or coleoptile is lying on its side, these hormones move to the lower surface in response to gravity, stimulate growth, and cause upward growth of the organ. In a horizontally placed root, the same high level of hormones inhibits growth of the lower surface, resulting in downward curvature. However, other physiological gradients may play a part, e.g. pH, electrical potential, or water potential. These gradients may arise owing to the detection of gravity by large starch grains (called *statoliths* or *amyloplasts*). The statoliths sediment towards the lower wall of the cell exerting a pressure that in some way causes the gravitational response. This is termed the *statolith hypothesis*. *See also* tropism.

germination The first outward sign of growth of a reproductive body, such as a spore or pollen grain. The term is most commonly applied to seeds, in which germination involves the emergence of the radicle or coleoptile through the testa. Both external conditions (e.g. water availability, temperature, and light) and internal biochemical status must be appropriate before germination can occur. Seed germination may be either *epigeal*, in which the cotyledons appear above ground, or *hypogeal*, in which the cotyledons remain below ground.

gibberellic acid (GA_3) A common GIBBERELLIN and one of the first to be discovered. Together with GA_1 and GA_2 it was isolated from *Gibberella fujikuroi*, a fungus that infects rice seedlings causing abnormally tall growth.

gibberellin A plant hormone involved chiefly in shoot extension. Gibberellins are diterpenoids; their molecules have the gibbane skeleton. More than thirty have been isolated, the first and one of the most common being gibberellic acid, GA_3.

Gibberellins stimulate elongation of shoots of various plants, especially the extension to normal size of the short internodes of genetically dwarf pea or maize plants. Increased gibberellin levels can mimic or mediate the effect of long days. Thus they stimulate internode extension and flowering in long-day plants such as lettuce and spinach. They are also effective in inhibiting tuber development or breaking tuber dormancy, e.g. in the potato, and breaking bud dormancy in woody species. They have similar effects in substituting for chilling in some species with a vernalization requirement; e.g. causing bolting in biennials at the rosette stage.

Synthesis of α-amylase and certain other hydrolytic enzymes in barley aleurone layers is regulated by gibberellin produced by the embryo. This initiates germination by mobilizing endosperm food reserves. Gibberellins may be produced in both shoots and roots and travel in both xylem and phloem.

Gibbs function (**Gibbs free energy**) Symbol: G A thermodynamic function defined by

$$G = H - TS$$

where H is the enthalpy, T the thermodynamic temperature, and S the entropy. It is useful for specifying the conditions of chemical equilibrium for reactions for constant temperature and pressure (G is a minimum). *See also* free energy.

giga- Symbol: G A prefix denoting 10^9. For example, 1 gigahertz (GHz) = 10^9 hertz (Hz).

Gilbert, Walter (1932–) American molecular biologist. He was awarded the Nobel Prize for chemistry in 1980 jointly with F. Sanger for their contributions concerning the determination of base sequences in nucleic acids. The prize was shared with P. Berg.

Gilman, Alfred Goodman (1941–) American pharmacologist. He was awarded the Nobel Prize for physiology or medicine in 1994 jointly with M. Rodbell for their discovery of G-proteins.

gland **1.** (*Zoology*) An organ that synthesizes a specific chemical substance and secretes this either through a duct into a tubular organ or onto the surface of the body or directly into the bloodstream.
2. (*Botany*) A specialized cell or group of cells concerned with the secretion of various substances produced as by-products of plant metabolism. The secretions may pass to the exterior or be contained in cavities or canals in the plant body. Ethereal oils, tannins, and resins are usually retained by the plant and give aromatic plants their characteristic scents. The hydathodes of leaves exude a watery solution onto the surface of the leaf in the process termed guttation, and the nectaries of flowers exude sugary substances to attract insects. Glandular hairs develop from the epidermis of many plants, e.g. stinging nettle and geranium.

GLC (**gas–liquid chromatography**) *See* gas chromatography.

globulin One of a group of proteins that are insoluble in water but will dissolve in neutral solutions of certain salts. They generally contain glycine and coagulate when heated. Three types of globulin are found in blood: *alpha* (α), *beta* (β), and *gamma* (γ). α and β globulins are made in the liver and are used to transport nonprotein material. γ globulins are made in reticuloendothelial tissues, lymphocytes, and plasma cells and most of them have antibody activity (*see* immunoglobulin).

glucagon A polypeptide hormone produced by the A-cells of the islets of Langerhans of the pancreas in response to somatotropin. Its action opposes that of insulin, causing an increase in blood glucose by promoting the breakdown of glycogen to glucose and the synthesis of glucose by gluconeogenesis in the liver. *Compare* insulin.

glucan *See* glycan.

glucocorticoid A type of steroid hormone produced by the adrenal cortex. Glucocorticoids (e.g. corticosterone and hydrocortisone) accelerate the formation of glucose from noncarbohydrate precursors (gluconeogenesis) and the breakdown of glycogen. They also inhibit inflammation by depressing T-cell activity. *See also* corticosteroid.

gluconeogenesis A metabolic process occurring in the liver by which glucose or other carbohydrates can be manufactured from lactic acid, glycerol, or noncarbohydrate precursors, such as amino acids. It is stimulated by the hormone glucagon. During prolonged fasting, it is the route by which amino acids derived from the breakdown of muscle proteins are converted into usable energy. The effect of gluconeogenesis is the reverse of glycolysis but it is not brought about by simply reversing the reactions of glycolysis; different reactions are required to bypass the irreversible steps of glycolysis.

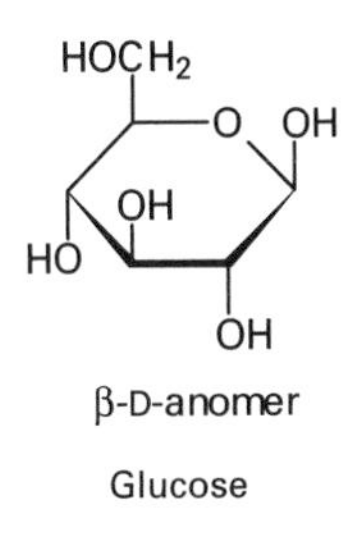

Glucose

glucose A monosaccharide with the formula $C_6H_{12}O_6$. The D-isomer occurs widely in nature and is sometimes called dextrose. It is an important source of energy and is stored in animals as glycogen and in plants as starch and cellulose. Brain and red blood cells are particularly dependent on glucose as a source of fuel and during starvation other body tissues and organs switch to different energy sources, e.g. ketone bodies. *See* glycolysis; gluconeogenesis; sugar.

glutamic acid *See* amino acids.

glutamine *See* amino acids.

glutathione A tripeptide of cysteine, glutamic acid, and glycine, widely distributed in living tissues. It takes part in many oxidation–reduction reactions, due to the reactive thiol group (–SH) being easily oxidized to the disulfide (–S–S–), and acts as an antioxidant, as well as a coenzyme to several enzymes.

gluten A mixture of proteins found in wheat flour. It is composed mainly of two proteins (gliaden and glutelin), the proteins being present in almost equal quantities. Certain people are sensitive to gluten (celiac disease) and must have a gluten-free diet.

glycan A polysaccharide made of more than 10 monosaccharide residues. A *homoglycan* is made up of a single type of sugar unit (i.e. >95%). As a class the glycans serve both as structural units (e.g. cellulose in plants and chitin in invertebrates) and energy stores (e.g. starch in plants and glycogen in animals). The most common homoglycans are made up of D-glucose units and called *glucans*.

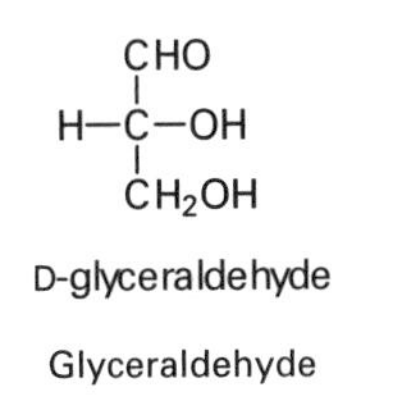

Glyceraldehyde

glyceraldehyde The simplest chiral aldose sugar, used as a standard for configuration.

glycerol (**glycerin; propane-1,2,3-triol**) An alcohol with three OH groups. Glycerol is biologically important as the alcohol involved in lipid formation (these particular lipids being called *glycerides*). *See* glyceride; lipid.

glyceride (**acylglycerol**) An ester of glycerol and one or more fatty acids. They may be mono-, di-, or triglycerides according to the number of –OH groups esterified. The fat stores of the body consist mainly of triglycerides. These can form a source of energy when carbohydrate levels are low, being broken down by lipases into fatty acids, which can enter metabolic pathways. *See also* lipid.

glycerol (**glycerin; propane-1,2,3-triol**) An alcohol with three OH groups. Glycerol is biologically important as the alcohol involved in lipid formation (these particular lipids being called *glycerides*). *See* glyceride; lipid.

glycine *See* amino acids.

glycogen (**animal starch**) A polysaccharide that is the main carbohydrate store of animals. It is composed of many glucose units linked in a similar way to starch.

Glycogen is readily hydrolyzed in a stepwise manner to glucose itself. It is stored largely in the liver and in muscle but is found widely distributed. After a meal, most of the glucose contained in food is absorbed via the intestine and blood and converted to glycogen in the liver (*glycogenesis*). The concentration of glucose in the blood is then normally regulated by conversion of glycogen back to glucose (*glycogenolysis*). The liver can store about 100 grams of glycogen.

glycogenesis *See* glycogen.

glycogenolysis *See* glycogen.

glycolipid A sugar-containing lipid with one or more sugar residues attached to a lipid by a glycosidic link (or ester link in prokaryotes). Glycolipids play an important structural role in cell membranes, where the sugar residues are always extracellular. Animal glycolipids are derived from *sphingosine*, an amino alcohol with a long unsaturated hydrocarbon chain. In glycolipids, the amino group of sphingosine is joined to a fatty acid chain by an amide bond and the primary hydroxyl group is linked to a sugar residue. The simplest glycolipid (in animal cells) is *cerebroside*, which has one sugar residue (either glucose or galactose). Glycolipids with branched chains of sugar residues are known as gangliosides. The fatty acid chain and sphingosine chain are hydrophobic while the sugar residues are hydrophilic, making glycolipids amphipathic.

glycolysis (**Embden–Meyerhof pathway; glycolytic pathway**) The conversion of glucose into pyruvate, with the release of some energy in the form of ATP. Glycolysis occurs in cell cytoplasm. It yields two molecules of ATP and two of $NADH_2$ per molecule of glucose. In anaerobic conditions, breakdown proceeds no further and pyruvate is converted into ethanol or lactic acid for storage or elimination. In aerobic conditions, glycolysis is followed by the KREBS CYCLE. The rate of glycolysis is controlled by the enzyme phosphofructokinase, which catalyzes an essentialy irreversible reaction. There are two other irreversible reactions catalyzed by hexokinase and pyruvate kinase.

glycoprotein A conjugated protein formed by the combination of a protein with carbohydrate side chains. Certain antigens, enzymes, and hormones are glycoproteins.

glycosaminoglycan (GAG) One of a group of compounds, sometimes called *mucopolysaccharides*, consisting of long unbranched chains of repeating disaccharide sugars, one of the two sugar residues being an amino sugar – either *N*-acetylglucosamine or *N*-acetylgalactosamine. These compounds are present in connective tissue; they include heparin and hyaluronic acid. Most glycosaminoglycans are linked to protein to form proteoglycans (sometimes called *mucoproteins*). *See also* glycoprotein.

glycoside A derivative of a pyranose sugar (e.g. glucose) in which there is a group attached to the carbon atom that is joined to the –CHO group. In a glycoside the C–OH is replaced by C–OR. The linkage –O– is a *glycosidic link*; it is the link joining monosaccharides in polysaccharides.

glyoxylate cycle A modification of the Krebs cycle occurring in some microorganisms, algae, and higher plants in regions where fats are being rapidly metabolized, e.g. in germinating fat-rich seeds. Acetyl groups formed from the fatty acids are passed into the glyoxylate cycle, with the eventual formation of mainly carbohydrates.

Goldstein, Joseph Leonard (1940–) American physician and molecular geneticist notable for his discovery of cellular receptors for low-density lipoproteins and their role in the removal of cholesterol from the bloodstream. He was awarded the Nobel Prize for physiology or medicine in 1985 jointly with M. S. Brown.

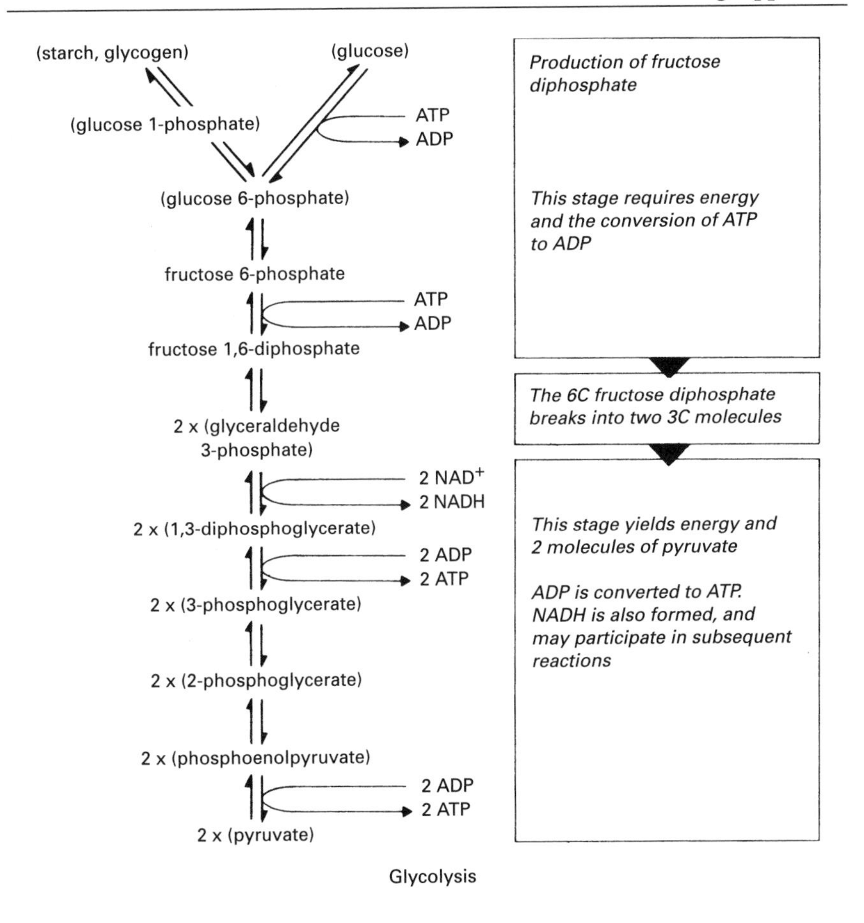

Glycolysis

Golgi, Camillo (1843–1926) Italian anatomist, cytologist, neurologist, and pathologist noted for his studies of the central nervous system and, in particular, for his discovery of the cellular organelle named for him. He was awarded the Nobel Prize for physiology or medicine in 1906 jointly with S. Ramón y Cajal for their work on the structure of the nervous system.

Golgi apparatus (**Golgi body; Golgi complex**) An organelle of eukaryotic cells discovered by Camillo Golgi in 1898. It is associated with the endoplasmic reticulum (ER) but lacks ribosomes. It consists of stacks of flattened membrane-bounded sacs (cisternae) and has two distinct faces: the *cis* which is closest to the ER and receives proteins from the rough ER in transfer vesicles and the *trans* which exports proteins to their various destinations in different vesicles. The region between the *cis* face and *trans* face is called the medial region. Secretory cells are rich in Golgi apparatus. The cisternae are either spread randomly (as dictyosomes) as in plant cells or form a single network as in most animal cells. The Golgi apparatus alters the carbohydrate units of glycoproteins and sorts proteins for transport to their cellular locations, e.g. the plasma membrane, lysosomes, or secretory vesicles.

The Golgi apparatus is also involved in various other processes, including: formation of zymogen granules; synthesis and transport of secretory polysaccharides, e.g. cellulose in cell plate formation or secondary wall formation and mucus in goblet cells; packaging of hormones in nerve cells carrying out neurosecretion. *See* lysosome; zymogen granule. *See* cell.

gonad The reproductive organ of animals. It produces the sex cells (gametes) and sometimes hormones. The female gonad, the ovary, produces ova; the male gonad, the testis, produces spermatozoa. Some invertebrates have both male and female gonads. *See also* ovotestis.

gonadotropic hormone *See* gonadotropin.

gonadotropin (**gonadotropic hormone**) A hormone that acts on the gonads (ovary and testis). The *pituitary gonadotrophins* are follicle-stimulating hormone, luteinizing hormone, and prolactin. They are involved in the initiation of puberty, regulation of the menstrual cycle, and lactation in females, and in the control of spermatogenesis in males. They are used in the treatment of infertility. In women such treatment may lead to multiple pregnancies if not carefully controlled.

G protein Any of a class of proteins that play a crucial role in relaying signals from certain types of cell-surface receptors (*G protein-coupled receptors*) to activate signal pathways inside the cell. G proteins are involved in signal transduction from receptors for various hormones and neurotransmitters, as well as light-activated receptors in the eye and odor receptors in the nose. Essentially, binding of the ligand (hormone, neurotransmitter, etc.) to the exterior region of the receptor changes its interior (cytosolic) region, which activates an associated G protein. The activated G protein in turn might activate or inhibit a further enzyme that produces a second messenger, or change the permeability of certain ion channels and alter membrane potential. For example, in epinephrine and glucagon receptors, the activated G protein binds to and activates the enzyme adenylyl cyclase, which catalyzes the synthesis of the second messenger cyclic AMP (cAMP).

graft The transplantation of an organ or tissue in plants and animals.

In plants, grafting is an important horticultural technique in which part (the scion) of one individual is united with another of the same or different species. Usually the shoot or bud of the scion is grafted onto the lower part of the stock. Incompatibility between species is much less likely to occur in plants than in animals.

In animals, a graft is a transplantation of an organ or tissue, either on the same individual or on different individuals (i.e. from a donor to a recipient). Antibody mechanisms of the recipient recognize a graft of dissimilar tissue and tend to cause its rejection. The closer the relationship between donor and recipient, the greater the chance of a successful graft. Grafts may be from one place to another on the same individual (*autograft*) or between different individuals. A graft between individuals of the same species is a *homograft* (or *allograft*); between individuals that are genetically identical (as between identical twins) it is an *isograft* (or *syngraft*); and between different species it is a *heterograft* (or *xenograft*).

Gram's stain A stain containing crystal violet and safranin used for bacteria and the basis for the division into Gram positive and Gram negative bacteria. The former retain the deep purple color of crystal violet; the latter are counterstained red with safranin.

Granit, Ragnar Arthur (1900–91) Finnish-born Swedish neurophysiologist. He was awarded the Nobel Prize for physiology or medicine in 1967 jointly with H. K. Hartline and G. Wald for their discoveries concerning the physiological and chemical visual processes in the eye.

granulocyte A white blood cell (leukocyte) that has granules in the cytoplasm. Granulocytes are sometimes called *polymorphonuclear leukocytes* (*polymorphs*) because the nucleus is lobed. The three types of granulocytes are neutrophils (70% of all leukocytes), eosinophils (1.5%), and basophils (0.5%).

granum A stack of membranes (resembling a pile of coins) in a chloroplast. With the light microscope these stacks are just visible as grains (grana). *See* chloroplast.

grape sugar *See* dextrose.

gravitropism *See* geotropism.

gray Symbol: Gy The SI unit of absorbed energy dose per unit mass resulting from the passage of ionizing radiation through living tissue. One gray is an energy absorption of one joule per kilogram of mass.

Greengard, Paul (1925–) American biochemist noted for his work on signal transduction in the nervous system. He shared the 2000 Nobel Prize for physiology or medicine with A. Carlsson and E. R. Kandel.

growth hormone (**somatotropin**) A hormone that controls growth. Deficiency leads to dwarfism; excessive secretion leads to gigantism. It is a polypeptide that is produced by the anterior pituitary gland and acts on the cells of the body, particularly those of bone, to stimulate metabolism and growth. Its actions *in vivo* are thought to be largely mediated by *somatomedins*, notably polypeptides resembling insulin and called *insulin-like growth factors*. Somatotropin can be used to boost meat and milk production. Unlike anabolic steroids (also used for this purpose), residual somatotropin present in foods is rendered metabolically inactive during digestion.

GSC Gas–solid chromatography. *See* gas chromatography.

GTP (**guanosine triphosphate**) A nucleoside triphosphate occurring in all cells as a coenzyme for various key processes. Often it provides energy by undergoing hydrolysis to GDP (guanosine diphosphate) and a phosphate group, a reaction catalyzed by an enzyme or other component having *GTPase* activity. In protein synthesis, GTP is essential for the assembly of ribosomes and elongation of the polypeptide chain. It is also required for the assembly of microtubules, for protein transport within cells, and for the relaying of messages to various cell components in signal transduction.

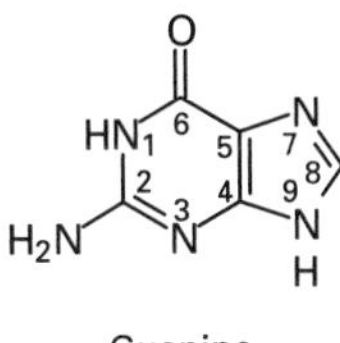

Guanine

guanine A nitrogenous base found in DNA and RNA. Guanine has a purine ring structure.

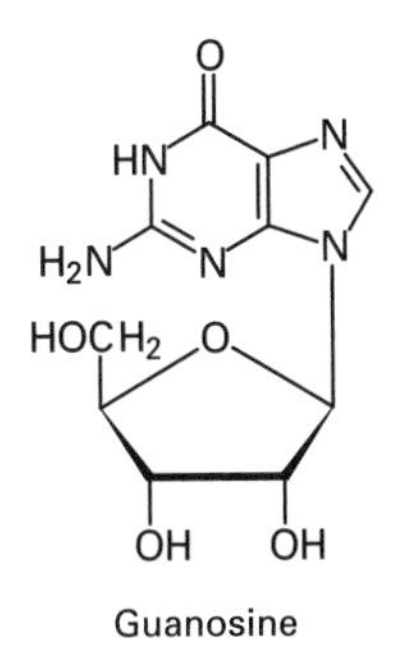

Guanosine

guanosine (**guanine nucleoside**) A nucleoside present in DNA and RNA and consisting of guanine linked to D-ribose via a β-glycosidic bond.

guanosine triphosphate *See* GTP.

guard cell A specialized kidney-shaped epidermal cell, a pair of which encircle each epidermal pore (*stoma*) in plants, and control the opening and closing of the stomatal aperture. Adjoining the guard cells, and assisting them in their actions, are *subsidiary cells*. Changes in the aperture are effected through changes in turgidity of the guard cells. The wall of the guard cell bordering the pore is heavily thickened whereas the opposite wall is comparatively thin. Thus when the guard cell is turgid the thin wall becomes distended, bulging out away from the pore, and causing the thickened wall, which cannot distend, to be drawn outwards with it. This results in opening of the aperture between the guard cells. When osmotic pressure of the guard cells drops, the guard cells shrink, and the pore closes.

The osmotic pressure inside the guard cells is regulated by light-dependent changes in the cytosolic concentration of dissolved solutes. During daylight, starch stored in the guard cell chloroplasts is converted to malate (the anion of malic acid) and hydrogen ions (H^+). ATP-powered proton pumps in the plasma membrane remove H^+ from the cell, creating an H^+ gradient across the plasma membrane. The consequent increased negative charge inside the cell triggers the opening of potassium channels in the plasma membrane and the influx of potassium ions (K^+), followed by the entry of chloride ions (Cl^-) via chloride channels. Hence the cytosolic ion concentration rises, drawing water into the cell by osmosis. During darkness, these events are reversed. The H^+ gradient across the plasma membrane falls, and channels open to transport K^+ and Cl^- from the cell. Hence water also leaves the cell, and the elastic walls of the guard cells cause them to shrink.

Guillemin, Roger Charles Louis (1924–) French-born American physician who worked on peptide hormone production in the brain. He was awarded the Nobel Prize for physiology or medicine in 1977 jointly with A. V. Schally. The prize was shared with R. S. Yalow.

gum One of a group of substances that swell in water to form gels or sticky solutions. Similar compounds that produce slimy solutions are called *mucilages*. Gums and mucilages are not distinguishable chemically. Most are heterosaccharides, being large, complex, flexible, and often highly-branched molecules.

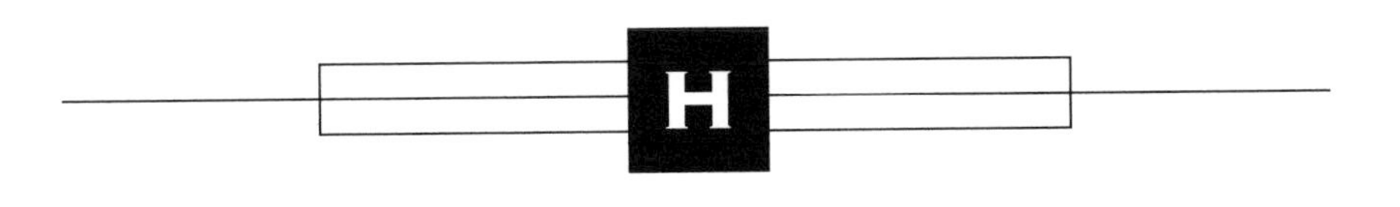

haem *See* heme.

hairpin A region of secondary structure of a linear molecule (e.g. a nucleotide) in which the chain of constituent groups is folded back on itself like a hairpin and held in place by chemical bonds (e.g. hydrogen bonds). Hairpins in nucleic acids occur as a result of bonding between quite widely separated stretches of complementary base sequence within single-stranded DNA and RNA molecules.

haploid (**monoploid**) A cell or organism containing only one representative from each of the pairs of homologous chromosomes found in the normal diploid cell. Haploid chromosomes are thus unpaired and the haploid chromosome number (n) is half the diploid number ($2n$). Meiosis, which usually precedes gamete formation, halves the chromosome number to produce haploid gametes. The diploid condition is restored when the nuclei of two gametes fuse to give the zygote. In man there are 46 chromosomes in 23 pairs and thus the haploid egg and sperm each contain 23 chromosomes. Gametes may develop without fertilization, or meiosis may substantially precede gamete formation. This is especially true in plants, and leads to the formation of haploid organisms, or haploid phases in the life cycles of organisms. Various multiples of the haploid number, e.g. the tetraploid ($4n$), hexaploid ($6n$), and octaploid ($8n$) conditions, are common in some plant groups, especially in certain cultivated plants.

Harden, (Sir) Arthur (1865–1940) British biochemist who discovered (with W. J. Young) that fermentation of sugar by yeast juice involved a separable substance (later called cozymase), required the presence of phosphate, and resulted in the intermediate formation of two hexose phosphates. He was awarded the Nobel Prize for chemistry in 1929, the prize being shared with H. von Euler.

hardening (of oils) A process for converting liquid vegetable oils into solid or semi-solid fats. Oils are glycerides of mainly unsaturated carboxylic acids. Hydrogenation of the double carbon–carbon bonds using a nickel catalyst produces saturated glycerides, which have higher melting points. Hydrogenation of vegetable oils is used in the manufacture of margarine.

Hartline, Haldan Keffer (1903–83) American neurophysiologist. He was awarded the Nobel Prize for physiology or medicine in 1967 jointly with R. Granit and G. Wald for their discoveries concerning the physiological and chemical visual processes in the eye.

Hartwell, Leyland H. (1939–) American cell biologist who was awarded the 2001 Nobel Prize for physiology or medicine jointly with R. T. Hunt and P. M. Nurse for discoveries of the key regulators of the cell cycle.

Hassel, Odd (1897–1981) Norwegian physical chemist who worked on the concept of conformation and its application in chemistry. He was awarded the Nobel Prize for chemistry in 1969, the prize being shared with D. H. R. Barton.

Hatch–Slack pathway *See* C_4 plant.

Haworth, (Sir) Walter Norman (1883–1950) British organic chemist noted for his structural studies of sugars and polysaccharides. Haworth made the first chemical synthesis of ascorbic acid. He was awarded the Nobel Prize for chemistry in 1937. The prize was shared with P. Karrer.

HCG (**human chorionic gonadotropin**) *See* chorionic gonadotropin.

hecto- Symbol: h A prefix denoting 10^2. For example, 1 hectometer (hm) = 10^2 meters (m).

hedgehog protein Any of a family of proteins that play key roles in the induction of patterning and tissue formation during embryonic development, in both invertebrates and vertebrates. In the fruit fly *Drosophila*, for example, Hedgehog protein is important in establishing the polarity (anterior–posterior orientation) of segments in the early embryo, and in patterning of adult body appendages, such as the wings. In vertebrates homologous proteins, such as *Sonic Hedgehog* in birds and mammals, are involved in the development of various embryonic and fetal structures, including the notochord, neural tube, gut, limbs, and heart. Hedgehog proteins are generally tethered to the exterior of the cell producing them. They bind to receptors on neighboring cells to activate or regulate the expression of other developmental genes, thus exerting a very localized influence.

helicase An enzyme that moves along double-stranded DNA to unwind the strands ahead of DNA replication. It melts the hydrogen bonds that link complementary bases between the two strands.

hematoxylin *See* staining.

heme (**haeme**) An iron-containing porphyrin that is the prosthetic group in HEMOGLOBIN, myoglobin, and some cytochromes.

hemerythrin A red oxygen-carrying blood pigment similar to hemoglobin in containing iron and found in various invertebrates.

hemicellulose One of a group of substances that make up the amorphous matrix of plant cell walls together with pectic substances (and occasionally, in mature cells, with lignin, gums, and mucilages). They are heteropolysaccharides, i.e. polysaccharides built from more than one type of sugar, mainly the hexoses (mannose and galactose) and the pentoses (xylose and arabinose). Galacturonic and glucuronic acids are also constituents. They vary greatly in composition between species. In some seeds (e.g. the endosperm of dates) hemicelluloses are a food reserve.

hemin The hydrochloride form of heme. Hemin is the crystalline form in which heme can be isolated and studied in the laboratory. The iron present is the trivalent state (iron(III)). Hemin can be made to crystallize by heating hemoglobin gently with acetic acid and sodium chloride. A variety of crystal forms are known.

hemizygous Describing genetic material that has no homologous counterpart and is thus unpaired in the diploid state. Both single genes and chromosome segments may be hemizygous; for example, the X chromosome in the heterogametic sex, and whole chromosomes in aneuploids.

hemocyanin A blue copper-containing blood pigment found in many mollusks and arthropods. Hemocyanin is the second most abundant blood pigment after hemo-

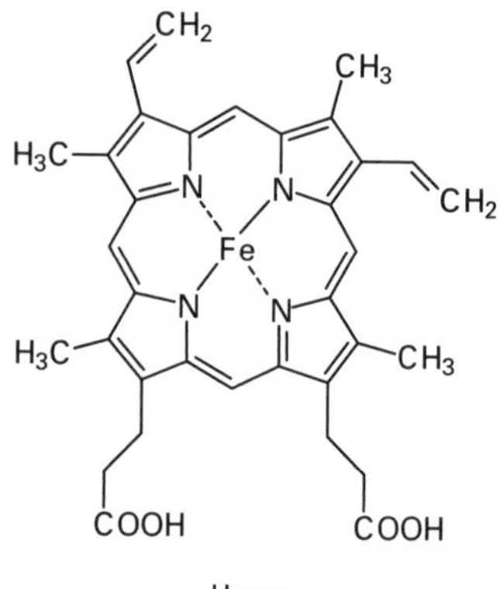

Heme

globin and functions similarly in acting as an oxygen-carrier in the blood.

hemoglobin The pigment of the red blood cells (erythrocytes) that is responsible for the transport of oxygen from the lungs to the tissues. Its molecular weight is 64 500 and consists of four polypeptide chains (globins), each being linked to an iron-containing heme group. There are variations in the polypeptide chains, giving rise to different types of hemoglobins in different species. The most important property of hemoglobin is its ability to combine reversibly with one molecule of oxygen per iron atom to form *oxyhemoglobin*, which has a bright red color. A molecule of hemoglobin can therefore carry four molecules of oxygen. The iron is present in the divalent state (iron(II)) and this remains unchanged with the binding of oxygen. When oxygen binds to the iron atom, the iron atom moves and causes structural changes in hemoglobin. Oxygen molecules, diffusing across the red cell membrane, are very readily attached to hemoglobin in the lungs and equally readily detached in the tissues. This is the mechanism by which blood transports oxygen through the body.

The binding of oxygen depends on the oxygen partial pressure; high pressure favors formation of oxyhemoglobin and low pressure favors release of oxygen. It also depends on pH. The affinity of oxygen decreases as the pH is lowered (more acid, as a result of dissolved carbon dioxide). This dependence is known as the *Bohr effect*. Carbon dioxide can also combine with hemoglobin at its amino groups. Oxygen binds cooperatively to hemoglobin: the binding of one oxygen molecule enhances the binding of further oxygen molecules.

hemolysis The release of hemoglobin from red corpuscles due to rupture of the cell membrane. It may be caused by such factors as toxins and incompatible blood transfusion.

hemophilia An inherited sex-linked condition caused by an abnormal gene on the X chromosome resulting in a deficiency of clotting Factor VIII. Hemophiliacs bleed profusely after the slightest wound or injury. As the abnormal gene is on the X chromosome, a female is affected only if she is homozygous for the condition, but a male is affected if his single X chromosome carries the gene. In this type of inheritance (X-linked recessive inheritance) the condition is transmitted either by an affected male or by a heterozygous female (the carrier). Treatment is by the administration of Factor VIII in a concentrated form.

hemopoiesis (**hematopoiesis**) The formation of blood cells. In the fetus hemopoiesis occurs in the spleen and liver; in the adult it occurs in the bone marrow (erythrocytes and polymorph white cells) and lymphoid tissue (lymphocytes and monocytes).

Hench, Philip Showalter (1896–1965) American physician who showed (with E. C. Kendall) that rheumatoid arthritis could be treated with an adrenal hormone (later known as cortisone). He was awarded the Nobel Prize for physiology or medicine in 1950 jointly with E. C. Kendall and T. Reichstein for their discoveries relating to the hormones of the adrenal cortex.

heparin A substance that prevents blood clotting by neutralizing prothrombin and stopping the action of thrombin. It is present in tissues of mammals, and is secreted by some blood-sucking animals.

hepatic portal system A venous pathway comprising the hepatic portal vein, which carries blood rich in absorbed food materials, such as glucose and amino acids, from the intestine to the liver. There the materials may be stored, converted, or released to the general circulation via the hepatic vein.

Hershey, Alfred Day (1908–97) American geneticist. He was awarded the Nobel Prize for physiology or medicine in 1969 jointly with M. Delbrück and S. E. Luria for their discoveries concerning the replication mechanism and the genetic structure of viruses.

hetero atom *See* heterocyclic compound.

heterochromatin *See* chromatin.

heterocyclic compound A compound that has a ring containing more than one type of atom. Commonly, heterocyclic compounds are organic compounds with at least one atom in the ring that is not a carbon atom. Pyridine and glucose are examples. The noncarbon atom is called a *hetero atom*. *Compare* homocyclic compound.

heterogeneous Relating to more than one phase. A heterogeneous mixture, for instance, contains two or more distinct phases.

heterogeneous nuclear RNA (**hnRNA**) An assortment of RNA molecules found associated with proteins to form *heterogeneous ribonucleoprotein* (hnRNP) particles in the nuclei of eukaryotic cells. hnRNA consists of newly transcribed pre-messenger RNA (pre-mRNA) plus other nuclear RNAs. The proteins bind to different regions of the pre-mRNA, and to each other. They are thought to facilitate subsequent processing of the pre-mRNA by the enzymes of the spliceosome to produce the mature mRNA transcript, for example, by preventing base pairing between complementary regions of the RNA. They might also function in transporting the mRNA out of the nucleus.

heterogeneous nuclear ribonucleoprotein (hnRNP) *See* heterogeneous nuclear RNA.

heterograft (**xenograft**) A type of graft from one organism to another of a different species. *See* graft.

heterolytic fission (**heterolysis**) The breaking of a covalent bond so that both electrons of the bond remain with one fragment. A positive ion and a negative ion are produced:

$$RX \rightarrow R^+ + X^-$$

Compare homolytic fission.

heterotrophism A type of nutrition in which the principal source of carbon is organic, i.e. the organism cannot make all its own organic requirements from inorganic starting materials. Most heterotrophic organisms are chemotrophic (i.e. show *chemoheterotrophism*); these comprise all animals and fungi and most bacteria. A few heterotrophic organisms are phototrophic (i.e. show *photoheterotrophism*). The nonsulfur purple bacteria, for instance, require organic molecules such as ethanol and acetate. *Compare* autotrophism. *See* chemotrophism; phototrophism.

heterozygous Having two different alleles at a given locus. Usually only one of these, the dominant allele, is expressed in the phenotype. On selfing or crossing heterozygotes some recessives may appear, giving viable offspring. Selfing heterozygotes halves the heterozygosity, and thus outbreeding maintains heterozygosity and produces a more adaptable population. *Compare* homozygous.

Hevesy, György (Károly) (1885–1966) Hungarian-born Swedish chemist and biochemist who was the first to recognize the possibility of using natural radioactive elements as indicators or tracers in chemical and biological systems. Hevesy was also the discoverer of hafnium. He was awarded the Nobel Prize for chemistry in 1943.

hexadecanoic acid *See* palmitic acid.

hexanoic acid (**caproic acid**) An oily carboxylic acid, $CH_3(CH_2)_4COOH$, found (as glycerides) in cow's milk and some vegetable oils.

hexose A sugar that has six carbon atoms in its molecules. *See* sugar.

hexose monophosphate shunt *See* pentose phosphate pathway.

Hill, (Sir) Archibald Vivian (1886–1977) British physiologist and biophysicist. He was awarded the Nobel Prize for physiology or medicine in 1922 for his work on the production of heat in muscle. The prize was shared with O. F. Meyerhof.

Hill, Robert (Robin) (1899–1991) British plant biochemist noted for his studies of the mechanism of photosynthesis.

Hill reaction The reaction, first demonstrated by Robert Hill in 1937, by which isolated illuminated chloroplasts bring about the reduction of certain substances with accompanying evolution of oxygen. For example, the blue dye dichlorophenol indophenol (DCPIP) may be reduced to a colorless substance. The reaction involves part of the normal light reaction of photosynthesis. Electrons from water involved in noncyclic photophosphorylation are used to reduce the added substance. It provided support for the idea that a light reaction preceded reduction of carbon dioxide in photosynthesis.

histamine An amine formed from the amino acid histidine by decarboxylation and produced mainly by the mast cells in connective tissue as a response to injury or allergic reaction. It causes contraction of smooth muscle, stimulates gastric secretion of hydrochloric acid and pepsin, and dilates blood vessels, which lowers blood pressure and produces inflammation, itching, or allergic symptoms (such as sneezing).

histidine *See* amino acids.

histochemistry The location of particular chemical compounds within tissues by the use of specific staining techniques, for example phloroglucinol to stain lignin.

histocompatibility The extent to which an organism's immune system will tolerate tissue grafts from another organism. Normally, a graft from an unrelated individual is recognized as foreign by the recipient's white blood cells because the marker molecules (self-antigens) on the surface of the foreign cells differ from the recipient's marker molecules. Hence, the white cells are stimulated to mount an immune response against the foreign tissue. Only certain close relatives share the same self-antigens, and can tolerate grafts of each other's tissues. The most important of these self-antigens are proteins coded by a complex cluster of genes called the major histocompatibility complex (MHC). Two classes of these proteins are important in histocompatibility and immune responses: class I MHC proteins, which occur on most body cells and 'present' viral antigens to cytotoxic T-cells; and class II MHC proteins, which occur only on certain immune system cells, such as macrophages and B-cells, and are essential for activating T-helper cells and consequently antibody secretion by B-cells. *See* B-cell; major histocompatibility complex; T-cell.

histology The study of tissues and cells at microscopic level.

histone One of a group of relatively small proteins found in chromosomes, where they organize and package the DNA. When hydrolyzed, they yield a large proportion of basic amino acids. They dissolve readily in water, dilute acids, and alkalis but do not coagulate readily on heating.

Hitchings, George Herbert (1905–98) American biochemist and pharmacologist noted for his introduction of a range of widely used synthetic drugs designed as antimetabolites, including the antifolate bactericidal agent co-trimoxazole, the immunosuppressants mercaptopurine and azathioprine, and the xanthine-oxidase inhibitor allopurinol. He was awarded the Nobel Prize for physiology or medicine in 1988 jointly with J. W. Black and G. B. Elion.

HIV (human immunodeficiency virus) A retrovirus that causes AIDS in humans by infecting and ultimately destroying certain cells, T-helper cells and macrophages, that are vital for immunity against infections. The viral particles are spherical and contain genes in the form of RNA. The virus recognizes a suitable host cell by its surface markers, binding to the cell surface protein CD4 and to one or more chemokine receptors, e.g. CCR5 before fusing with the cell membrane. Infected T-cells can bud off new virus particles from their surface. However, they can also fuse with uninfected cells carrying the same surface markers, forming multinucleate cells that are ineffective as immune cells. As increasing numbers of T-helper cells become infected, the effectiveness of the immune system diminishes and the patient becomes more and more prone to infections. There are two types of HIV: HIV-1, which occurs worldwide; and HIV-2, which is found mainly in Africa. *See* AIDS; retrovirus; T-cell.

HLA system (**human leukocyte-associated antigen system**) A group of histocompatibility antigens found on the surface of human body cells and encoded by the major histocompatibility complex of genes. It is the most important system of self-antigens in humans, and is crucial in determining the compatibility of tissue grafts between different persons, for example. *See* major histocompatibility complex.

hnRNA *See* heterogeneous nuclear RNA.

hnRNP *See* heterogeneous nuclear RNA.

Hodgkin, (Sir) Alan Lloyd (1914–98) British physiologist noted for his work on the condition of nerve impulses. He was awarded the Nobel Prize for physiology or medicine in 1963 jointly with J. C. Eccles and A. F. Huxley.

Hodgkin, Dorothy (Mary) (1910–94) British x-ray crystallographer noted for her determination of the molecular structures of benzylpenicillin and cyanocobalamin and, in particular, of the structure of crystalline zinc insulin. She was awarded the Nobel Prize for chemistry in 1964.

Holley, Robert William (1922–93) American biochemist and molecular biologist noted for his determination of the structure of alanine transfer RNA (the first determination of the sequence of a nucleic acid). He was awarded the Nobel Prize for physiology or medicine jointly with H. G. Khorana and M. W. Nirenberg (1968) for their interpretation of the genetic code and its function in protein synthesis.

holoenzyme A catalytically active complex made up of an apoenzyme and a coenzyme. The former is responsible for the specificity of the holoenzyme whilst the latter determines the nature of the reaction.

holophytic The type of nutrition in which complex organic molecules are synthesized from inorganic molecules using light energy. It is another term for *photoautotrophic.*

holozoic (**heterotrophic**) Designating organisms that feed on other organisms or solid organic matter, i.e. most animals and insectivorous plants. *Compare* holophytic.

homeobox A segment of DNA found in many so-called HOMEOTIC GENES concerned with controlling the development of organisms. It consists of 180 base pairs, and the sequence of bases is remarkably similar across a wide range of species, from yeasts to human beings. This suggests it arose early in evolutionary time and has been little changed since. The sequence encodes the amino acids of a peptide sequence that enables the parent protein to bind to DNA. This is consistent with the suggested role of homeotic proteins as genetic switches, binding to genes to control their expression. *See* differentiation.

homeostasis The maintenance of a constant internal environment by an organism. It enables cells to function more efficiently. Any deviation from this balance results in reflex activity of the nervous and hormone systems, which tend to negate the effect. The degree to which homeostasis is achieved by a particular group, independent of the environment, is a measure of evolutionary advancement.

homeotic gene (***Hox* gene**) Any of a class of genes that are crucial in determining the differentiation of tissues in different parts of the body during development. They encode proteins that regulate the expression of other genes by binding to DNA. This binding capability can be pinpointed to a characteristic base sequence known as a HOMEOBOX. Homeotic genes have been intensively studied in the fruit fly *Drosophila*. They come into play when the basic pattern of body segments has been established in the fly embryo, and direct the development of particular groups of cells in each segment. In *Drosophila* there are two major clusters of homeotic genes: the *antennapedia* complex controls development of the head and front thoracic segments, while the *bithorax* complex governs the fate of cells in more posterior segments. The physical order of the genes in these complexes corresponds to the order in which they are expressed from anterior to posterior in the developing embryo. This, together with studies of homeotic mutants, has prompted the theory that differentiation in each segment requires expression of the homeotic gene regulating that particular segment in combination with the gene for the preceding segment. A similar ordered arrangement of homeotic genes is seen in other species, including humans where there are four clusters of homeotic genes located on separate chromosomes. *See also* differentiation.

homocyclic compound A compound containing a ring made up of the same atoms. Benzene is an example of a homocyclic compound. *Compare* heterocyclic compound.

homogeneous Relating to a single phase. A homogeneous mixture, for instance, consists of only one phase.

homoglycan *See* glycan.

homograft (**allograft**) A type of graft between individuals of the same species. *See* graft.

homoiothermy The maintenance of the body temperature at a constant level, irrespective of environmental conditions. Birds and mammals are homoiothermic ('warm-blooded'). *Compare* poikilothermy.

homologous chromosomes Chromosomes that pair at meiosis. Each carries the same genes as the other member of the pair but not necessarily the same alleles for a given gene. During the formation of the germ cells only one member of each pair of homologs is passed on to the gametes. At fertilization each parent contributes one homolog of each pair, thus restoring the diploid chromosome number in the zygote. With the exception of the sex chromosomes, for example in mammals the Y chromosome is much smaller than the X chromosome, the members of each homologous pair are similar to one another in size and shape.

homologous series A group of organic compounds possessing the same functional group and having a regular structural pattern so that each member of the series differs from the next one by a fixed number of atoms. The members of a homologous series can be represented by a general formula.

homolytic fission (**homolysis**) The breaking of a covalent bond so that one electron from the bond is left on each fragment. Two free radicals result:

$$RR' \rightarrow R\bullet + R'\bullet$$

Compare heterolytic fission.

homozygous Having identical alleles for any specified gene or genes. A homozygote breeds true for the character in question if it is selfed or crossed with a similar homozygote. An organism homozygous at every locus produces offspring identical to itself on selfing or when crossed with a genetically identical organism. Homozygosity is obtained by inbreeding, and homozygous populations may be well adapted to a certain environment, but slow to adapt to changing environments. *Compare* heterozygous.

Hopkins, (Sir) Frederick Gowland (1861–1947) British biochemist noted for his discoveries of essential amino acids and vitamins. He also isolated glutathione and tryptophan. He was awarded the Nobel Prize for physiology or medicine in 1929. The prize was shared with Christiaan Eijkman.

hormone 1. (*Zoology*) A chemical messenger liberated by a certain type of gland (endocrine gland) or group of cells and transported in the blood to a specific (target) organ or tissue where it acts to control growth, metabolism, sexual reproduction, and other body processes. Hormones may be steroids, polypeptides, or amines. They are recognized by specific molecules or receptors in the cells of target organs. These receptors are usually proteins located at the membrane (e.g. for insulin, glucagon, and epinephrine) or in the cytoplasm (e.g. for estrogens and progesterone). The hormones exert their effects on enzymes or nucleic acids.
2. (*Botany*) *See* plant hormone.

Houssay, Bernardo Alberto (1887–1971) Argentinian physiologist and endocrinologist noted for his researches on diabetes and on the adrenal, pituitary, and thyroid glands. He was awarded the Nobel Prize for physiology or medicine in 1947 for his discovery of the part played by the hormone of the anterior pituitary lobe in the metabolism of sugar. The prize was shared with C. F. and G. T. Cori.

HPLC High-performance liquid chromatography; a sensitive analytical technique, similar to gas–liquid chromatography but using a liquid carrier. The carrier is specifically choosen for the particular substance to be detected.

Huber, Robert (1937–) German biochemist who worked on the structure of proteins. He was awarded the Nobel Prize for chemistry in 1988 with J. Deisenhofer and H. Michel.

Human Genome Project An international project launched in 1989 with the aim of mapping and sequencing the entire human genome. The target date for completion was 2005 but the first draft of the human genome sequence was published in 2001. It covers about 97% of the human genome. The speed of the project was enabled by cooperation between publicly and privately funded organizations, although this has been (and still is) a major source of controversy over who 'owns' the sequence. While sequencing the human genome is a major achievement, it is not the end because the function of many of the genes that make up the human genome are still uncertain. It is not even known how many genes there are. Working out the function of every human gene and its product is perhaps an even more daunting project.

The results are being used in the diagnosis of a wide range of diseases. These include not only hereditary disorders, such as Huntington's disease, but also many common ailments with a genetic component, such as heart disease and breast cancer. Sequencing the genes responsible for the development of these diseases enables the design of DNA probes that will identify susceptible individuals, so allowing preventive measures or routine check-ups. However, such knowledge has profound ethical and commercial implications. *See* chromosome map.

human immunodeficiency virus *See* HIV.

Hunt, R. Timothy (1943–) British cell biologist who was awarded the 2001 Nobel Prize for physiology or medicine jointly with L. H. Hartwell and P. M. Nurse for discoveries of the key regulators of the cell cycle.

Huxley, (Sir) Andrew Fielding (1917–) British physiologist noted for his work on the contraction of muscle and the conduction of nerve impulses. He was awarded the Nobel Prize for physiology or medicine in 1963 jointly with J. C. Eccles and A. L. Hodgkin.

Huxley, Hugh Esmor Huxley (1924–) British biophysicist noted for his study of the relation of ATP hydrolysis to the mechanics of muscle contraction.

hyaloplasm **(cell matrix; ground substance)** *See* cytoplasm.

hyaluronic acid A type of organic acid that has the properties of a lubricant. It is found, for example, in the synovial fluid of joints and vitreous humor of the eye.

hydrazone A type of organic compound containing the $C:NNH_2$ group, formed by the reaction between an aldehyde or ketone and hydrazine (N_2H_4). Derivatives of hydrazine are often used to produce crystalline products, which have sharp melting points that can be used to characterize the original aldehyde or ketone. Phenylhydrazine ($C_6H_5NH.NH_2$), for instance, produces *phenylhydrazones*.

hydrocarbon Any compound containing only the elements carbon and hydrogen. Examples are the alkanes, alkenes, alkynes, and aromatics such as benzene and naphthalene.

hydrocortisone **(cortisol)** A steroid hormone produced by the adrenal cortex having glucocorticoid activity. *See* glucocorticoid.

hydrogen An essential element in living tissues. It enters plants, with oxygen, as water and is used in building up complex reduced compounds such as carbohydrates and fats. Water itself is an important medium, making up 70–80% of the weight of organisms, in which chemical reactions of the cell can take place. Hydrogenated compounds, particularly fats, are rich in energy and on breakdown release energy for driving living processes.

hydrogenation The reaction of a compound with hydrogen. In organic chemistry, hydrogenation refers to the addition of hydrogen to multiple bonds, usually with the aid of a catalyst. Unsaturated natural liquid vegetable oils can be hydrogenated to form saturated semisolid fats – a reaction used in making types of margarine.

hydrogen bond An intermolecular bond in which two atoms of different molecules share a hydrogen atom. Hydrogen bonding takes place between an electronegative atom, e.g. oxygen, nitrogen, or fluorine atoms and a hydrogen atom joined to another electronegative atom. In biology the electronegative atom is usually nitrogen or oxygen. The attraction is due to electrostatic forces. Hydrogen bonding is responsible for the properties of water and is important in many biological systems for holding together the structure of large molecules, such as proteins and DNA.

hydrolase An enzyme that catalyzes a hydrolysis reaction. Digestive enzymes are an example. Hydrolases play an important part in rendering insoluble food material into a soluble form, which can then be transported in solution.

hydrolysis A reaction between a compound and water. For example,

$$CH_3COOC_2H_5 + H_2O \leftrightarrow CH_3COOH + C_2H_5OH.$$

hydrophilic *See* lyophilic.

hydrophobic *See* lyophobic.

hydroxybenzoate *See* salicylate.

2-hydroxybutanedioic acid *See* malic acid.

5-hydroxytryptamine *See* serotonin.

hyperplasia Enlargement of a tissue due to an increase in the number of its cells. For example, if part of the liver is removed, the remaining part may undergo hyperplasia in order to regenerate. *Compare* hypertrophy.

hyperpolarization An increase in the polarity of the potential difference across a membrane of a nerve or muscle cell, i.e. an increase in the resting potential. It is caused by the pumping of ions across the membrane so that differential concentrations are created on either side – the inside becoming more negative. As a result, a stronger stimulus is needed to evoke a response.

hypertonic Designating a solution with an osmotic pressure greater than that of a specified other solution, the latter being hypotonic. When separated by a semipermeable membrane (e.g. a cell membrane) water moves by osmosis into the hypertonic solution from the hypotonic solution. *Compare* hypotonic; isotonic.

hypertrophy Enlargement of a tissue or organ due to an increase in the size of its cells or fibers. An example is the enlargement of muscles as a result of exercise. *Compare* hyperplasia.

hypostasis The situation in which the expression of one gene (the *hypostatic gene*), is prevented in the presence of another, nonallelic, gene (the *epistatic gene*). *See* epistasis.

hypothalamus Part of the vertebrate forebrain that is concerned primarily with regulating the physiological state of the body. Blood temperature and chemical composition are monitored by the hypothalamus, which also regulates body temperature, drinking, eating, water excretion, and other metabolic functions, largely by its influence on the release of hormones by the pituitary gland. Heart rate, breathing rate, blood pressure, and sleep patterns are also controlled by the hypothalamus via its connections to centers in the cerebral cortex and the medulla oblongata. The hypothalamus connects directly to the pituitary gland via the infundibulum.

hypotonic Designating a solution with an osmotic pressure less than that of a specified other solution, the latter being hypertonic. When separated by a semipermeable membrane (e.g. a cell membrane) water is lost by osmosis from the hypotonic to the hypertonic solution. *Compare* hypertonic; isotonic.

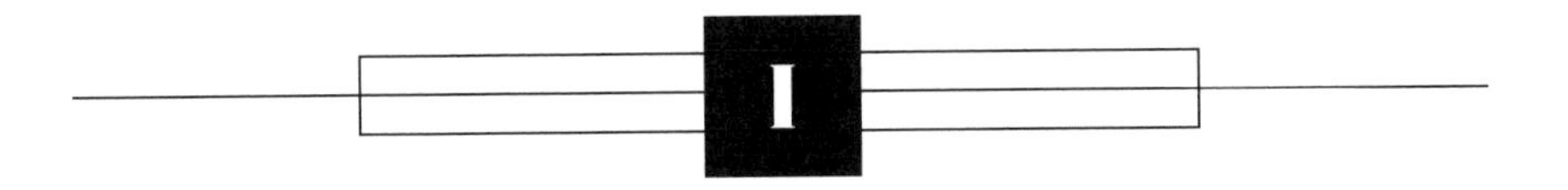

IAA (indole acetic acid) A naturally occurring auxin. *See* auxin.

ICSH (interstitial-cell-stimulating hormone) *See* luteinizing hormone.

idiogram *See* karyogram.

IFN *See* interferon.

Ignarro, Louis J. (1941–) American biochemist who was awarded the 1998 Nobel Prize for physiology or medicine jointly with R. F. Furchgott and F. Murad for work on the action of nitric oxide as a signaling molecule in the cardiovascular system.

imbibition The phenomenon in which a substance absorbs a liquid and swells, but does not necessarily dissolve in the liquid. The process is reversible, the substance contracting on drying. Water is imbibed by many biological substances: cellulose, hemicelluloses, pectic substances, lignin (all plant cell wall constituents); starch; certain proteins; etc.

imide An organic compound, general formula R-CONHCO-R′, where R and R′ are alkyl or aryl radicals. The group -CONHCO- is known as the *imido group*.

imido group *See* imide.

imine A compound containing the group -NH-, where the nitrogen atom is not joined to carbonyl groups or to other hydrogen atoms. The group is the *imino group*

imino group *See* imine.

immortal cells Normal cells have a finite life span, but they can acquire the ability to grow and divide indefinitely. Such cells are known as *immortal cells*. Cancer cells are immortal. Artificial immortal *cell lines* have been created e.g. from human tumor cells for use in research and biotechnology.

immune clearance (immune elimination) The rapid removal of antigen introduced into the body of an immune individual, as a result of its complexing with antibody.

immune response A response to the introduction of antigen into the body involving production of specific antibodies or lymphocytes, which combine with the antigen. It is the basic mechanism of active immunity, and of hypersensitivity, e.g. allergies. *See* immunity.

immunity The ability of plants and animals to withstand harmful infective agents and toxins. Immunity arises partly from a number of nonspecific mechanisms, such as inflammation and phagocytosis or an impervious skin (*nonspecific immunity*). However, in vertebrates it is largely the result of specific mechanisms, whereby certain substances (ANTIBODIES) or lymphocytes present in the body combine with an introduced foreign substance (antigen) – *specifically acquired immunity*. Specifically acquired immunity includes *passive immunity*, where the antibody has been derived from another individual (e.g. from the mother to offspring), and *active immunity*, where the antibody is produced following stimulation with antigen (e.g. by vaccination or by exposure to infection). *Cell-mediated immunity* is achieved by T-cells, which detect and destroy virus-infected body cells (*see* T-cell). *Humoral immunity* involves the circulation of free antibody in the blood. The immune system distinguishes between 'self' and 'nonself' and the loss of this ability leads to so-called *autoimmune diseases*, in which the immune system attacks its own body. *See also* immunoglobulin; lymphocyte.

immunization The process of making an animal resistant to infection or harmful agents. *See* immunity.

immunoassay Any of various techniques for measuring biological substances that depend on the substance acting antigenically and binding with a specific antibody. Most also require the addition of labeled antigen or antibody, bearing a radioisotope, fluorescent molecule, or enzyme, for example, to reveal the extent of the antigen–antibody binding, and enable the amount or concentration of the test sample to be established. *See* ELISA; radioimmunoassay.

immunoelectrophoresis *See* electrophoresis.

immunoglobulin (Ig) Any of a group of proteins that take part in the immune responses of higher animals. They are produced by B-lymphocytes (*see* B-cell), and act mainly as antibodies by binding to foreign antigens, such as bacteria, viruses, or toxins. Each consists of four polypeptide chains, two heavy chains and two light chains, arranged to form a Y-shaped structure. The two arms of the 'Y' each bear an antigen-binding site. The antigen-binding site has a unique sequence and is specific. There are five classes, distinguished by the structure of their heavy chains.
Immunoglobulin A (IgA) is found mainly in secretions such as saliva and tears, and its main role is to neutralize viruses and bacteria on the external mucous surfaces of the body.
Immunoglobulin D (IgD) occurs on the surface of lymphocytes and is thought to control the activation and suppression of B-cell activity.
Immunoglobulin E (IgE) is normally present in very low concentrations in blood and connective tissue underlying epithelia, but its level is raised in allergies. It binds to mast cells, and in the presence of antigen causes histamine release from mast cells, with consequent inflammation and the other common symptoms of allergies.
Immunoglobulin G (IgG) is the main immunoglobulin of blood and tissue fluid. It binds to microorganisms, enhancing their engulfment by phagocytic cells, and neutralizes viruses and bacterial toxins. It also activates complement, leading to lysis of target cells, and can cross the placenta, affording protection to the fetus.
Immunoglobulin M (IgM) is a star-shaped molecule comprising five of the basic Y-shaped units. It occurs within blood vessels and is most prominent early in the immune response, mopping up microorganisms or other antigens with its formidable array of binding sites.

immunological tolerance The failure of the antibody response to an antigen, usually one to which the animal has been exposed previously.

implantation (nidation) The attachment of the developing mammalian egg to the wall of the uterus. The human egg enters the uterus from the Fallopian tube, where it has been fertilized four days earlier. Cells on its outer surface destroy the cells of the uterine wall and invade the mother's tissues, anchoring the growing embryo and making way for the development of the placenta. Implantation occurs at the blastocyst stage of embryonic development. In women, implantation prevents the next menstrual period from occurring, probably by the action of human chorionic gonadotropin (HCG), which maintains the secretion of progesterone by the corpus luteum during pregnancy.

incompatibility **1.** The rejection of grafts, transfusions, or transplants between animals or plants of different genetic composition.
2. A mechanism in flowering plants that prevents fertilization and development of

an embryo following pollination by the same or a genetically identical individual. It is due to interaction between genes in the pollen grain and those in the stigma, in such a way that the pollen is either unable to grow or grows more slowly on the stigma. It results in self-sterility, thus preventing inbreeding.

incomplete dominance *See* co-dominance.

independent assortment The law, formulated by Mendel, that genes segregate independently at meiosis so that any one combination of alleles is as likely to appear in the offspring as any other combination. The work of T. H. Morgan later showed that genes are linked together on chromosomes and so tend to be inherited in groups. The law of independent assortment therefore only applies to genes on different chromosomes. *See* linkage; Mendel's laws.

indicator A substance used to test for acidity or alkalinity of a solution by a color change. Examples are litmus and phenolphthalein. A *universal indicator* shows a range of color changes over a wide range from acid to alkaline, and can be used to estimate the pH.

indigo A blue organic dye that occurs (as a glucoside) in plants of the genus *Indigofera*. It is a derivative of indole, and is now made synthetically.

indole (**benzpyrrole**) A colorless solid organic compound, whose molecules consist of a benzene ring fused to a pyrrole ring. It occurs in coal tar and various plants, and is the basis of indigo and of several plant hormones.

indole acetic acid (**IAA**) A naturally occurring auxin. *See* auxin.

inducible enzyme *See* adaptive enzyme.

inflammation A defensive reaction of animal tissues to injury, infection, or irritation, characterized by redness, swelling, heat, and pain. HISTAMINE released by basophils and mast cells dilates capillaries and causes them to leak plasma (including proteins) and leukocytes. The effects are enhanced by various other substances, including prostaglandins and kinins. Fibrin forms a clot behind the site of injury, sealing it off, and phagocytic cells (neutrophils and macrophages) engulf infecting microorganisms, thus destroying them.

inhibin A glycoprotein involved in the differentiation and growth of certain cells. It inhibits secretion of follicle-stimulating hormone by the pituitary gland.

inhibition **1.** (in the nervous system) The reduction or complete prevention of activation of an effector by means of inhibitory nerve impulses. An example is the inhibition of the reflex tonic contraction of antagonistic muscles when a voluntary skeletal muscle is to be contracted.

Inhibitory synapses release a neurotransmitter that opens channels in the postsynaptic membrane that are permeable to potassium ions but not to sodium ions. Thus there is an outflow of positive potassium ions from the postsynaptic cell, increasing the polarization of the membrane and rendering depolarization, and therefore formation of an action potential, less likely. Inhibitory synapses play an important part in central-nervous-system control of motor activity.

2. (of enzyme action) The action of a substance (an *inhibitor*) that interferes with and negatively affects the action of an enzyme. There are three types, depending on how the inhibitor interacts with the enzyme: in competitive inhibition, the inhibitor binds to the active site and prevents substrate binding; in noncompetitive inhibition the inhibitor binds to a site other than the active site and does not affect substrate binding; in uncompetitive inhibition, the inhibitor binds the enzyme–substrate complex but not the free enzyme.

inhibitor *See* inhibition.

initiation complex **1.** A complex assembly of proteins and RNAs that is formed at

the start of translation during protein synthesis at ribosomes. The small ribosomal subunit associates with initiation factors, GTP, and the initiator transfer RNA to form a *preinitiation complex*. This then becomes positioned at the start site on a messenger RNA (mRNA) molecule, and the large ribosomal subunit binds to the small subunit, forming the complete initiation complex and enabling translation to proceed.
2. A large enzyme complex (holoenzyme) formed at the start of transcription of eukaryotic genes. It comprises RNA polymerase II and as many as 60–70 polypeptides, including transcription factors, and binds to the promoter site.

initiation factor Any of various proteins that enable the formation and correct positioning of the INITIATION COMPLEX that precedes translation of messenger RNA (mRNA) during protein synthesis at ribosomes. These factors are designated IF1, IF2, etc. in prokaryotes, and eIF1, eIF2, etc. in eukaryotes.

inositol An optically active cyclic sugar alcohol. It is a component of the vitamin B complex and is required for growth in certain animals and microorganisms. The stereoisomer *myo-inositol* is a precursor of phosphatidyl inositol, an important constituent of animal membranes, muscle, and brain and is involved in mediating the actions of several hormones.

inositol 1,4,5-trisphosphate (IP_3) An intracellular signaling molecule, derived from phosphatidylinositol, that acts as a second messenger in various cell signaling pathways. It is water soluble and can diffuse through the cell cytoplasm to its target. For example, it can bind to IP_3-sensitive calcium channels in the endoplasmic reticulum (ER). Opening of the latter releases calcium ions from the ER, which mediates various cellular processes. Hormone-induced cell responses often involve such an IP_3-triggered rise in intracellular calcium ions. IP_3 is hydrolyzed, and thus inactivated, within a second of its formation.

insulin A hormone that controls the metabolism of glucose. Lack of insulin results in DIABETES, but excess insulin leads to coma. Insulin is a polypeptide produced by the B-cells of the islets of Langerhans of the pancreas. Its secretion is stimulated by high blood levels of glucose and amino acids after a meal. Glucose uptake is then stimulated by the action of insulin on various tissues (e.g. muscles, liver, and fat). It also stimulates glycogen and fat synthesis. Insulin is used therapeutically in the treatment of diabetes mellitus.

insulin-like growth factor (IGF) Any of a group of polypeptides that are structurally related to insulin and act as growth-promoting agents. They stimulate cell division and protein synthesis, and also promote glycogen formation in muscle and fat deposition. Their production is itself stimulated by growth hormone (somatotropin), and they include IGF-I (somatomedin C) and IGF-II (somatomedin A).

intercellular Describing materials found and processes occurring between cells. *Compare* intracellular.

interferon (IFN) Any of a group of proteins produced by animal cells in response to infection by viruses or other agents. They act to inhibit viral replication within the cell; some also have antibacterial and anticancer properties. They are produced commercially for various therapeutic applications, using recombinant DNA technology.

interleukin (IL) Any of a class of cytokines that act as chemical messengers between various types of white blood cells (leukocytes). All are proteins or glycoproteins. For example, interleukin 2 (IL-2) stimulates the growth of T-cells; IL-4 activates B-cells, among other functions; while IL-7 stimulates the proliferation of leukocyte precursor cells. *See* cytokine. *Compare* lymphokine.

intermediate A transient chemical entity in a complex reaction. *See also* precursor.

intermediate bonding A form of covalent bond that also has an ionic or electrovalent character. *See* polar bond.

intermolecular forces Forces of attraction between molecules rather than forces within the molecule (chemical bonding). If these intermolecular forces are weak the material will be gaseous and as their strength progressively increases materials become progressively liquids and solids. The intermolecular forces are divided into H-bonding forces and van der Waals forces, and the major component is the electrostatic interaction of dipoles. *See* hydrogen bond; van der Waals force.

internal energy (**intrinsic energy**) Symbol: U The quantity of energy possessed by a given substance or system. The internal energy is the sum of the kinetic and potential energies of the atoms or molecules of the system. In practice, changes in internal energy are important in chemical reactions. *See* enthalpy.

internal environment The medium surrounding the body cells of multicellular animals, i.e. the intercellular fluid. In vertebrates its composition is kept relatively constant by the mechanisms of homeostasis.

interphase The stage in the cell cycle when the nucleus is not in a state of division. Interphase is divisible into various stages each characterized by a differing physiological activity. *See* cell cycle.

interstitial-cell-stimulating hormone (**ICSH**) *See* luteinizing hormone.

intracellular Describing the material enclosed and processes occurring within a cell membrane. *Compare* intercellular.

intrinsic energy *See* internal energy.

intrinsic factor A glycoprotein secreted by the stomach lining that is required for the absorption of vitamin B_{12} (cobalamin). It binds the vitamin in the small intestine and is then itself bound to a specific receptor on the intestinal mucosa, where the vitamin is released and transported across the mucosal cell into the bloodstream. Deficiency of intrinsic factor leads to malabsorption of vitamin B_{12} and hence pernicious anemia.

intron A noncoding DNA sequence that occurs between coding sequences (exons) in many eukaryote genes. Mature RNA does not contain introns, these being removed during the transcription process (splicing). Some introns are spliced by an autocatalytic process in which the RNA acts as its own enzyme (*see* ribozyme). However, in the case of mRNA in the nucleus the process is regulated by a complex of proteins called a *spliceosome*. Different mRNAs can be produced from a DNA transcript by the splicing together of different exons – a process known as *alternative splicing*.

The function and evolution of introns is a highly contentious subject. One view is that by acting as 'spacers' for exons in mRNA they enable *exon shuffling* – recombination or rearrangement of exons – and hence the rapid evolution of proteins with different combinations of functional groups. Another view is that they represent selfish DNA, which confers no advantage on the host. It is now known that introns also occur in some bacteria, particularly in some archaebacteria and blue-green bacteria, and even in certain viruses. *Compare* exon.

inulin A polysaccharide food reserve of some higher plants, particularly the Compositae, e.g. *Dahlia* root tubers. It is a polymer of fructose.

inversion *See* chromosome mutation.

invert sugar *See* sucrose.

in vitro Literally 'in glass'; describing experiments or techniques performed in laboratory apparatus rather than in the living organism. Cell tissue cultures and *in vitro* fertilization (to produce 'test-tube babies') are examples. *Compare in vivo*.

in vivo Literally 'in life'; describing processes that occur within the living organism. *Compare in vitro.*

involuntary muscle *See* smooth muscle.

iodine A trace element essential in animal diets mainly as a constituent of the thyroid hormones. Deficiency of iodine causes the thyroid gland to enlarge, giving the condition known as goiter. Iodine is not essential to plant growth although it is accumulated in large amounts by the brown algae.

ion An atom or molecule that has a negative or positive charge as a result of losing or gaining one or more electrons.

ion channel A channel through a cell membrane that allows the passage through the membrane of specific inorganic ions. There are several different types of channel, characterized by how they are controlled: *ligand gated* channels are opened by an agonist binding to a receptor; *voltage gated* channels are controlled by the membrane potential; *mechanically gated* channels are opened by mechanical stress.

ion exchange A process that takes place in certain insoluble materials, which contain ions capable of exchanging with ions in the surrounding medium. Zeolites, the first ion exchange materials, were used for water softening. These have largely been replaced by synthetic resins made of an inert backbone material, such as polyphenylethene, to which ionic groups are weakly attached. If the ions exchanged are positive, the resin is a cationic resin. An anionic resin exchanges negative ions. When all available ions have been exchanged (e.g. sodium ions replacing calcium ions) the material can be regenerated by passing concentrated solutions (e.g. sodium chloride) through it. The calcium ions are then replaced by sodium ions. Ion-exchange techniques are used for a vast range of purification and analytical purposes.

ionic bond *See* electrovalent bond.

iron An essential nutrient for animal and plant growth. It is contained in the protein HEMOGLOBIN, which gives the color to red blood cells and is responsible for oxygen transport from the lungs. Iron deficiency leads to anemia. Iron is also found in other porphyrins and in cytochromes, which are important components of the electron-transport chain. It is also required as a cofactor for certain enzymes, e.g. catalase and peroxidase. Certain iron-containing proteins are essential for the fixation of nitrogen by bacteria.

islets of Langerhans Clusters of endocrine cells within the pancreas (described by Langerhans in 1869) that produce various hormones, notably insulin, which controls blood-sugar level. The islets consist of three types of cells: A-cells, which secrete glucagon; B-cells, which secrete insulin; and D-cells, which secrete somatostatin.

isoantigen A type of antigen that induces antibody production in genetically different individuals of the same species but not in the individual itself.

isoenzyme (**isozyme**) An ENZYME that occurs in different structural forms within a single species. The isomeric forms all have the same molecular weight but differing structural configurations and properties. Large numbers of different enzymes are known to have isomeric forms; for example, lactate dehydrogenase has five forms. Variations in the isoenzyme constitution of individuals can be distinguished by electrophoresis.

isograft (**syngraft**) A type of graft between individuals that are genetically identical. *See* graft.

isoleucine *See* amino acids.

isomer Any one of the possible compounds arising from the different ways of grouping the atoms in a given molecular formula. For example, methoxymethane and ethanol are isomers of C_2H_6O. Types of isomers are classified according to the type of isomerism exhibited. *See* isomerism.

isomerase An enzyme that catalyses the interconversion of isomers. They are classified by the type of isomers they act on. A cis-trans isomerase interconverts a cis C=C and a trans C=C e.g. in the unsaturated hydrocarbon chains of fatty acids. A racemase interconverts enantiomers. An epimerase interconverts the configuration of chiral centres of diastereoisomers, e.g. phosphopentose epimerase interconverts rubulose-5-phosphate and xylulose-5-phosphate in the pentose phosphate pathway. An intramolecular oxidoreductase interconverts oxidised and reduced isomers e.g. glucose-6-phosphate isomerase interconverts glucose-6-phosphate and fructose-6-phosphate in the glycolytic pathway. A mutase catalyses the transfer of chemical groups within a molecule. See isomer.

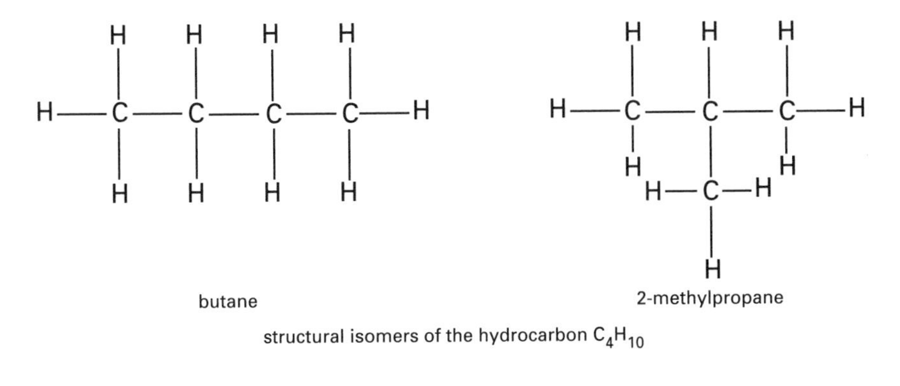

structural isomers of the hydrocarbon C_4H_{10}

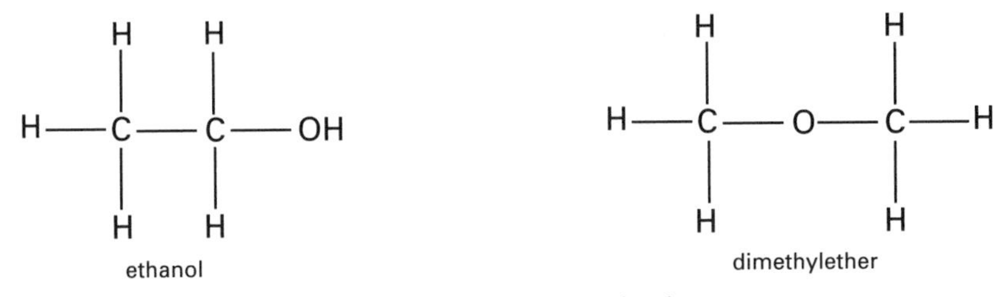

structural isomers differing in functional groups

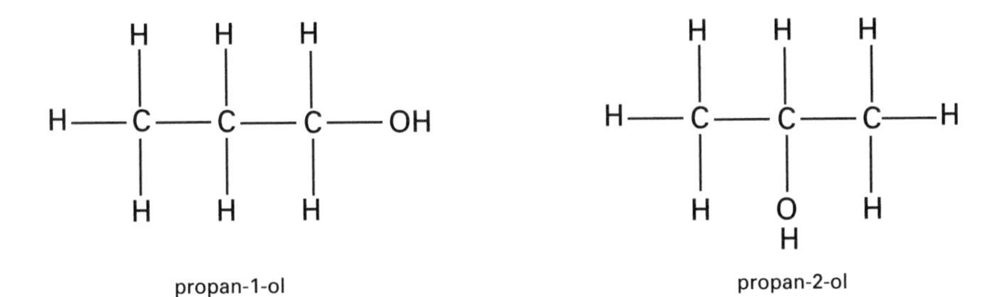

structural isomers differing in the position of the functional group

Isomer

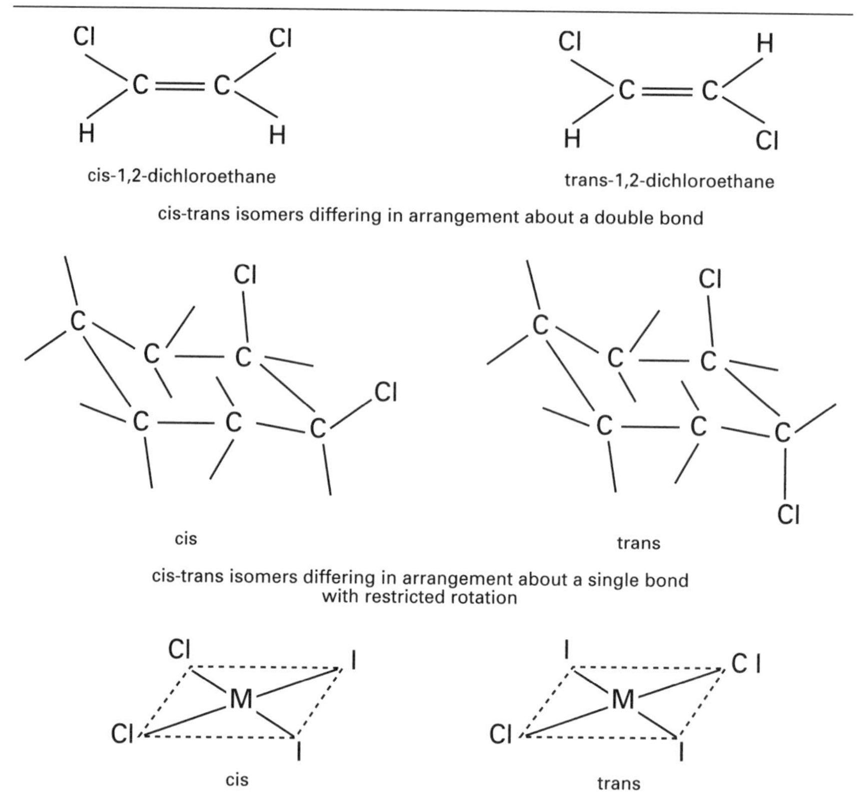

Isomer

isomerism The existence of two or more chemical compounds with the same molecular formulae but different structural formulae or different spatial arrangements of atoms. The different forms are known as *isomers.* For example, the compound C_4H_{10} may be butane (with a straight chain of carbon atoms) or 2-methyl propane ($CH_3CH(CH_3)CH_3$, with a branched chain).

Structural isomerism is the type of isomerism in which the structural formulae of the compounds differ. There are two main types. In one the isomers are different types of compound. An example is the compounds ethanol (C_2H_5OH) and methoxymethane (CH_3OCH_3), both having the formula C_2H_6O. In the other type of structural isomerism, the isomers differ because of the position of a functional group. For example, the primary alcohol propan-1-ol (C_3H_7OH) and the secondary alcohol propan-2-ol ($CH_3CH(OH)CH_3$) are isomers.

Stereoisomerism occurs when two compounds with the same molecular formulae and the same groups differ only in the arrangement of the groups in space. There are two types of stereoisomerism.

Cis-trans isomerism occurs when there is restricted rotation about a bond between two atoms. Groups attached to each atom may be on the same side of the bond (the *cis isomer*) or opposite sides (the *trans isomer*). *Cis-trans* isomerism also occurs in some inorganic complex compounds, when two groups may be at adjacent (*cis*) or opposite (*trans*) positions. *Cis-trans* isomerism was formerly called *geometrical isomerism.* Sometimes the terms *syn* and *anti* are used, which correspond to *cis* and *trans*, respectively.

Optical isomerism occurs when the compound has no plane of symmetry and can exist in left- and right-handed forms that are mirror images of each other. Such molecules have an asymmetric atom – i.e. one attached to four different groups – called a *chiral center*. *See also* optical activity.

isotonic Designating a solution with an osmotic pressure or concentration equal to that of a specified other solution, usually taken to be within a cell. It therefore neither gains nor loses water by osmosis. *Compare* hypertonic; hypotonic.

isotope One of two or more atoms of the same element that differ in atomic mass, having different numbers of neutrons. For example ^{16}O and ^{18}O are isotopes of oxygen, both with eight protons, but ^{16}O has eight neutrons and ^{18}O has ten neutrons. A natural sample of most elements consists of a mixture of isotopes. Many isotopes are radioactive and can be used for labeling purposes. The isotopes of an element differ in their physical properties and can therefore be separated by techniques such as fractional distillation, diffusion, and electrolysis.

Isotopes are also used in kinetic studies. For example, if the bond between two atoms X–Y is broken in the rate-determining step, and Y is replaced by a heavier isotope of the element, Y*, then the reaction rate will be slightly lower with the Y* present. This *kinetic isotope effect* is particularly noticeable with hydrogen and deuterium compounds, because of the large relative difference in mass.

isozyme *See* isoenzyme.

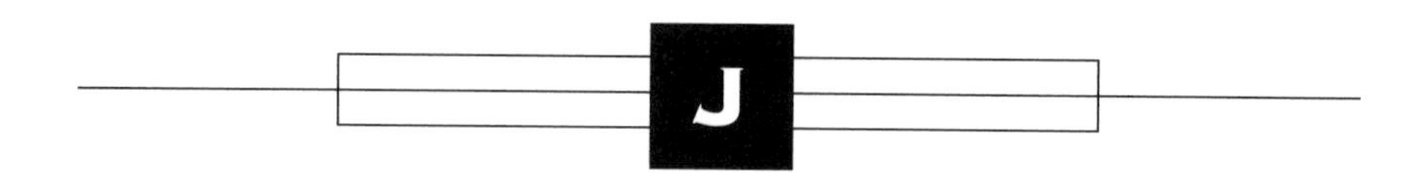

Jacob, François (1920–) French molecular biologist who postulated the existence of messenger RNA. He also investigated genetic regulatory mechanisms in protein synthesis. He was awarded the Nobel Prize for physiology or medicine in 1965 jointly with J. L. Monod and A. M. Lwoff.

Jerne, Niels Kaj (1911–94) Danish microbiologist and immunologist who studied antibody generation. He was awarded the Nobel Prize for physiology or medicine in 1984 jointly with G. J. F. Köhler and C. Milstein.

jumping gene *See* transposon.

junk DNA Repetitive sequences of eukaryotic DNA that apparently have no useful function. *See* selfish DNA.

juvenile hormone (**neotonin**) A hormone secreted by endocrine glands associated with the brain in insects that prevents metamorphosis into the adult form and maintains the presence of larval characteristics. The exact mechanism of juvenile hormone action is not clear, but it appears to modify the effect of the molting hormone, ecdysterone.

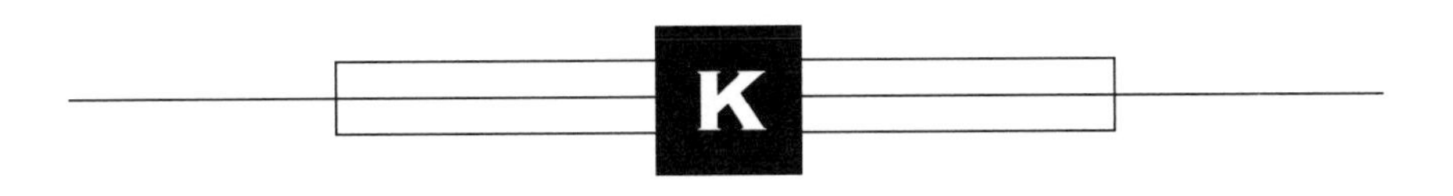

kairomone A chemical messenger emitted by an individual of a species and causing a response in an individual of another species. This may be detrimental to the producer of the kairomone, for example many parasites are attracted to their hosts by an excreted kairomone. *See also* pheromone.

Kandel, Eric R. (1929–) American biochemist noted for his work on signal transduction in the nervous system. He shared the 2000 Nobel Prize for physiology or medicine with P. Greengard and A. Carlsson.

Karrer, Paul (1889–1971) Swiss organic chemist noted for the synthesis of vitamins A, B_2, and E. He was awarded the Nobel Prize for chemistry in 1937. The prize was shared with W. N. Haworth.

karyogram (**idiogram**) The formalized layout of the karyotype of a species, often with the chromosomes arranged in a certain numerical sequence.

karyokinesis *See* mitosis.

karyotype The physical appearance of the chromosome complement of a given species. A species can be characterized by its karyotype since the number, size, and shape of chromosomes vary greatly between species but are fairly constant within species.

Katz, (Sir) Bernard (1911–) German-born British biophysicist and neurophysiologist who worked on the physiology of muscle and nerve, especially the mechanism of release of acetylcholine by neural impulses. He was awarded the Nobel Prize for physiology or medicine in 1970 jointly with J. Axelrod and U. S. von Euler.

kb *See* kilobase.

kelvin Symbol: K The SI base unit of thermodynamic temperature. It is defined as the fraction 1/273.16 of the thermodynamic temperature of the triple point of water. Zero kelvin (0 K) is absolute zero. One kelvin is the same as one degree on the Celsius scale of temperature.

Kendall, Edward Calvin (1886–1972) American chemist who first isolated the thyroid hormone thyroxine (1915) and also isolated a number of hormones of the adrenal cortex and studied their action. He was awarded the Nobel Prize for physiology or medicine in 1950 jointly with P. S. Hench and T. Reichstein.

Kendrew, (Sir) John Cowdery (1917–97) British biophysicist and molecular biologist who determined the three-dimensional structure of the myoglobin molecule by x-ray diffraction. He was awarded the Nobel Prize for chemistry in 1962; the prize was shared with M. F. Perutz for their studies of the structures of globular proteins.

keratin One of a group of fibrous insoluble sulfur-containing proteins (scleroproteins) found in ectodermal cells of animals, as in hair, horns, and nails. Leather is almost pure keratin. There are two types: α keratins and β keratins. The former have a coiled structure, whereas the latter have a beta pleated sheet structure. *Cytokeratins* form part of the filamentous cytoskeleton of cells.

keratinization (**cornification**) A process occurring in vertebrate epidermis and epidermal structures in which KERATIN replaces the cytoplasm of a cell. For example, the cornified outer layer of the epidermis of the skin consists of dead horny cells. Hairs and nails also consist of keratinized cells.

ketal A type of organic compound formed by addition of an alcohol to a ketone. Addition of one molecule gives a *hemiketal*; two molecules give a full ketal. *See* acetal.

keto–enol tautomerism An equilibrium between two isomers in which one isomer is a ketone (the *keto form*) and the other an enol. The equilibrium is by transfer of a hydrogen atom between the oxygen atom of the carbonyl group and an adjacent carbon atom.

ketogenesis The formation of ketone bodies.

ketohexose A ketose SUGAR with six carbon atoms.

ketone A type of organic compound with the general formula RCOR, having two alkyl or aryl groups bound to a carbonyl group. They are formed by oxidizing secondary alcohols (just as aldehydes are made from primary alcohols). Simple examples are propanone (acetone, CH_3COCH_3) and butanone (methyl ethyl ketone, $CH_3COC_2H_5$).

The chemical reactions of ketones are similar in many ways to those of aldehydes.

ketone body One of a group of organic substances formed in fat metabolism, mainly in the liver. Examples are acetoacetic acid and acetone. Ketone bodies are the major fuel source for resting skeletal muscle. If the body has little or no carbohydrate as a respiratory substrate, *ketosis* occurs, in which more ketone bodies are produced than the body can use.

ketopentose A ketose SUGAR with five carbon atoms.

ketose A SUGAR containing a ketone (=CO) or potential ketone group.

Khorana, Har Gobind (1922–) Indian organic chemist and biochemist. He was awarded the Nobel Prize for physiology or medicine in 1968 jointly with R. W. Holley and M. W. Nirenberg for their interpretation of the genetic code and its function in protein synthesis.

kidney One of a pair of major excretory organs of vertebrates, which may also function in osmoregulation. They are made up of excretory units (*see* nephron), which are responsible for the filtration and selective reabsorption of materials (water, mineral salts, glucose, etc.) and the production of waste. In mammals, the kidneys are red-brown oval structures, which are attached to the dorsal side of the abdominal cavity. They receive oxygenated blood by the renal artery and are drained of deoxygenated blood by the renal vein. A collecting duct, the ureter, conveys excess water, salts, and nitrogenous compounds (urea and uric acid) as urine from each kidney to the bladder and hence to the exterior.

kidney tubule (**uriniferous tubule**) A long narrow tube forming the part of the excretory unit (nephron) of the vertebrate kidney that is responsible for selective reabsorption of useful substances. As filtrate leaves the Bowman's capsule it passes through a series of coiled loops (the proximal convoluted tubule), where glucose, amino acid, and some water are absorbed. It then passes through a long straight loop (loop of Henle) – the major site of water reabsorption – which may reach far down into the medulla before returning to the cortex. There is a second series of coils (the distal convoluted tubule), concerned with salt and water reabsorption. The remaining liquid enters a collecting duct as urine. Some animals (e.g. amphibians) have little or no loops of Henle and hence are unable to produce concentrated urine.

killer cell *See* natural killer cell; T-cell.

kilo- Symbol: k A prefix denoting 10^3. For example, 1 kilometer (km) = 10^3 meters (m).

kilobase (**kb**) 1000 base residues of a single-stranded polynucleotide or 1000 base pairs of a double-stranded polynucleotide.

kinase Any enzyme that transfers a phosphate group, usually from ATP.

kinetin (**6-furfurylaminopurine**) An artificial CYTOKININ found in extracts of denatured DNA; the first of the cytokinins to be isolated.

kinetochore *See* centromere.

kinin **1.** Any of a class of peptides that are formed from precursors in blood or other tissues and function as local hormones. They include the angiotensins, which constrict blood vessels and raise blood pressure; bradykinin, which dilates blood vessels and lowers blood pressure; and tachykinins, which have a similar effect and also stimulate salivation and tear production. *See* angiotensin.
2. A former name for cytokinin.

kinomere *See* centromere.

Klug, (Sir) Aaron (1926–) Lithuanian-born British molecular biologist distinguished for his determination of the structure of transfer RNA and for his work on three-dimensional structure of complexes of proteins and nucleic acids. He was awarded the Nobel Prize for chemistry in 1982 for his development of crystallographic electron microscopy and his work on the structures of biologically important nucleic acid–protein complexes.

Köhler, Georges Jean Franz (1946–95) German immunologist famous for his discovery of the use of cell fusion to produce a clonal cell population capable of producing unlimited quantities of a pure antibody. He was awarded the Nobel Prize for physiology or medicine in 1984 jointly with N. K. Jerne and C. Milstein.

Kornberg, Arthur (1918–) American biochemist who discovered the enzyme DNA polymerase and used it to synthesize segments of DNA molecules and viral DNA. He was awarded the Nobel Prize for physiology or medicine in 1959 jointly with S. Ochoa for their discovery of the mechanisms in the biological synthesis of ribonucleic acid and deoxyribonucleic acid.

Kossel, Karl Martin Leonhard Albrecht (1853–1927) German chemist noted for his discovery of adenine, cytosine, thymine, and uracil as breakdown products of nucleic acids. He also discovered histone and histidine. He was awarded the Nobel Prize for physiology or medicine in 1910 in recognition of his work on cell chemistry.

Krebs, Edwin Gerhard (1918–) American biochemist. He was awarded the Nobel Prize for physiology or medicine in 1992 jointly with Edmond H. Fischer for their discoveries concerning reversible protein phosphorylation as a biological regulatory mechanism.

Krebs, (Sir) Hans Adolf (1900–81) German-born British biochemist renowned for his work on metabolism and, in particular, for the discovery of the cyclic metabolic pathways named after him. He was awarded the Nobel Prize for physiology or medicine in 1953 for his discovery of the TCA cycle. The prize was shared with F. A. Lipmann.

Krebs cycle (**tricarboxylic acid cycle; TCA cycle; citric acid cycle**) A complex and almost universal cycle of reactions in which the acetyl group of acetyl CoA is oxidized to carbon dioxide and water, with the production of large amounts of energy. It is the final common pathway for the oxidation of carbohydrates, fatty acids, and amino acids. It requires oxygen, and in eukaryotes occurs in the mitochondrial matrix.

2-carbon acetate reacts with 4-carbon oxaloacetate to form 6-carbon citrate, which is then decarboxylated to reconsti-

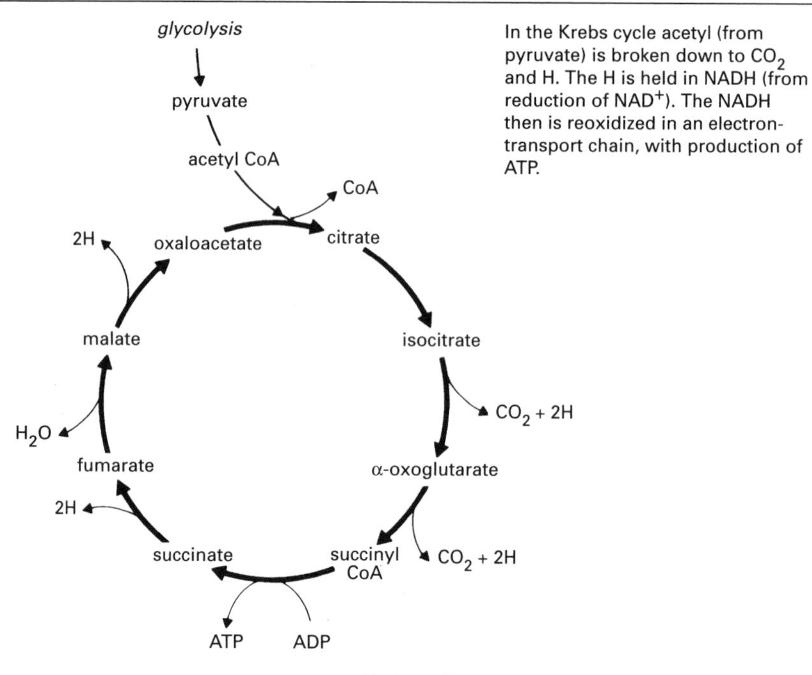

Krebs cycle

tute oxaloacetate. Some ATP is produced by direct coupling with cycle reactions, but most production is coupled to the electron-transport chain via the generation of reduced coenzymes, NADH and $FADH_2$. *See* electron-transport chain.

Kuhn, Richard Johann (1900–67) Austrian-born German chemist and biochemist. He is noted for his work on carotenoids and vitamins, in particular for his synthesis of vitamins A and B_2. He also made the first synthesis of the coenzyme flavin mononucleotide. He was awarded the Nobel Prize for chemistry in 1938.

labeling The technique of using isotopes (usually radioactive isotopes) or other recognizable chemical groups to investigate biochemical reactions. For instance, a compound can be synthesized with one of the atoms replaced by a radioactive isotope of the element. The radioactivity can then be used to follow the course of reactions involving this compound.

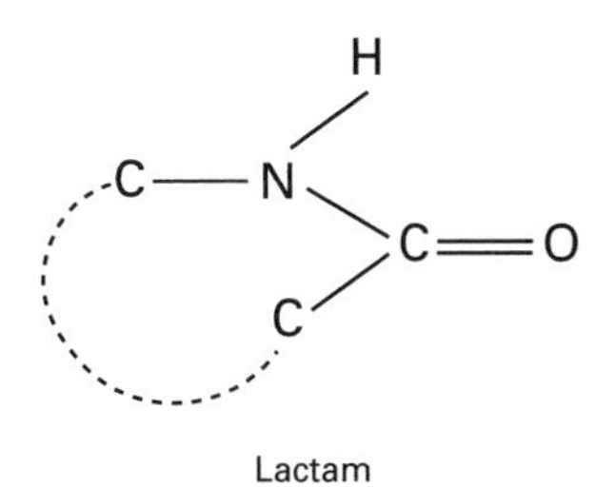

Lactam

lactam A type of organic compound containing the –NH.CO– group as part of a ring in the molecule. Lactams can be regarded as formed from a straight-chain compound that has an amine group ($-NH_2$) at one end of the molecule and a carboxylic acid group (–COOH) at the other; i.e. from an amino acid. The reaction of the amine group with the carboxylic acid group, with elimination of water, leads to the cyclic lactam, which is thus an internal amide.

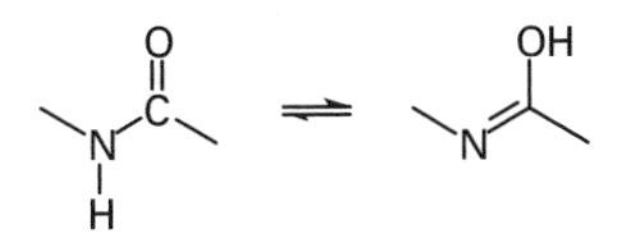

Lactam–lactim tautomerism

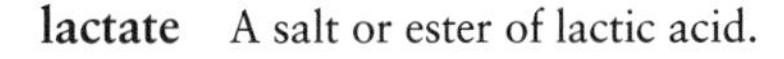

lactate A salt or ester of lactic acid.

lactation The secretion of milk from the mammary glands of female mammals, which occurs after parturition and is initiated by hormone activity, particularly by an increase in the level of prolactin. Another hormone, oxytocin, stimulates the ejection of milk. During pregnancy, estrogen and progesterone cause an increase in the amount of milk-producing tissue in the breasts, but inhibit prolactin. The levels of both these hormones fall after birth allowing prolactin to act. The sucking action of the young stimulates the continued production of prolactin and oxytocin, so that lactation continues for a prolonged period. *See also* colostrum.

lacteal A microscopic blind-ending tube containing lymph, found in each villus of the lining of the small intestine. Digested fat is absorbed into the lacteals and forms an emulsion, which makes the lymph look milky. The lacteals are connected to the lymphatic capillaries and larger lymph vessels of the intestine. Through them the fat from the lacteals reaches the blood system via the thoracic duct, which opens into the jugular vein.

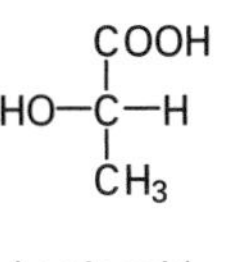

Lactic acid

lactic acid (2-hydroxypropanoic acid) A syrupy liquid, $CH_3CH(OH)COOH$, occurring in sour milk as a result of fermentation by lactobacilli. It is produced (L-form only) during anaerobic respiration in animals as the end product of glycolysis.

lactic acid bacteria A group of bacteria that ferment carbohydrates in the presence or absence of oxygen, with lactic acid always a major end product. Lactic acid bacteria are involved in the formation of yoghurt, cheese, sauerkraut, and silage. They have a high tolerance of acid conditions.

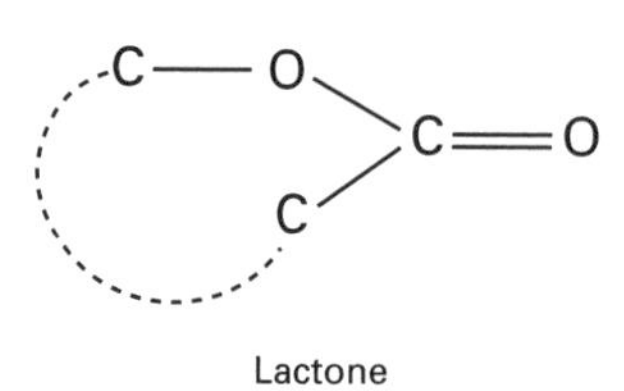

Lactone

lactone A type of organic compound containing the group –O.CO– as part of a ring in the molecule. A lactone can be regarded as formed from a compound with an alcohol (–OH) group on one end of the chain and a carboxylic acid (–COOH) group on the other. The lactone then results from reaction of the –OH group with the –COOH group; i.e. it is an internal ester.

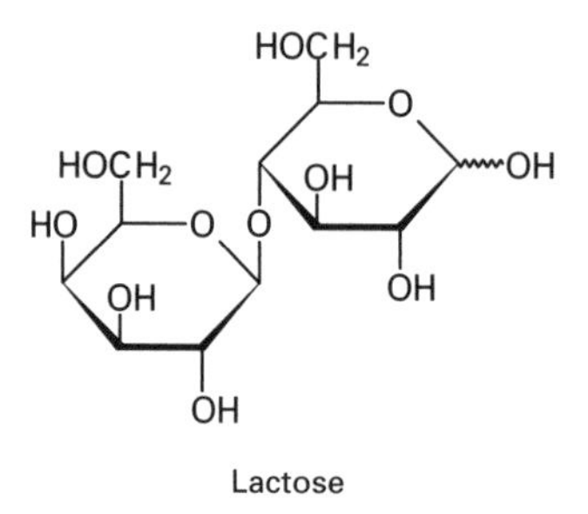

Lactose

lactose (milk sugar) A sugar found in milk. It is a disaccharide composed of glucose and galactose units.

laminarin The chief carbohydrate food reserve of the brown algae. It is a soluble polymer of glucose.

lampbrush chromosome An extended chromosome structure found in the oocytes of certain animals, notably amphibians, during the prophase of meiosis. In those species that show a great increase in nuclear and cytoplasmic volume during prophase, the lampbrush chromosomes may measure up to 1 mm in length and 0.02 mm in width. Such chromosomes consist of two central strands along which fine loops extend laterally. The loops are sites of active transcription.

Landsteiner, Karl (1868–1943) Austrian-born American pathologist and immunologist distinguished for his discoveries of the human blood groups. He designated factors A, B, and O (1901), factors M and N (1927), and factor Rh (1940). He was awarded the Nobel Prize for physiology or medicine in 1930.

Langerhans, Paul (1847–88) German physician and anatomist who first described the ISLETS OF LANGERHANS.

lanolin A yellowish viscous substance obtained from wool fat. It contains cholesterol and terpene compounds, and is used in cosmetics, in ointments, and in treating leather.

latent virus A virus that can remain inactive in its host cell for a considerable period after initial infection. The viral nucleic acid becomes integrated in the host chromosome and multiplies with it. Eventual replication inside the host cell may be triggered by such factors as radiation and chemicals. An example of a latent virus is herpes simplex. *See also* provirus.

latex A liquid found in some flowering plants contained in special cells or vessels called laticifers (or laticiferous vessels). It is a complex variable substance that may contain terpenes (e.g. rubber), resins, tannins, waxes, alkaloids, sugar, starch, enzymes, crystals, etc. It is often milky in appearance (e.g. dandelion and lettuce) but may be colorless, orange, or brown. Its function is obscure, but may be involved in wound healing as well as a repository for excretory substances. Commercial rubber comes from the latex of the rubber plants *Ficus elastica* and *Hevea brasiliensis.*

Opium comes from alkaloids found in the latex of the opium poppy.

Law of Independent Assortment *See* Mendel's laws.

Law of Segregation *See* Mendel's laws.

LD_{50} Median lethal dose, i.e. the dose of toxin at which 50% of exposed animals are killed. It is used as a standard measure of toxicity.

LDL *See* low-density lipoprotein.

L-DOPA (L-3,4-dihydroxyphenylalanine) An intermediate in the synthesis of dopamine, norepinephrine, and epinephrine and in the conversion of tyrosine to melanin pigments. L-DOPA is used to treat Parkinson's disease, a primary cause of which is a deficiency of dopamine in the brain cells. Dopamine itself cannot be administered because it cannot pass from the blood to the brain, so L-DOPA is taken orally, passes via the bloodstream to the brain, and is converted by decarboxylation to dopamine.

lecithin (phosphatidylcholine) One of a group of phospholipids that contain glycerol, fatty acid, phosphoric acid, and choline and are found widely in higher plants and animals, particularly as a component of cell membranes.

Lederberg, Joshua (1925–) American geneticist. He was awarded the Nobel Prize for physiology or medicine in 1958 for his discoveries concerning genetic recombination and the organization of the genetic material of bacteria. The prize was shared with G. W. Beadle and E. L. Tatum.

Leloir, Luis Federico (1906–87) Argentinian biochemist who first isolated uridine diphosphate glucose and discovered the role of sugar nucleotides in interconversion of sugars and in polysaccharide formation. He was awarded the Nobel Prize for chemistry in 1970.

leptotene In meiosis, the stage in early prophase I when the chromosomes, already replicated, start to condense and appear as fine threads, although sister chromatids are not yet distinct. The spindle starts to form around the intact nucleus.

leucine *See* amino acids.

leukocyte (white blood cell) A type of blood cell that has a nucleus but no pigment. White cells are larger and less numerous than red cells (about 6000–8000 per cubic millimeter of blood). They are important in defending the body against disease because they devour bacteria and produce antibodies. They are all capable of ameboid movement and are divided into two groups, granulocytes and agranulocytes, according to the presence or absence of granules in the cytoplasm. Within these groups, there are several types of leukocytes. The most numerous are the neutrophils (70%) and lymphocytes (25%). Leukocytes have a very short lifespan and are continuously produced in the myeloid tissue of the red marrow.

leukosin One of the structurally stable scleroproteins found in wheat.

leukotrienes A group of substances that are produced in tissues to serve as local hormones. They are derived from polyunsaturated fatty acids, and all contain at least three double bonds. For example, leukotriene B promotes increased permeability of blood vessels and infiltration of certain types of white blood cells (basophils, neutrophils, and eosinophils), while leukotriene C_4 causes constriction of blood capillaries in various tissues, including the heart and brain.

Levi-Montalcini, Rita (1909–) Italian cell biologist. She was awarded the Nobel Prize for physiology or medicine in 1986 jointly with S. Cohen for their discoveries of growth factors.

levo-form *See* optical activity.

levorotatory Describing compounds that rotate the plane of polarized light to the left (anticlockwise as viewed facing the oncoming light). Levorotatory compounds are indicated by the symbol (–). *Compare* dextrorotatory. *See* optical activity.

Lewis, Edward B. (1918–) American biochemist who shared the 1995 Nobel Prize for physiology or medicine with C. Nüsslein-Volhard and E. F. Wieschaus for work on the genetic control of early embryonic development.

Lewis acid A substance that can accept an electron pair to form a coordinate bond; a *Lewis base* is an electron-pair donor. In this model, the neutralization reaction is seen as the acquisition of a stable octet by the acid, for example:

$$Cl_3B + :NH_3 \rightarrow Cl_3B:NH_3$$

Metal ions in coordination compounds are also electron-pair acceptors and therefore Lewis acids. The definition includes the 'traditional' Brønsted acids since H^+ is an electron acceptor, but in common usage the terms Lewis acid and Lewis base are reserved for systems without acidic hydrogen atoms.

Lewis base *See* Lewis acid.

L-form *See* optical activity.

LH *See* luteinizing hormone.

Libby, Willard Frank (1908–80) American chemist. He was awarded the Nobel Prize for chemistry in 1960 for the invention of radiocarbon dating.

ligament A tough band or capsule of connective tissue that connects two bones together at a joint. It has a high proportion of elastic fibers and white collagen fibers combining strength with elasticity. The fibers of a ligament penetrate the tissue of the bones, making a very strong connection.

ligand **1.** An ion, atom, or chemical group that is attached to a central metal atom and supplies a pair of electrons to it, forming a type of covalent bond called a coordinate bond.
2. Any molecule that binds specifically to a site on a cell receptor molecule, causing a change in the shape of the receptor and triggering some response in the cell.

ligase An enzyme that catalyzes the bond formation between two substrates at the expense of the breakdown of ATP or some other nucleotide triphosphate. The degree of bond formation by ligases is proportional to the amount of ATP available in the cell at a particular time.

light green *See* staining.

light microscope *See* microscope.

light reactions The light-dependent reactions of PHOTOSYNTHESIS that convert light energy into the chemical energy of NADPH and ATP.

lignin One of the main structural materials of vascular plants. With cellulose it is one of the main constituents of wood. Lignified tissues include sclerenchyma and xylem. Lignin is deposited during secondary thickening of cell walls. The degree of lignification varies from slight in protoxylem to heavy in sclerenchyma and some xylem vessels, but values of 25–30% lignin and 50% cellulose are average. It is a complex variable polymer, derived from sugars via aromatic alcohols. Phenylpropane (C_6–C_3) units are linked in various ways by oxidation reactions during polymerization. Lignin is characteristically stained yellow by aniline sulfate or chloride, and red by phloroglucinol with hydrochloric acid.

limiting factor Any factor in the environment that alone governs the behavior of an organism or system by being above or below a certain level. In general, the behavior of a system depends on a number of different factors; under certain conditions, one of these can limit the behavior. For instance, plant growth is limited by low temperature and increases with rising temperature to an optimum, beyond which growth rate decreases.

limiting step *See* rate-determining step.

linkage The occurrence of genes together on the same chromosome so that they tend to be inherited together and not independently. Groups of linked genes are termed *linkage groups* and the number of linkage groups of a particular organism is equal to its haploid chromosome number. Linkage groups can be broken up by crossing over at MEIOSIS to give new combinations of genes. Two genes close together on a chromosome are more strongly linked, i.e. there is less chance of a cross over between them, than two genes further apart on the chromosome. Linked genes are symbolized Ab...Y/aB...y, indicating that Ab...Y are on one homolog while aB...y are on the other homolog.

linkage map *See* chromosome map.

linoleic acid A common unsaturated fatty acid, $C_{17}H_{31}COOH$, occurring as glycerides in linseed oil, cottonseed oil and other vegetable oils. It is an essential nutrient in the diet of mammals. *See* essential fatty acids.

linolenic acid An unsaturated fatty acid occurring commonly in plants as the glyceryl ester, for example in linseed oil and poppy-seed oil. The biological function of linolenic acid is similar to that of linoleic acid and administration of linolenic acid is also used to cure fat deficiency in animals. It is an essential nutrient in the diet of mammals. *See* essential fatty acids.

linseed oil An oil extracted from the seeds of flax (linseed). It hardens on exposure to air (it is a drying oil) because it contains linoleic acid, and is used in enamels, paints, putty, and varnishes. *See* linoleic acid.

lipase Any of various enzymes that catalyze the hydrolysis of fats to fatty acids and glycerol. Lipases are present in the pancreatic juice of vertebrates.

lipid A collective term used to describe a group of substances in cells characterized by their solubility in organic solvents such as ether and benzene, and their absence of solubility in water. The group is rather heterogeneous in terms of both function and structure. They encompass the following broad bands of biological roles: (1) basic structural units of cellular membranes and cytologically distinct subcellular bodies such as chloroplasts and mitochondria; (2) compartmentalizing units for metabolically active proteins localized in membranes; (3) a store of chemical energy and carbon skeletons; and (4) primary transport systems of nonpolar material through biological fluids. There are also the more physiologically specific lipid hormones, e.g. the steroid hormones and lipid vitamins.

The simple lipids include *neutral lipids* or glycerides, which are esters of glycerol and fatty acids, and the waxes, which are esters of long chain monohydric alcohols and fatty acids.

Compound lipids have one of the fatty acid parts replaced, such that complete hydrolysis gives only two fatty acids; the PHOSPHOLIPIDS are particularly important examples.

Lipmann, Fritz Albert (1899–1986) German-born American biochemist. He was awarded the Nobel Prize for physiology or medicine in 1953 for his discovery of coenzyme A and its importance for intermediary metabolism. The prize was shared with H. A. Krebs.

lipoic acid A sulfur-containing fatty acid found in a wide variety of natural materials. It is an essential as a coenzyme for certain dehydrogenase enzymes, notably pyruvate dehydrogenase, which catalyzes the dehydrogenation of pyruvic acid to form acetyl-CoA. Lipoic acid is classified with the water-soluble B vitamins and has not yet been demonstrated to be required in the diet of higher animals.

lipolysis The splitting of the component fatty acids from a lipid; i.e. part of the process of catabolism of lipid molecules.

Lipolysis is effected in the body, largely in the gut, by the lipase enzymes.

lipolytic hormone *See* lipotropin.

lipopolysaccharide A conjugated polysaccharide in which the noncarbohydrate part is a lipid. Lipopolysaccharides are a constituent of the cell walls of certain bacteria.

lipoprotein Any conjugated protein formed by the combination of a protein with a lipid. In the blood of humans and other mammals, cholesterol, triglycerides, and phospholipids associate with various plasma proteins to form lipoproteins. These are particles with diameters in the region of 7.5–70 nm, and are placed in several classes. The largest lipoproteins in this range are the *very low-density lipoproteins* (*VLDLs*), which are formed in the liver and contain up to about 20% cholesterol. *Low-density lipoproteins* (*LDLs*) are formed in plasma from VLDLs and contain over 50% cholesterol. LDLs transport cholesterol from the liver to peripheral tissues, and an excess of LDLs in the blood is a factor in the development of fatty arterial deposits and cardiovascular disease. *High-density lipoproteins* (HDLs), with about 20% cholesterol, are the smallest of the plasma lipoproteins, and apparently function in transporting cholesterol from tissues to the liver.

The largest of all plasma lipoproteins are the *chylomicrons*, which act as vehicles for the absorption of fat from the intestine. Measuring up to 100 nm in diameter, they are formed in the intestinal mucosa and contain mostly triglycerides, with relatively small amounts of cholesterol and protein. They enter the lacteals of the intestinal villi and are conveyed via the lympatic system to the bloodstream.

liposome A microscopic spherical sac consisting of a lipid envelope enclosing fluid. They occur naturally in the cytoplasm of some cells and can be made *in vitro* by adding an aqueous solution to a gel of complex lipids. Liposomes have a wall consisting of a double layer of lipids similar to that of cell membranes. They are used experimentally as models of cells or cell organelles, and also in medicine to deliver toxic drugs to target tissues in the body.

lipotropin (**lipotropic hormone; lipolytic hormone; LPH**) Either of two polypeptide hormones produced in mammals by the pituitary gland, that stimulate lipolysis. Both occur in human blood but their precise physiological role remains uncertain; they may simply be by-products of the formation of corticotropin.

liter Symbol: l A unit of volume now defined as 10^{-3} meter^3; i.e. 1000 cm^3. The milliliter (ml) is thus the same as the cubic centimeter (cm^3). However, the name is not recommended for precise measurements since the liter was formerly defined as the volume of one kilogram of pure water at 4°C and standard pressure. On this definition, one liter is the same as 1000.028 cubic centimeters.

litmus A natural pigment that changes color when in contact with acids and alkalis; above a pH of 8.3 it is blue and below a pH of 4.5 it is red. Thus it gives a rough indication of the acidity or basicity of a solution; because of its rather broad range of color change it is not used for precise work. Litmus is used both in solution and as litmus paper.

liver A large dark red organ, made up of several lobes, lying close to the stomach in vertebrates. It accounts for one fifth of the whole contents of the abdomen, and all the blood in the body passes through the liver every two minutes. Its main function is to regulate the chemical composition of the blood and to act as a depot where the nutrients obtained by digestion or from body stores are converted into the materials that fulfill the body's metabolic requirements. It is supplied with oxygenated blood by the hepatic artery, but receives 80% of its blood in the hepatic portal vein from the intestine. After the digestion of food, this blood is rich in glucose and amino acids. The liver removes the surplus glucose (over

0.1%) and stores it as glycogen until it is needed. It changes the surplus amino acids to urea in a process of deamination. Its other functions are:
1. The production of bile, stored in the gall bladder and then passed into the duodenum along the bile duct.
2. The removal of damaged red corpuscles from the blood.
3. The storage of iron.
4. The manufacture of vitamin A from carotene, and the storage of vitamins A and D.
5. The manufacture of some of the proteins of blood plasma.
6. The manufacture of blood-clotting factors, prothrombin and fibrinogen.
7. The removal of poisons from the blood (detoxication).
8. The conversion of fats into compounds suitable for oxidation during starvation, and the conversion of carbohydrates to fat when there is too much to be stored as glycogen in the liver.

The microscopic structure of the liver insures that every cell is in direct contact with the blood (there are no endothelial linings in the capillaries) so that diffusion of molecules in and out of the cells is very rapid.

locus The position of a gene on a chromosome. Alleles of the same gene occupy the equivalent locus on homologous chromosomes.

Loewi, Otto (1873–1961) German-born American physiologist. He shared the Nobel Prize for physiology or medicine in 1936 with H. H. Dale for their work on the chemical transmission of nerve impulses.

loop of Henle *See* kidney tubule.

low-density lipoprotein (LDL) A spherical particle, typically about 20–25 nm in diameter, that is found in blood plasma and transports cholesterol to tissue cells. It is bounded by a single layer of phospholipid and free cholesterol, which encloses a core of cholesterol esterified to a long-chain fatty acid. Embedded in the surface layer is a single large protein, called apo-B, which assists in binding of the LDL to cell-surface receptors. LDLs are taken into cells by receptor-mediated endocytosis. The cholesterol is incorporated into cell membranes or stored as lipid droplets. High concentrations of LDLs in the blood have been associated with an increased risk of atherosclerosis ('hardening of the arteries').

LPH Lipotropic (lipolytic) hormone. *See* lipotropin.

LSD *See* lysergic acid diethylamide.

LTH Luteotropic hormone. *See* prolactin.

luciferase An enzyme that catalyzes the oxidation of luciferin to produce the light of bioluminescent reactions, such as occur in fireflies and some bacteria (*see* bioluminescence). It is also used as a reporter molecule in certain antibody assays of proteins or other molecules. The luciferase is linked to a specific antibody, which binds to a target molecule. When ATP and luciferin are added, the appearance and amount of light emitted corresponds to the presence and amount of the target molecule.

luciferin Any substance that acts as substrate for an enzyme (*luciferase*) that catalyzes a light-emitting reaction in living organisms (bioluminescence). Luciferins vary widely in different organisms.

Luria, Salvador Edward (1912–91) Italian-born American physician and biologist. He was awarded the Nobel Prize for physiology or medicine in 1969, jointly with M. Delbrück and A. D. Hershey, for discoveries concerning the replication mechanism and the genetic structure of viruses.

lutein The commonest of the xanthophyll pigments. It is found in green leaves and certain algae, e.g. the Rhodophyceae. *See* photosynthetic pigments.

luteinizing hormone (LH; **interstitial-cell-stimulating hormone;** ICSH) A glycoprotein hormone secreted by the anterior pituitary lobe under regulation of the hypothalamus. In female mammals it stimulates secretion of estrogen, ovulation, and formation of corpora lutea. In male mammals it stimulates interstitial cells in the testes to secrete androgens. *See also* gonadotropin.

luteotropic hormone (**luteotrophic hormone; LTH**) *See* prolactin.

Lwoff, André Michel (1902–94) French microbiologist. He was awarded the Nobel Prize for physiology or medicine in 1965, jointly with F. Jacob and J. L. Monod, for work on the genetic control of enzyme and virus synthesis.

lyase An enzyme that catalyzes the separation of two parts of a molecule with the formation of a double bond in one of them. For example, fumerase catalyzes the interconversion of maleic acid and fumaric acid.

lymph The fluid contained within the vessels of the lymphatic system. It is derived from tissue fluid that is drained from intercellular spaces, and is similar to plasma but with a lower protein concentration and contains cells (mainly lymphocytes), bacteria, etc. It is colorless except in the region of the small intestine where absorbed fat gives the lymph a milky appearance. *See* lacteal.

lymphatic system A series of vessels (lymphatic vessels) and associated lymph nodes that transports lymph from the tissue fluids into the bloodstream and the heart. Tissue fluid not returned to the circulation via blood capillaries is drained as lymph into the blind-ending thin-walled lymphatic capillaries, which also occur between cells. These join to form larger vessels, which eventually unite into major ducts (the right lymphatic duct and the thoracic duct) and empty into the large veins entering the heart. The flow of lymph is achieved by muscular and respiratory movements in mammals and by the pumping of *lymph hearts* in other vertebrates. It is unidirectional: some vessels contain valves to prevent any backflow. The lymphatic system is also the main route by which fats reach the bloodstream from the intestine.

lymphatic tissue *See* lymphoid tissue.

lymph node (**lymph gland**) One of a large number of flat oval structures distributed along the lymphatic vessels and clustered in certain regions, such as the neck, armpits, and groin. Lymph nodes are composed of lymphatic tissue and contain special white blood cells (lymphocytes), some of which produce antibodies. They act as defense posts against the spread of infection, their lymphocytes engulfing bacteria and other foreign materials from the lymph; the nodes may become inflamed and enlarged as a result.

lymphocyte A type of white blood cell (leukocyte) with a very large nucleus, rich in DNA, and a small amount of clear cytoplasm. They comprise 25% of all leukocytes and are divided into two classes: B-lymphocytes or B-cells, and T-lymphocytes, or T-cells. Lymphocytes are made in myeloid tissue in red bone marrow, lymph nodes, thymus, tonsils, and spleen. During infection, antigens stimulate B-cells in the lymphoid tissue to multiply rapidly, and the resulting lymphocytes, called *plasma cells*, are released into the bloodstream to produce the appropriate antibody: the humoral immune response T-cells are involved in the cell-mediated immune response, which destroys cells infected by viruses. *See also* antibody; B-cell; immunity; leukocyte; plasma cells; T-cell.

lymphoid tissue (**lymphatic tissue**) Tissue in which lymphocytes are produced, found in the lymph nodes, tonsils, spleen, and thymus. It consists of a delicate network of cells through which lymph flows continuously. Lymphocytes have a life span of only a few days and must be constantly replaced. When an antigen enters lymphoid tissue, it is 'recognized' by one particular type of lymphocyte, which then multiplies rapidly; the resulting plasma cells circulate in the blood, producing the necessary antibody for that antigen. Lymphoid tissue also contains numerous macrophages, which ingest foreign particles, especially bacteria, hence the lymph nodes act as filters to remove bacteria from the lymph. *See also* antibody; macrophage; plasma cells.

lymphokine Any of various cytokines that are released by activated lymphocytes. They serve as chemical signals to other cell types involved in the cell-mediated immune response, such as macrophages, neutrophils, and basophils. Examples include *macrophage migration inhibition factor*, which inhibits the movement of macrophages, and *tumor necrosis factor*, which stimulates a range of responses in body cells. Soluble lymphokines that act as signals for other lymphocytes are known as interleukins. *See* cytokine; interleukin.

Lynen, Feodor (1911–79) German biochemist. He was awarded the Nobel Prize for physiology or medicine in 1964, jointly with K. E. Bloch, for discoveries concerning the mechanism and regulation of the cholesterol and fatty acid metabolism.

lyophilic Solvent attracting. When the solvent is water, the word *hydrophilic* is often used. The terms are applied to:
1. Ions or groups on a molecule. In aqueous or other polar solutions ions or polar groups are lyophilic. For example, the $-COO^-$ group on a soap is the lyophilic (hydrophilic) part of the molecule.
2. The disperse phase in colloids. In lyophilic colloids the dispersed particles have an affinity for the solvent, and the colloids are generally stable. *Compare* lyophobic.

lyophobic Solvent repelling. When the solvent is water, the word *hydrophobic* is used. The terms are applied to:
1. Ions or groups on a molecule. In aqueous or other polar solvents, the lyophobic group will be nonpolar. For example, the hydrocarbon group on a soap molecule is the lyophobic (hydrophobic) part.
2. The disperse phase in colloids. In lyophobic colloids the dispersed particles are not solvated and the colloid is easily solvated. Gold and sulfur sols are examples. *Compare* lyophilic.

lysergic acid diethylamide (**LSD**) A synthetic organic compound that has physiological effects similar to those produced by alkaloids in certain fungi. Even small quantities, if ingested, produce hallucinations and extreme mental disturbances. The initials LSD come from the German form of the chemical's name, *Lysergic-Saure-Diathylamide.*

lysine *See* amino acids.

lysis (**degeneration**) The death and subsequent breakdown of a cell. Under normal conditions such cells are engulfed by phagocytes and degraded by their lysosomes. Only under rare conditions are such cells degraded from within by lysosomes. During some morphological changes (e.g. regression of the tadpole tail) internal lysosomal action does not initiate degradation but simply participates together with phagocytotic action.

lysosome An organelle of plant and animal cells that contains a range of digestive enzymes whose destructive potential necessitates their separation from the rest of the cytoplasm. They have many important functions, e.g. contributing enzymes to food vacuoles, as in *Amoeba*, or to similar vacuoles formed in white blood cells during phagocytosis. They may be involved in destruction of cells and tissues during development, e.g. loss of tadpole tails. Lysosomes are bounded by a single membrane and have homogeneous contents that often appear uniformly gray with the electron microscope. They are usually spherical and about 0.5 μm in diameter, although lysosomal compartments may range from small Golgi vesicles to large plant vacuoles. Lysosomes may be formed directly from endoplasmic reticulum or by budding off from the Golgi apparatus containing processed proteins derived from the endoplasmic reticulum. These *primary lyso-*

somes carry the enzymes (hydrolases) to the material to be digested, which has also become membrane delimited during endocytosis or autophagy. The structures fuse forming *secondary lysosomes*. The final structure after digestion is called a *residual body*. Its contents may be excreted from the cell by exocytosis.

The material digested may be of extracellular or intracellular origin. Examples of the former are substances found in vacuoles formed by phagocytosis, e.g. bacteria in white blood cells, or pinocytosis, e.g. thyroglobulin taken up from the lumen of the thyroid gland for hydrolysis to thyroxine; examples of the latter include the phenomena of autophagy and autolysis.

A special type of lysosome is the Golgi-derived acrosome of sperm heads which, on attachment of the sperm to the egg, releases enzymes to dissolve the vitelline membrane. Sometimes lysosomal enzymes are released by exocytosis for extracellular digestion as in the replacement of cartilage with bone during ossification. *See* autolysis.

lysozyme An enzyme present in saliva, tears, egg white, and mucus. It also destroys bacteria by hydrolysis of the cell walls. *See also* lysis.

McClintock, Barbara (1902–92) American geneticist noted for her studies on chromosome breakage and reunion and her proposal of the existence of transposable elements. She was awarded the Nobel Prize for physiology or medicine in 1983.

Macleod, John James Richard (1876–1935) British physiologist and biochemist. He shared the Nobel Prize for physiology or medicine in 1923 with F. G. Banting for the discovery of insulin.

macromolecule A very large molecule, usually a polymer, having a very high molecular weight. Proteins, polysaccharides, and nucleic acids are examples.

macronutrient A nutrient required in more than trace amounts by an organism. *See* essential element. *Compare* micronutrient.

macrophage A large ameboid cell that can engulf, ingest, and destroy bacteria, damaged cells, and worn-out red blood cells. This process is called *phagocytosis* and is an important part of nonspecific immunity. Macrophages are found free ('wandering') in the tissues, in the blood (as monocytes), in connective tissue (as histiocytes), in the lining of the blood sinusoids of the liver (as Kupffer cells), and in lymphoid tissue. They make up the reticuloendothelial system.

magnesium An element essential for plant and animal growth. In animals it is found in bones and teeth. As magnesium carbonate it is found in large quantities in the skeletons of certain marine organisms, and is found in smaller quantities in the muscles and nerves of higher animals. A magnesium ion, Mg^{2+}, it is contained in the chlorophyll molecule and is thus essential for photosynthesis. Mg^{2+} is an essential cofactor for most enzymes that use ATP. High concentrations of magnesium ions are needed to maintain ribosome structure.

major histocompatibility complex (MHC) A large cluster of related genes found in mammals and encoding cell-surface proteins (MHC proteins) that play several vital roles in the immune system. They form the markers (i.e. 'self' antigens) of the MHC system of self-recognition, which prevents cells of the immune system (lymphocytes) attacking self tissue, and they serve to 'present' foreign antigens to lymphocytes. In humans, the MHC complex consists of numerous genes located on chromosome 6; the genes and the antigens they encode are called the *HLA system*.

MHC genes generally fall into three classes, with protein products having different roles. *Class I MHC proteins* are found on virtually all cells. Within the cell they attach themselves to processed foreign antigen, for example from an invading virus, and migrate to the cell surface where they are recognized by cytotoxic T-CELLS, which are stimulated to destroy the infected cell. *Class II MHC proteins* occur only on certain cells of the immune system, notably macrophages and B-CELLS. They combine with foreign antigen such as debris from bacteria phagocytosed by the macrophage, and are recognized by another set of T-cells called T-helper cells. The latter stimulate the macrophages to destroy the foreign antigens they contain, and also activate antibody secretion by B-cells. *Class III MHC proteins* are components of the complement system.

A large number of alleles exist for the MHC genes, which creates immense genetic variability in the MHC proteins. Hence, only close relatives are likely to have similar class I MHC proteins, i.e. show some degree of histocompatibility. As a consequence, a graft from such a relative is more likely to be tolerated by the immune system. Tissue grafts from unrelated individuals have different class I MHC proteins, and are recognized as foreign and killed by the recipient's cytotoxic T-cells, causing rejection of the graft.

maleic acid *See* butenedioic acid.

malic acid (2-hydroxybutanedioic acid) A colorless crystalline carboxylic acid, $HCOOCH_2CH(OH)COOH$, that occurs in acid fruits such as grapes and gooseberries. In biological processes malate ion is an important part of the Krebs cycle.

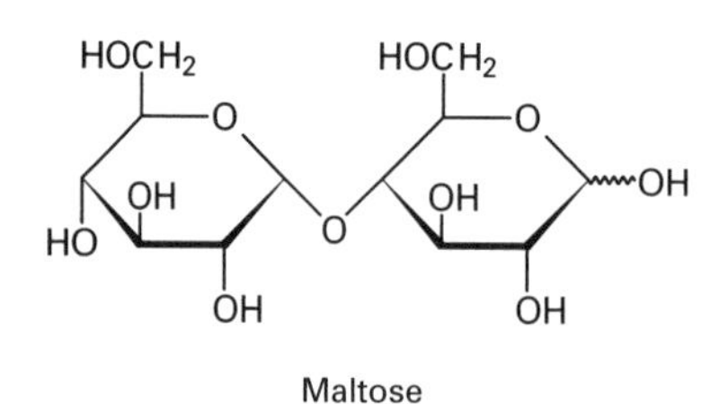

Maltose

maltose A sugar found in germinating cereal seeds. It is a disaccharide composed of two glucose units. Maltose is an important intermediate in the enzyme hydrolysis of starch. It is further hydrolyzed to glucose.

mammary gland The milk-producing gland in female mammals.

manganese *See* trace element.

mannitol A soluble sugar alcohol (carbohydrate) found widely in plants and forming a characteristic food reserve of the brown algae. It is a hexahydric alcohol, $HOCH_2(CHOH)_4CH_2OH$, i.e. each of the six carbon atoms has an alcohol (hydroxyl) group attached. In medicine it is used as a diuretic to treat fluid retention.

mannose A simple sugar found in many polysaccharides. It is an aldohexose, isomeric with glucose.

marker gene A gene of known location and function that can be used to establish the relative positions and functions of other genes. During gene transfer, a marker gene may be linked to the transferred gene to determine whether or not the transfer has been successful. *See* chromosome map; genetic engineering.

Martin, Archer John Porter (1910–) British biochemist and physical chemist. He shared the Nobel Prize for chemistry in 1952 with R. L. M. Synge for their invention of partition chromatography.

mast cell A type of white blood cell (leukocyte) with granular cytoplasm found within connective tissue, e.g. beneath the skin and around blood vessels. Mast cells bind certain immunoglobulin antibodies (IgE) to their cell membrane, and when a specific antigen is encountered this triggers the mast cell to release histamine, heparin, and serotonin. These substances increase vascular permeability at the site, causing inflammation and attracting other types of immune cell. Mast cells are responsible for causing the symptoms of various allergies and other hypersensitivity reactions.

Mb *See* megabase.

mechanism A step-by-step description of the events taking place in a chemical reaction. It is a theoretical framework accounting for the fate of bonding electrons and illustrates which bonds are broken and which are formed.

Medawar, (Sir) Peter Brian (1915–87) British zoologist and immunologist. He shared the Nobel Prize for physiology or medicine in 1960 with F. M. Burnet for their discovery of acquired immunological tolerance.

mega- Symbol: M A prefix denoting 10^6. For example, 1 megahertz (MHz) = 10^6 hertz (Hz).

megabase 1 000 000 base residues of a single-stranded polynucleotide or 1000 base pairs of a double-stranded polynucleotide.

meiosis The process of cell division leading to the production of daughter nuclei with half the genetic complement of the parent cell. Cells formed by meiosis give rise to gametes and fertilization restores the correct chromosome complement.

Meiosis consists of two divisions during which the chromosomes replicate only once. Like mitosis the stages prophase, metaphase, and anaphase can be recognized. The first prophase (prophase I) is longer and is subdivided into *leptotene*, *zygotene*, *pachytene*, and *diplotene*. During prophase homologous chromosomes attract each other and become paired forming bivalents. At the end of prophase genetic material may be exchanged between the chromatids of homologous chromosomes. Meiosis also differs from mitosis in that after anaphase, instead of nuclear membranes forming, there is a second division, which may be divided into metaphase II and anaphase II. The centromeres divide and pull homologous chromatids to opposite poles. The second division ends with the formation of four haploid cells.

melanin One of a group of pigments found in animals and plants, derived from the amino acid tyrosine. The colors range from black through brown to yellow, orange, or red. In animals melanin occurs in *melanophores* (pigment cells) in the skin, usually below the epidermis. Melanin gives color to the skin, hair, and eyes of animals and causes color in various seedlings and roots of plants. The absence of the enzyme tyrosinase in animals leads to a condition known as *albinism*, in which no pigment develops in the eyes, skin, or hair.

melanism The possession of a dark appearance due to the presence of a dark brown pigment, melanin.

melanocyte-stimulating hormone (MSH) A peptide hormone produced by the anterior pituitary gland in vertebrates. It has a marked action on pigmentation in the skin of amphibians and reptiles, but its physiological role in mammals is unclear. *Compare* melatonin.

melatonin A hormone, produced by the pineal gland, that produces marked lightening of the skin in embryonic fish and larval amphibians. It has been shown to be involved in the perception of photoperiod in mammals; melatonin synthesis is inhibited by daylight. In animals with seasonal breeding, such as sheep and deer, injections of melatonin can be used to control the breeding cycle. In humans, melatonin has been implicated in the condition of 'winter depression'.

membrane A structure consisting mainly of lipid and protein (lipoprotein) surrounding all living cells as the *plasma membrane*, or *plasmalemma*, and also found surrounding organelles within cells. Membranes function as selectively permeable barriers, controlling passage of substances between the cell and its organelles, and the environment, either actively or passively. Membranes are typically 7.5–10 nm in thickness with two regular layers of lipid molecules (a *bilayer*) containing various types of protein molecules. Some proteins penetrate through the membrane, others are associated with one side; some float freely over the surface while others remain stationary. Membrane proteins mediate specific functions e.g. transport, communication, and energy transduction. Small molecules and ions pass passively through protein channels (*see* ion channels) or are actively transported across by carrier proteins. Larger molecules or particles enter or leave cells by endocytosis or exocytosis, respectively.

The lipids are mostly PHOSPHOLIPIDS. These are amphipathic polar molecules, i.e. one end (the phosphate end) is *hydrophilic* (water-loving) and faces outwards, while the other end (two fatty acid tails) is *hydrophobic* (water-hating) and faces inwards. Membranes also contain glycolipids, which are also amphipathic polar molecules, and cholesterol, a neutral sterol. The particular types of carbohydrates, lipids, and proteins determine the characteristics of the membrane, affecting, for example, cell–cell recognition (as in embryonic development and immune mechanisms), fluidity, permeability, and hormone recognition. Some membranes contain efficient arrangements of molecules involved in certain metabolic processes, e.g. electron transport and phosphorylation (ATP production) in mitochondria and chloroplasts. *See* osmosis; freeze fracturing; active transport; facilitated diffusion; glycolipid.

memory cell *See* B-cell.

menaquinone *See* vitamin K.

Mendel, Gregor Johann (1822–84) Austrian monk and botanist who did early work on genetics.

Mendelism The theory of inheritance according to which characteristics are determined by particulate 'factors', or genes, that are transmitted by the germ cells. It is the basis of classical genetics, and is founded on the work of Mendel in the 1860s. *See* Mendel's laws.

Mendel's laws Two laws formulated by Mendel to explain the pattern of inheritance he observed in plant crosses. The first law, the *Law of Segregation*, states that any character exists as two factors, both of which are found in the somatic cells but only one of which is passed on to any one gamete. The second law, the *Law of Independent Assortment*, states that the distribution of such factors to the gametes is random; if a number of pairs of factors is considered each pair segregates independently.

Today Mendel's 'characters' are termed genes and their different forms (factors) are called alleles. It is known that a diploid cell contains two alleles of a given gene, each of which is located on one of a pair of homologous chromosomes. Only one homolog of each pair is passed on to a gamete. Thus the Law of Segregation still holds true. Mendel envisaged his factors as discrete particles but it is now known that they are grouped together on chromosomes. The Law of Independent Assortment therefore only applies to pairs of alleles found on different chromosomes.

Menten, Maude Leonora (1879–1960) Canadian physician and biochemist, noted for her collaboration with Leonor Michaelis in the formulation of Michaelis–Menten kinetics of enzyme action.

Merrifield, Robert Bruce (1921–) American chemist and biochemist. He was awarded the Nobel Prize for chemistry in 1984 for his development of methods of chemical synthesis on a solid matrix.

mesoderm The germ layer in an early animal embryo from which the muscles, reproductive organs, circulatory system, and, in vertebrates, the skeleton and excretory system usually develop. At gastrulation the mesoderm comes to lie between ectoderm on the outside and endoderm lining the gut. Either side of the notochord and neural tube, blocks of mesoderm called *somites* develop, arranged in a series of paired segments running from front to rear. These spread laterally and ventrally, and give rise to epithelial sheets of cells and wandering cells known as mesenchyme. In most animals the coelom divides the mesoderm into an outer *somatopleure* under the skin and an inner *splanchnopleure* around the gut.

***meso*-form** *See* optical activity.

mesophilic Describing microorganisms that have an optimum temperature for growth between 25–45°C) for growth. *Compare* thermophilic; psychrophilic.

mesophyll Specialized tissue located between the epidermal layers of the leaf. Veins, supported by sclerenchyma and collenchyma, are embedded in the mesophyll. *Palisade mesophyll* consists of cylindrical cells, at right angles to the upper epidermis, with many chloroplasts and small intercellular spaces. It is the main photosynthesizing layer in the plant. *Spongy mesophyll*, adjacent to the lower epidermis, comprises interconnecting irregularly shaped cells with few chloroplasts and large intercellular spaces that communicate with the atmosphere through pores (stomata) allowing gas exchange between the cells and the atmosphere. The distribution of mesophyll tissue varies in different leaves depending on the environment in which the plant lives.

messenger RNA (**mRNA**) The form of RNA that transfers the information necessary for protein synthesis from the DNA in the nucleus to the ribosomes in the cytoplasm. One strand of the double helix of DNA acts as a template along which complementary RNA nucleotides become aligned. These form a polynucleotide identical to the other DNA strand, except that the thymine bases are replaced by uracil. This polynucleotide is called HETEROGENEOUS NUCLEAR RNA (hnRNA) and it contains both coding and noncoding sequences (*see* exon; intron). This process is termed *transcription*. The introns are then removed by *splicing* to form mRNA. The new mRNA molecule thus has a copy of the genetic code, which directs the formation of proteins in the ribosomes (*translation*). *Compare* transfer RNA.

meta- Designating a benzene compound with substituents in the 1,3 positions. The position on a benzene ring that is two carbon atoms away from a substituent is the meta position. This was used in the systematic naming of benzene derivatives. For example, meta-dinitrobenzene (or *m*-dinitrobenzene) is 1,3-dinitrobenzene.

metabolism The chemical reactions that take place in cells. The molecules taking part in these reactions are termed *metabolites*. Some metabolites are synthesized within the organism, while others have to be taken in as food. It is metabolic reactions, particularly those that produce energy, that keep cells alive. Metabolic reactions characteristically occur in small steps, comprising a *metabolic pathway*. These pathways are tightly regulated by various systems including control of enzyme action by allosteric molecules or phosphorylation and control of protein synthesis (which regulates the amount of enzyme). The regulatory systems operate in a coordinated manner. Metabolic reactions involve the breaking down of molecules to provide energy (CATABOLISM) and the building up of more complex molecules and structures from simpler molecules (ANABOLISM).

metabolite A substance that takes part in a metabolic reaction, either as reactant or product. Metabolites are thus intermediates in metabolic pathways. Some are synthesized within the organism itself, whereas others have to be taken in as food. *See also* metabolism.

metalloporphyrin *See* porphyrins.

metamorphosis A phase in the life history of many animals during which there is a rapid transformation from the larval to the adult form. Metamorphosis is widespread among invertebrates, especially marine organisms and arthropods, and is typical of the amphibians. It is normally under hormone control and usually involves widespread lysosome-mediated destruction of larval tissues.

metaphase The stage in mitosis and meiosis when the chromosomes become aligned along the equator of the nuclear spindle.

methanal (**formaldehyde**) A colorless gaseous aldehyde, HCOH.

methane A gaseous alkane, CH_4. Natural gas is about 99% methane and this provides an important starting material for the organic-chemicals industry. Methane is the first member of the homologous series of alkanes.

methanoate (**formate**) A salt or ester of methanoic acid.

methanoic acid (**formic acid**) A liquid carboxylic acid, HCOOH, occurring naturally in ants and nettles.

methionine *See* amino acids.

methoxy group The group CH_3O–.

methylene blue *See* staining.

methyl alcohol See methanol.

methyl orange An acid–base indicator that is red in solutions below a pH of 3 and yellow above a pH of 4.4. As the transition range is clearly on the acid side, methyl orange is suitable for the titration of an acid with a moderately weak base, such as sodium carbonate.

methyl red An acid–base indicator that is red in solutions below a pH of 4.2 and yellow above a pH of 6.3. It is often used for the same types of titration as methyl orange but the transition range of methyl red is nearer neutral (pH7) than that of methyl orange. The two molecules are structurally similar.

methyl salicylate (**oil of wintergreen; methyl 2-hydroxybenzoate**) The methyl ester of salicylic acid, which occurs in certain plants. It is absorbed through the skin and used medicinally to relieve rheumatic symptoms. It is also used in perfumes and as a flavoring agent in various foods.

Meyerhof, Otto Fritz (1884–1951) German physiologist and biochemist. He was awarded the Nobel Prize for physiology or medicine in 1922 for his discovery of the relationship between the consumption of oxygen and the metabolism of lactic acid in the muscle. The prize was shared with A. V. Hill.

MHC *See* major histocompatibility complex.

Michaelis, Leonor (1875–1949) German physician and biochemist; noted for his development of the theory of enzyme–substrate combination.

Michaelis constant Symbol: K_m For an enzyme-catalyzed reaction obeying MICHAELIS KINETICS under steady-state conditions (when the reaction intermediates have reached a steady concentration):

$$[ES] = [E][S]/K_m$$

where [ES] is the concentration of enzyme–substrate complex, [E] is the concentration of enzyme, [S] is the concentration of substrate, and K_m is the Michaelis constant.

K_m gives the concentration of substrate at which half the active sites are filled and also gives an indication of the strength of the enzyme–substrate complex if the rate of product formation is much slower than the rate of dissociation of the enzyme–substrate complex into enzyme and substrate. If this is the case then a high K_m indicates weak binding of the complex and a low K_m indicates strong binding.

Michaelis kinetics (**Michaelis–Menten kinetics**) A simple and useful model of the kinetics of enzyme-catalyzed reactions. It assumes the formation of a specific enzyme–substrate complex. Many enzymes obey Michaelis kinetics and a plot of reaction velocity (V) against substrate concentration [S] gives a characteristic curve showing that the rate increases quickly at first and then levels off to a maximum value. When substrate concentration is low, the rate of reaction is almost proportional to substrate concentration. When substrate concentration is high, the rate is at a maximum, V_{max}, and independent of substrate concentration. The Michaelis constant K_m is the concentration of substrate at half the maximum rate and can be determined experimentally by measuring reaction rate at varying substrate concentrations. Different types of inhibition can also be distinguished in this way. Allosteric enzymes do not obey Michaelis kinetics.

Michel, Hartmut (1948–) German biophysicist. He was awarded the Nobel Prize for chemistry in 1988 jointly with J. Deisenhofer and R. Huber.

micro- Symbol: μ A prefix denoting 10^{-6}. For example, 1 micrometer (μm) = 10^{-6} meter (m).

microbiology The study of microscopic organisms (e.g. bacteria and viruses), including their interactions with other organisms and with the environment. Microbial biochemistry and genetics are important branches, due to the increasing use of microorganisms in biotechnology and genetic engineering.

microfilament A minute filament, about 6 nm wide, found in eukaryotic cells and having roles in cell motion and shape. Microfilaments are made of two helically twisted strands of globular subunits of the protein actin, almost identical to the actin of muscle. They can undergo rapid extension or shortening by subunit assembly or disassembly, or form complex three-dimensional networks. Microfilaments can associate with myosin-like protein (as in muscle) and are part of the machinery of cell contraction. They often occur in sheets or bundles just below the plasma membrane and at the interface of moving and stationary cytoplasm.

micrograph A photograph taken with the aid of a microscope. *Photomicrographs* and *electron micrographs* are produced using optical microscopes and electron microscopes respectively.

micrometer Symbol: μm A unit of length equal to 10^{-6} meter (one millionth of a meter). It is often used in measurements of cell diameter, sizes of bacteria, etc. Formerly, it was called the *micron*.

micron *See* micrometer.

micronutrient A nutrient required in trace amounts by an organism. For example, a plant can obtain sufficient of the essential trace element manganese from a solution containing 0.5 parts per million of manganese. Micronutrients include trace elements and vitamins. *See* deficiency disease.

microscope An instrument designed to magnify objects and thus increase the resolution with which one can view them. *Resolution* is the ability to distinguish between two separate adjacent objects. Radiation (light or electrons) is focused through the specimen by a *condenser lens*. The resulting image is magnified by further lenses. Since radiation must pass through the specimen, it is usual to cut larger specimens into thin slices of material (sections) with a microtome. Biological material has little contrast and is therefore often stained. If very thin sections are required the material is preserved and embedded in a supporting medium.

The *light microscope* uses light as a source of radiation. With a *compound microscope* the image is magnified by two lenses, an *objective lens* near the specimen, and an *eyepiece*, where the image is viewed, at the opposite end of a tube. Its maximum magnification is limited by the wavelength of light. Much greater resolution became possible with the introduction of the *electron microscope*, which uses electrons as a source of radiation, because electrons have much shorter wavelengths than light. However, only dead material can be observed because the specimen must be in a vacuum and electrons eventually heat and destroy the material.

Electron microscopes are of two main types, the *transmission electron microscope* and the *scanning electron microscope*. The former produces an image by passing electrons through the specimen. With the scanning microscope electrons scan the surfaces of specimens rather as a screen is scanned in a TV tube, allowing surfaces of objects to be seen with greater depth of field and giving a 3D appearance to the image. Scanning microscopes cannot operate at such high magnifications as transmission microscopes.

microsomes Fragments of endoplasmic reticulum and Golgi apparatus in the form of vesicles formed during homogenization of cells and isolated by high speed centrifugation. Microsomes from rough endoplasmic reticulum are coated with ribosomes and can carry out protein synthesis in the test tube.

microtome An instrument for cutting thin sections (slices a few micrometers thick) of biological material for microscopic examination. The specimen is usually embedded in wax for support and cut by a steel knife. Alternatively it is frozen and a *freezing microtome*, which keeps the specimen frozen while cutting, is used. For electron microscopy, extremely thin (20–100 nm) sections can be cut by an *ultramicrotome*. Here the specimen is embedded in resin or plastic for support and mounted in an arm that advances slowly, moving up and down, towards a glass or diamond knife. As sections are cut they float off on to the surface of water contained in a trough behind the knife.

microtubule A thin cylindrical unbranched tube of variable length found in eukaryotic cells, either singly or in groups, and a part of the cytoskeleton. Microtubules have a skeletal role, helping cells to maintain their shape if not spherical, e.g. nerve cells. They form part of the structure of centrioles, basal bodies, and undulipodia (cilia and flagella); and form the spindle during cell division, bringing about chromosome movement. Microtubules also help to orientate materials and structures in the cell, e.g. cellulose fibrils during the formation of plant cell walls.

The wall subunits are mainly of globular proteins called *tubulins*, arranged helically around the wall to give an outside diameter of about 25 nm. Longitudinally the subunits form 13 parallel rows. Various proteins bind to microtubules and give them particular properties; for example, dynein is a motor protein that uses ATP to generate force.

microvilli Elongated slender projections (~2 μm long and ~0.2 μm diameter) of the plasma membrane, found especially in secretory and absorptive cells. The closely packed arrangement of microvilli on the free surface of epithelial cells constitutes a *brush border*. Microvilli provide an increased surface area for the exchange of molecules. Numerous microvilli occur on the epithelial cells of the intestine and also in the kidney tubules. In addition to their presence in secretory and absorptive cells, they are commonly observed in many other cells although they may not be permanent structures. Observations confirm that microvilli adhere to solid surfaces and appear to act as a signal for the cell to send out more extensive lamellar projections (*lamellipodia*).

See epithelium; membrane.

middle lamella A thin cementing layer holding together neighboring plant cell walls. It consists mainly of pectin substances (e.g. calcium pectate). The middle lamella is laid down at the cell plate during cell division. *See* pectic substances.

milk **1.** A whitish opaque nutritious fluid that is produced by the mammary glands of female mammals for feeding their young. It contains carbohydrates (mainly lactose), fats, proteins (mainly casein and whey), certain mineral salts, vitamins, and water. It is rich in calcium and phosphorus, vitamins A, D, and riboflavin, but a poor source of iron, and vitamins C and K. The composition of milk differs between species. For example, cow's milk contains more protein, calcium, phosphorus, and riboflavin than human milk, but less nicotinic acid and lactose.
2. In plants, any of various milklike fluids, such as coconut milk or the latex of certain flowering plants.

milk sugar *See* lactose.

milli- Symbol: m A prefix denoting 10^{-3}. For example, 1 millimeter (mm) = 10^{-3} meter (m).

Milstein, César (1927–2002) Argentinian-born British biochemist and immunologist. He was awarded the Nobel Prize for physiology or medicine in 1984 jointly with N. K. Jerne and G. J. F. Köhler for theories concerning the specificity in development and control of the immune system and the discovery of the principle for production of monoclonal antibodies.

mineral acid An inorganic acid, especially an acid used commercially in large quantities. Examples are hydrochloric, nitric, and sulfuric acids.

mineralocorticoid A type of steroid hormone produced by the adrenal cortex. Mineralocorticoids (e.g. aldosterone and deoxycorticosterone) control salt and water balance by their action on the kidney. *See also* corticosteroid.

minisatellite *See* genetic fingerprinting.

miscible Denoting combinations of substances that, when mixed, give rise to only one phase; i.e. substances that dissolve in each other.

Mitchell, Peter Dennis (1920–92) British biochemist. He was awarded the Nobel Prize for chemistry in 1978 for his contribution to the understanding of biological energy transfer.

mitochondrion An organelle of all plant and animal cells chiefly associated with aerobic respiration. Cells with high rates of aerobic respiration, e.g. flight muscles, have many mitochondria. It is surrounded by two membranes separated by an intermembrane space; the inner membrane forms fingerlike processes called *cristae*, which project into the gel-like *matrix*. Mitochondria are typically sausage-shaped, but may assume a variety of forms, including irregular branching shapes. The diameter is always about 0.5–1.0 μm. They contain the enzymes and cofactors of aerobic respiration and therefore are most numerous in active cells (up to several thousand per cell). They may be randomly distributed or functionally associated with other organelles, for example with the contractile fibrils of muscle cells.

The reactions of the Krebs cycle and fatty acid oxidation take place in the matrix and those of electron transport coupled to oxidative phosphorylation (i.e. the respiratory chain) on the inner membrane. Within the membrane the components of the respiratory chain are highly organized. The matrix is also involved in amino acid metabolism via Krebs cycle acids and transaminase enzymes.

It is possible that mitochondria, like chloroplasts, may be the descendents of once independent organisms that early in evolution invaded eukaryotic cells, leading to an extreme form of symbiosis. Mitochondria possess DNA (*mitochondrial DNA* or *mtDNA*) that is replicated and expressed within the mitochondria. *See* endosymbiont theory.

mitosis (karyokinesis) The ordered process by which the cell nucleus and cytoplasm divide in two during the division of body (i.e. nongermline) cells. The chromosomes replicate prior to mitosis and are then separated during mitosis in such a way that each daughter cell inherits a genetic complement identical to that of the parent cell. Although mitosis is a continuous process it is divided into four phases; prophase, metaphase, anaphase, and telophase. *Compare* meiosis; endomitosis.

moiety In a complex molecule, either of two distinct parts; e.g. a steroid and a saccharide forming a cardiac glycoside.

molar **1.** Denoting a physical quantity divided by the amount of substance. In almost all cases the amount of substance will be in moles. For example, volume (V) divided by the number of moles (n) is molar volume $V_m = v/n$.
2. A *molar solution* contains one mole of solute per cubic decimeter of solvent.

molarity A measure of the concentration of solutions based upon the number of molecules or ions present, rather than on the mass of solute, in any particular volume of solution. The molarity (M) is the number of moles of solute in one cubic decimeter (liter).

mole Symbol: mol The SI base unit of amount of substance, defined as the amount of substance that contains as many elementary entities as there are atoms in 0.012 kilogram of ^{12}C. The elementary entities may be atoms, molecules, ions, electrons, photons, etc., and they must be specified. The amount of substance is proportional to the number of entities, the constant of proportionality being the Avogadro number. One mole contains $6.022\ 045 \times 10^{23}$ entities. One mole of an element with relative atomic mass *A* has a mass of *A* grams (this mass was formerly called one *gram-atom* of the element).

molecular cloning *See* gene cloning.

molecular formula The formula of a compound showing the number and types of the atoms present in a molecule, but not the arrangement of the atoms. For example, C_2H_6O represents the molecular formula both of ethanol (C_2H_5OH) and methoxymethane (CH_3OCH_3).

molecular sieve A substance through which molecules of a limited range of sizes can pass, enabling volatile mixtures to be separated. Zeolites and other metal aluminum silicates can be manufactured with pores of constant dimensions in their molecular structure. When a sample is passed through a column packed with granules of this material, some of the molecules enter these pores and become trapped. The remainder of the mixture passes through the interstices in the column. The trapped molecules can be recovered by heating. Molecular-sieve chromatography is widely used in biochemistry laboratories. A modified form of molecular sieve is used in GEL FILTRATION.

molecular systematics A branch of biology that compares functionally equivalent macromolecules from different organisms as a basis for classification. Sequences of amino acids in proteins (e.g. enzymes) or of nucleotides in nucleic acids (e.g. ribosomal RNA) are determined using automated techniques, and compared statistically using sophisticated computer programs. Essentially, how closely two organisms are related in evolutionary terms is reflected in the degree of similarity of their macromolecules.

molecular weight *See* relative molecular mass.

molecule A particle formed by the combination of atoms in a whole-number ratio. A molecule of an element (combining atoms are the same, e.g. O_2) or of a compound (different combining atoms, e.g. HCl) retains the properties of that element or compound. Thus, any quantity of a compound is a collection of many identical molecules. Molecular sizes are characteristically 10^{-10} to 10^{-9} m.

Many molecules of natural products are so large that they are regarded as giant molecules (macromolecules); they may contain thousands of atoms and have complex structural formulae that require very advanced techniques to identify.

Molisch's test *See* alpha-naphthol test.

molting **1.** *See* ecdysis.
2. The seasonal loss of hair or feathers by mammals or birds respectively.

molybdenum *See* trace element.

monoclonal antibody A specific antibody produced by a cell clone (i.e. one of many identical cells derived from a single parent). The parent cell is obtained by the artificial fusion of a normal antibody-producing mouse spleen cell with a cell from cancerous lymphoid tissue of a mouse. This hybrid cell, or *hybridoma*, multiplies rapidly *in vitro* and yields large amounts of antibody, which comprises only a single species of immunoglobulin molecule. Monoclonal antibodies can identify a specific antigen within a mixture and are now used widely as reagents in IMMUNOASSAYS.

monocyte The largest type of white blood cell (LEUKOCYTE). It has nongranular cytoplasm and a large kidney-shaped nucleus. Monocytes are actively phagocytic, devouring foreign particles (such as bacteria). They make up 4–5% of all leukocytes.

Monod, Jacques Lucien (1910–76) French microbiologist and molecular biologist. He was awarded the Nobel Prize for physiology or medicine in 1965 jointly with F. Jacob and A. M. Lwoff for their discoveries concerning genetic control of enzyme and virus synthesis.

monomer The molecule, group, (or compound) from which a dimer, trimer, or polymer is formed.

monoploid *See* haploid.

monosaccharide A sugar that cannot be hydrolyzed to simpler carbohydrates of smaller carbon content. Glucose and fructose are examples. *See* sugar.

monosodium glutamate (MSG) A white crystalline solid compound, made from soya-bean protein. It is a sodium salt of glutamic acid (*see* amino acid) used as a flavor enhancer, particularly in Chinese cuisine. Monosodium glutamate can cause an allergic reaction in people who are ultrasensitive to it.

Moore, Stanford (1913–82) American biochemist. He was awarded the Nobel Prize for chemistry in 1972 jointly with W. H. Stein for their contribution to the understanding of the connection between chemical structure and catalytic activity of the active center of the ribonuclease molecule. The prize was shared with C. B. Anfinsen.

Morgan, Thomas Hunt (1866–1945) American geneticist and zoologist who developed *Drosphila* mutants and for techniques of gene mapping. He was awarded the Nobel Prize for physiology or medicine in 1933 for his discoveries concerning the role played by the chromosome in heredity.

morphine An alkaloid present in *opium* (from the poppy *Papaver somniferum*). It is used for the relief of severe pain. The drug *heroin* is a derivative.

morphogen *See* differentiation.

morphogenesis The development of form and structure.

morphology The study of the form of organisms. The term may be used synonymously with 'anatomy' although generally the study of external form is termed 'morphology' while the study of internal structures is termed 'anatomy'.

motor neurone A nerve cell (neurone) that transmits impulses from the brain or spinal cord to a muscle or other effector.

mRNA *See* messenger RNA.

MSG *See* monosodium glutamate.

MSH *See* melanocyte-stimulating hormone.

mtDNA (mitochondrial DNA) *See* mitochondrion.

mucilage *See* gum.

mucin The main constituent of mucus. It is a glycoprotein.

mucopolysaccharide *See* glycosaminoglycan.

mucoprotein *See* proteoglycan.

mucosa *See* mucous membrane.

mucous membrane The tissue, in vertebrates, that lines many tracts (e.g. the intestinal and respiratory tracts) that open to the exterior. It consists of surface epithelium containing goblet cells, which secrete mucus, and is underlaid by connective tissue.

mucus A slimy substance produced by goblet cells in mucous membranes of animals. It is viscous and insoluble, consisting mainly of glycoproteins. Its function is to protect and lubricate the surface on which it is secreted.

Muller, Hermann Joseph (1890–1967) American geneticist. He was awarded the Nobel Prize for physiology or medicine in 1946 for the discovery of the production of mutations by means of X-ray irradiation.

Mullis, Kary B. (1944–) American biochemist who invented the polymerase chain reaction. He was awarded the Nobel

Prize for chemistry in 1993. The prize was shared with M. Smith.

multicellular Consisting of many cells.

multiple allelism The existence of a series of alleles (three or more) for one gene. In humans, for example, there are three alleles (A, B, and O) governing blood type. Only two alleles of the series can be present in a diploid cell. Dominance relationships within an allelic series are often complicated.

Murad, Ferid (1936–) American biochemist who was awarded the 1998 Nobel Prize for physiology or medicine jointly with R. F. Furchgott and L. J. Ignarro for work on the action of nitric oxide as a signaling molecule in the cardiovascular system.

muscle Tissue consisting of elongated cells (*muscle fibers*) containing fibrils that are highly contractile.

mutagen Any physical or chemical agent that induces mutation or increases the rate of spontaneous mutation. Chemical mutagens include ethyl methanesulfonate, which causes changes in the base pairs of DNA molecules, and acridines, which cause base pair deletions or additions. Physical mutagens include ultraviolet light, x-rays, and gamma rays.

mutarotation A change in the optical rotation of a solution with time.

mutation (**gene mutation**) A change in one or more of the bases in DNA, which results in the formation of an abnormal protein. Mutations can be spontaneous, caused by the action of a mutagen, or result from errors in DNA replication. Mutations are inherited only if they occur in the cells that give rise to the gametes; somatic mutations may give rise to chimaeras and cancers. Mutations result in new allelic forms of a gene and hence new variations upon which natural selection can act. Most mutations are deleterious but are often retained in the population because they also tend to be recessive and can thus be carried in the genotype without affecting the viability of the organism. Although the natural rate of mutation is low, organisms could not survive without specific enzymes to repair damaged DNA. *See also* chromosome mutation; mutagen; polyploid.

mycelium A filamentous mass comprising the body of a fungus, each filament being called a *hypha*. The mycelium often forms a loose mesh as in *Mucor*, but the hyphae may become organized into definite structures, e.g. the fruiting body of a mushroom. The mycelium produces the reproductive organs of a fungus. The whole thallus of unicellular fungi may be thus employed, but only part of the thallus produces gametangia or sporangia in most species, the rest of the thallus being vegetative.

mycoprotein Any protein produced by a fungus or bacterium.

myelinated nerve fiber A nerve fiber that is surrounded by a fatty (myelin) sheath. Most nerves of vertebrates consist of thousands of medullated fibers, which appear white because of the fatty sheaths. *See* myelin sheath.

myelin sheath An insulating covering that surrounds the axon of a neurone. It is composed of the cell membranes of Schwann cells wound tightly in a spiral around the axon. The membranes consist of a fatty material (myelin). In between each Schwann cell is a short region of bare axon (*node of Ranvier*). Myelinated (medullated) axons occur in most vertebrate neurones but are less common in invertebrates. *See* Schwann cell.

myelocyte A cell in the myeloid tissue of red bone marrow. Myelocytes are formed by cell division of precursor cells (myeloblasts) and they change into granulocytes, which are released into the bloodstream. *See also* granulocyte.

myeloid tissue Tissue that manufactures white blood cells. It occurs in the red bone marrow, surrounding the blood vessels. It contains myeloblast cells, which divide continuously to give myelocytes (which develop into granulocytes), and lymphoblasts and monoblasts, which give rise to agranular leukocytes.

myofibril A very fine fiber (1–2 μm in diameter) many of which are embedded in the sarcoplasm of a muscle fiber. In SKELETAL MUSCLE these fibrils are striated, being divided along their length into a great number of SARCOMERES, which constitute the contractile apparatus of the muscle fiber.

myoglobin A conjugated protein found in muscles (sometimes referred to as 'muscle hemoglobin'). It is similar to HEMOGLOBIN in being a heme protein capable of binding oxygen but is structurally simpler, having only one polypeptide chain combined with the heme group. Each molecule of myoglobin can attach one molecule of oxygen.

myosin A contractile protein; the most abundant protein found in muscle. Filaments of myosin form the thick filaments observed in muscle myofibrils. In muscle contraction, myosin molecules combine with actin present in adjacent thin filaments to form actomyosin complexes. *See* sarcomere.

NAD (**nicotinamide adenine dinucleotide**) A derivative of nicotinic acid that acts as a coenzyme in electron-transfer reactions (e.g. the electron-transport chain). Its role is to carry hydrogen atoms; the reduced form is written NADH.

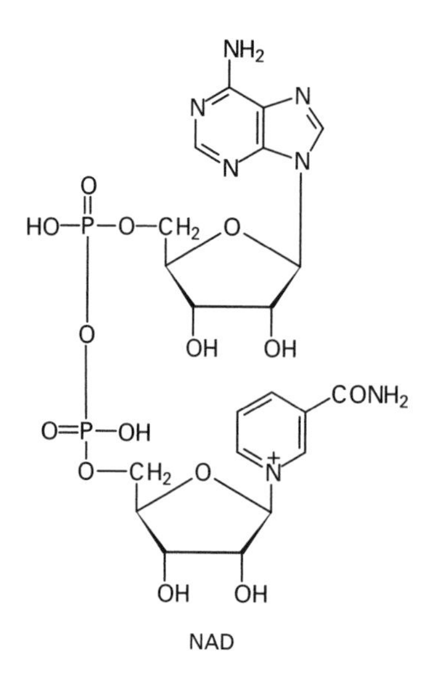

NAD

NADP Nicotinamide adenine dinucleotide phosphate; a coenzyme similar in its action to NAD. The reduced form is written NADPH, which acts as an electron donor in many synthetic reactions.

nano- Symbol: n A prefix denoting one thousand-millionth, or 10^{-9}. For example, 1 nanometer (nm) = 10^{-9} meter.

Nathans, Daniel (1928–) American molecular biologist. He was awarded the Nobel Prize for physiology or medicine in 1978 jointly with W. Arber and H. O. Smith for the discovery of restriction enzymes.

natural killer cell (**NK cell**) A large lymphocyte with prominent lysosomes that attaches to the surface of virus-infected cells and secretes substances (PERFORINS) that puncture the target cell's membrane. This allows the entry of other substances that trigger the death of the cell. Unlike cytotoxic T-CELLS, NK cells are not primed to attack specific target cells, and how they recognize their targets is unclear.

natural selection The process, which Darwin called the 'struggle for survival', by which organisms less adapted to their environment tend to perish, and better-adapted organisms tend to survive. According to Darwinism, natural selection acting on a

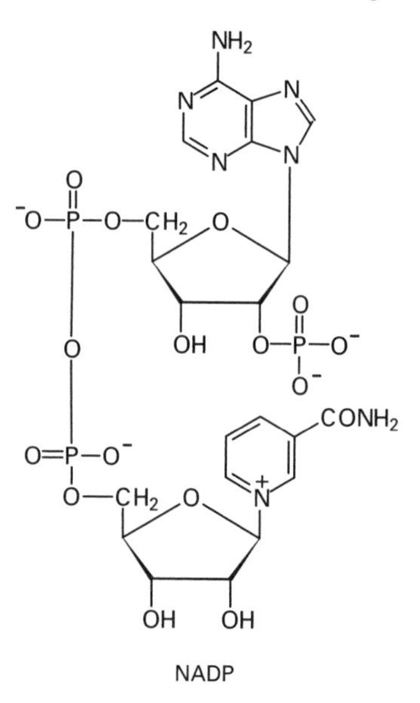

NADP

varied population results in evolution. *See* Darwinism; evolution.

nectar A sugar-containing fluid secreted by the nectaries in plants.

negative staining A method of preparation of material for electron microscopy used for studying three-dimensional and surface features, notably of viruses, macromolecules (e.g. enzyme complexes), and the cristae of mitochondria. A stain is used that covers the background and penetrates surface features of the specimen, but leaves the specimen itself unstained.

Neher, Erwin (1944–) German biophysicist. He was awarded the Nobel Prize for physiology or medicine in 1991 jointly with B. Sakmann for their work on the function of single ion channels in cells.

neo-Darwinism Darwin's theory of evolution through natural selection, modified and expanded by modern genetic studies arising from the work of Mendel and his successors. Such studies have answered many questions which Darwin's theory raised, but could not adequately explain because of lack of knowledge at the time it was formulated. Notably, modern genetics has revealed the source of variation on which natural selection operates, namely mutations of genes and chromosomes, and provided mathematical models of how alleles fluctuate in natural populations, thereby quantifying the process of evolution.

neotonin *See* juvenile hormone.

nerve A bundle of nerve fibers surrounded by a protective covering of connective tissue. *Mixed nerves*, such as the spinal nerves, contain both sensory and motor fibers. *See also* neurone.

nerve cell *See* neurone.

nerve cord An enclosed cylindrical tract of nerve fibers that forms a central route for the conduction of nerve impulses within the body. Vertebrates and other chordates have a single hollow nerve cord (the spinal cord) situated dorsally. Invertebrates generally have two or more nerve cords, each lacking a central cavity and with ganglia situated at intervals along its length.

nerve fiber The axon of a neurone. *See* neurone.

nerve impulse The signal transmitted along neurones. All nerve impulses are identical in form and strength and consist of changes in permeability of the axon membrane followed by flows of ions into and out of the cell, thereby producing potential changes that can be detected as the action potential passing along the axon. The energy required to pass the impulse is derived from the neurone itself, not from the stimulus.

In its passive state the axon has a resting potential of –70 mV inside the membrane, caused by sodium ions being pumped out of the cell. As the impulse passes, the membrane becomes transiently permeable to sodium ions, which flow into the cell. This causes a change in potential to about +30 mV – the action potential. In myelinated neurones this mechanism operates only at the nodes of Ranvier; the myelin sheath insulates the axon so that the action potential is conducted, more rapidly, from one node to the next (saltatory conduction). There is a refractory period following the passage of an impulse, during which another impulse cannot be transmitted and sodium is pumped back out of the neurone. In living organisms the impulse is triggered by local depolarization at a synapse or a receptor cell, but in isolated axons almost any disturbance of the membrane will set off an impulse. Since the impulse is an all-or-nothing event, the strength of the stimulus is signaled by the frequency and number of identical impulses.

nerve net A netlike layer of interconnecting nerve cells that is found in the body wall of certain groups of invertebrate animals; the most primitive type of nervous system. It occurs in cnidarians, echinoderms, and hemichordates.

nervous system A ramifying system of cells, found in all animals except sponges, that forms a communication system between receptors and effectors and allows varying degrees of coordination of information from different receptors and stored memory, producing integrated responses to stimuli. The system consists of neurones, supportive glial cells, and various fibrous tissues surrounding the softer matter. NERVE IMPULSES are transmitted through the neurones, which communicate with each other at specialized junctions, the synapses, which are essentially one-way and are the basis of all integration within the system. The impulse is electrochemical, consisting of a propagated change in the potential on either side of the neurone membrane, the action potential, which travels at between 1 and 120 m/s, depending on the animal and the type of neurone.

At its simplest, as found in the Cnidaria, the nervous system is merely a diffuse net with little concentration of function, but higher animals possess groups of neurones (ganglia), within which integration can take place. The major ganglion develops in the head, as the brain, and becomes increasingly important as a control center in more advanced types. The brain communicates with the body through the spinal cord, which is composed mostly of long axons transmitting impulses to and from the brain but also contains the circuits for body reflexes, and the peripheral nervous system, which contains sensory or motor neurones running from receptors or to effectors. *See also* autonomic nervous system; central nervous system

neuroendocrine systems The systems involving both nervous and endocrine factors that control functions of the body. Many examples are known, particularly involving the pituitary gland, from which hormones are secreted under direct nervous stimulation.

neurohormone A hormone that is produced by specialized nervous tissue. Examples are norepinephrine, serotonin, vasopressin, and oxytocin. *See* neuroendocrine systems.

neurohypophysis The posterior lobe of the pituitary gland in higher vertebrates. It is derived from a fold in the floor of the brain and stores and releases into the blood the hormones oxytocin and vasopressin. These are manufactured in the hypothalamus by neurosecretory cells that have their endings in the neurohypophysis.

neuromuscular junction The specialized region in which a nerve ending makes close contact with a muscle. Impulses arriving at the nerve ending cause it to release a chemical transmitter, which diffuses across the intervening gap and stimulates the muscle to contract.

neurone (**nerve cell; neuron**) A cell that is specialized for the transmission of nervous impulses. It consists of a *cell body*, which contains the nucleus and Nissl granules and has numerous branching extensions (*dendrites*), and a single long fine *axon* (nerve fiber), which has few branches and may be surrounded by a myelin sheath. Dendrites carry nervous impulses towards the cell body and the axon carries them away from the cell body. The end of the axon connects with another neurone at a synapse or with an effector (e.g. a muscle or gland).

Sensory neurones carry impulses from sense organs to the central nervous system and usually have rounded cell bodies; *motor neurones* carry impulses from the central nervous system to muscles and usually have star-shaped cell bodies. *Interneurones* relay impulses between sensory and motor neurones. *See also* nerve impulse; synapse.

neurotransmitter A chemical that is released from neurone endings to cause either excitation or inhibition of an adjacent neurone or muscle cell. It is stored in minute vesicles near the synapse and released when a nerve impulse arrives. It diffuses across the synaptic cleft and binds to specific receptors on the adjacent neurone. In mammals the main neurotransmitters are acetylcholine, found throughout the nervous system, and norepinephrine, occurring in the sympathetic nervous system. *See* synapse.

neutrophil A white blood cell (LEUKOCYTE) containing granules that do not stain with either acid or basic dyes. Neutrophils have a many-lobed nucleus and are therefore called *polymorphonuclear leukocytes* or *polymorphs*. Comprising about 70% of all leukocytes in humans and certain other mammals, they engulf and digest foreign particles, such as bacteria, using enzymes from their granules. This is the body's first line of defense against disease. They can pass out of capillaries by an ameboid process (*diapedesis*) and wander in the tissues, gathering in large numbers at the site of an infection, where they may die, forming pus. *Compare* macrophage.

niacin *See* nicotinic acid.

nicotinamide adenine dinucleotide *See* NAD.

nicotinic acid (**niacin**) One of the water-soluble B-group of vitamins. Its deficiency in man causes pellagra. Nicotinic acid functions as a constituent of two coenzymes, NAD and NADP, which operate as hydrogen and electron transfer agents and play a vital role in metabolism. *See also* vitamin B complex.

ninhydrin A reagent used to test for the presence of proteins and amino acids. An aqueous solution turns blue in the presence of alpha amino acids in solution. When dissolved in an organic solvent it is used as a developer to color amino acids on chromatograms. If a chromatogram treated with ninhydrin is heated strongly the amino acids appear as purple spots. Ninhydrin is carcinogenic.

Nirenberg, Marshall Warren (1927–) American biochemist and molecular biologist. He was awarded the Nobel Prize for physiology or medicine in 1968 jointly with R. W. Holley and H. G. Khorana for their interpretation of the genetic code and its function in protein synthesis.

Nissl granules Densely staining material found in the cell bodies of neurones. They consist of endoplasmic reticulum covered by ribosomes, plus many free ribosomes. The granules are stained by the same basic dyes that stain nuclei.

nitric oxide A colorless odorless gas, NO. Nitric oxide is a biogenic messenger with properties as a vasodilator. It also has anticoagulant activity. It is formed in most mammalian tissues from arginine, NADPH, and oxygen under the influence of the enzyme *nitric-oxide synthase*.

nitrification The conversion of ammonia to nitrite, and nitrite to nitrate, carried out by certain *nitrifying bacteria* in the soil. The chemosynthetic bacteria *Nitrosomonas* and *Nitrobacter* carry out the first and second stages respectively of this conversion. The process is important in the nitrogen cycle since nitrate is the only form in which nitrogen can be used directly by plants. *See* nitrogen cycle.

nitrifying bacteria *See* nitrification.

nitrogen An essential element found in all amino acids and therefore in all proteins, and in various other important organic compounds, e.g. nucleic acids. Gaseous nitrogen forms about 80% of the atmosphere but is unavailable in this form except to a few nitrogen-fixing bacteria. Nitrogen is therefore usually incorporated into plants as the nitrate ion, NO_3^-, absorbed in solution from the soil by roots. In animals, the nitrogen compounds urea and uric acid form the main excretory products. *See also* nitrogen cycle.

nitrogen cycle The circulation of nitrogen between organisms and the environment. Atmospheric gaseous nitrogen can only be used directly by certain nitrogen-fixing bacteria (e.g. *Clostridium*, *Nostoc*). They convert nitrogen to ammonia, ni-

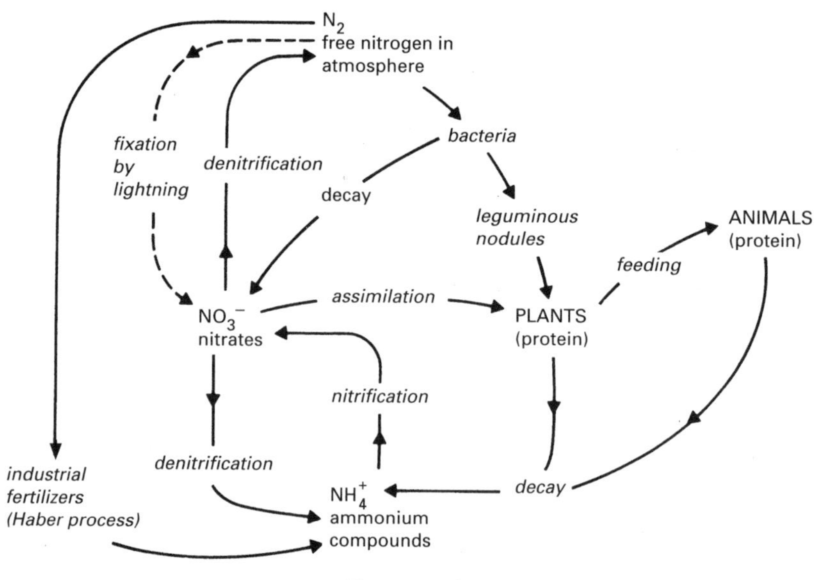

Nitrogen cycle

trites, and nitrates, which are released into the soil by excretion and decay. Some are free-living, while others form symbiotic associations with plants (*See* nitrogen fixation). Another method by which atmospheric nitrogen is fixed is by lightning. When plants and animals die, the organic nitrogen they contain is converted back into nitrate in the process termed *nitrification*. Apart from uptake by plants, nitrate may also be lost from the soil by *denitrification* and by leaching. The increasing use of nitrogen fertilizers in agriculture and the emission of nitrous oxides in car exhaust fumes are also important factors in the nitrogen cycle. *See* denitrification; nitrification; nitrogen fixation.

nitrogen fixation The formation of nitrogenous compounds from atmospheric nitrogen. In nature this may be achieved by electric discharge in the atmosphere (lightning) or by the activities of certain microorganisms. For example, the symbiotic bacteria, *Rhizobium* species, are associated with leguminous plants, forming the characteristic nodules on their roots. The bacteria contain the nitrogenase enzyme that catalyzes the fixation of molecular nitrogen to ammonium ions (a process which requires ATP), which the plant can assimilate. In return the legume supplies the bacteria with carbohydrate. Free-living bacteria that can fix nitrogen include members of the genera *Azotobacter* and *Clostridium*. Some sulfur bacteria (e.g. *Chlorobium*), some Cyanobacteria (e.g. *Anabaena*), and some yeast fungi have also been shown to fix nitrogen. In industry the most important method for fixing nitrogen is the Haber process, which is used to make ammonia from nitrogen and hydrogen.

NK cell *See* natural killer cell.

NMR *See* nuclear magnetic resonance.

node of Ranvier A region of bare axon that occurs at intervals of up to 2 mm along the length of myelinated nerve axons. *See also* myelin sheath; saltatory conduction.

noncoding strand The DNA strand of a double helix that is the template for transcription of RNA. Its sequence is therefore complementary to the RNA sequence (with T substituted for U). *Compare* coding strand.

nondisjunction The failure of homologous chromosomes to move to separate poles during anaphase I of meiosis, both homologs going to a single pole. This results in two of the four gametes formed at telophase missing a chromosome (i.e. being $n - 1$). If these fuse with normal haploid (n) gametes then the resulting zygote is *monosomic* (i.e. $2n - 1$). The other two gametes formed at telophase have an extra chromosome (i.e. are $n + 1$) and give a *trisomic* zygote (i.e. $2n + 1$) on fusion with a normal gamete. If two gametes deficient for the same chromosome fuse then *nullisomy* ($2n - 2$) will result, which is almost always lethal, and if two gametes with the same extra chromosomes fuse, *tetrasomy* ($2n + 2$) results. All these abnormal chromosome conditions are collectively referred to as *aneuploidy*. In humans the condition of Down's syndrome is due to trisomy of chromosome 21.

noradrenaline *See* norepinephrine.

norepinephrine (**noradrenaline**) A catecholamine, secreted as a hormone by the adrenal medulla, that regulates heart muscle, smooth muscle, and glands. It causes narrowing of arterioles and hence raises blood pressure. It is also secreted by nerve endings of the sympathetic nervous system in which it acts as a neurotransmitter. In the brain, levels of norepinephrine are related to mental function; lowered levels lead to mental depression.

Northrop, John Howard (1891–1987) American biochemist. He was awarded the Nobel Prize for chemistry in 1946 jointly with W. M. Stanley for their preparation of enzymes and virus proteins in a pure form. The prize was shared with J. B. Sumner.

nuclear magnetic resonance (**NMR**) A property of atomic nuclei utilized in the analysis of molecular structure. Small changes in the resonance of certain atomic nuclei (e.g. ^{1}H, ^{13}C) can be induced by irradiating them with radio waves in the presence of a strong magnetic field. This property is utilized in a form of spectroscopy called *NMR spectroscopy* to provide information about the composition and structure of complex molecules. It is also exploited in medicine as the basis of *NMR imaging*, which is used to detect tumors, etc. in the body.

nuclease *See* endonuclease; exonuclease.

nucleic acid hybridization The pairing of a single-stranded DNA or RNA molecule with another such strand, forming a DNA–DNA or RNA–DNA hybrid. In order to achieve hybridization the base sequences of the strands must be complementary. This phenomenon is exploited in many techniques, notably in gene probes, which are designed to bind to particular complementary base sequences among a mass of DNA fragments. Hybridization can also be used in comparative biology to assess the degree of similarity of base sequences between, say, corresponding genes from different organisms. The DNA forms double helices in homologous regions where it is heated and then cooled. *See* gene probe.

nucleic acids Organic acids whose molecules consist of chains of alternating sugar and phosphate units, with nitrogenous bases attached to the sugar units. They occur in the cells of all organisms. In DNA the sugar is deoxyribose; in RNA it is ribose. *See* DNA; RNA.

nucleoid The region of a bacterium containing DNA and not enclosed by membranes. It may be associated with the mesosome during cell divisions. *Compare* nucleus.

nucleolar organizer *See* nucleolus.

nucleolus A more or less spherical region found in nuclei of eukaryote cells, and easily visible with a light microscope. One to several per nucleus may occur. It is the site of ribosome manufacture and is thus most conspicuous in cells making large quantities of protein. Nucleoli disappear during cell division. The nucleolus synthesizes ribosomal RNA (rRNA) and is rich in

RNA (about 10%) and protein. It forms around particular loci of one or more chromosomes called *nucleolar organizers.* These loci contain numerous tandem repeats of the genes coding for ribosomal RNA.

nucleophile An electron-rich ion or molecule that takes part in an organic reaction. The nucleophile can be a negative ion (Br^-, CN^-) or a molecule with a lone pair of electrons (NH_3, H_2O). The nucleophile attacks positively charged parts of molecules, which usually arise from the presence of an electronegative atom elsewhere in the molecule. *Compare* electrophile.

nucleoprotein A compound consisting of a protein associated with a nucleic acid. Examples of nucleoproteins are the chromosomes, made up of DNA, some RNA, and histones (proteins), and the ribosomes (ribonucleoproteins), consisting of ribosomal RNA and proteins.

nucleoside A molecule consisting of a purine or pyrimidine base linked to a sugar, either ribose or deoxyribose. ADENOSINE, CYTIDINE, GUANOSINE, THYMIDINE, and URIDINE are common nucleosides.

nucleotide The compound formed by condensation of a nitrogenous base (a purine, pyrimidine, or pyridine) with a sugar (ribose or deoxyribose) and phosphoric acid. The coenzymes NAD and FAD are *dinucleotides* (consisting of two linked nucleotides) while the nucleic acids are *polynucleotides* (consisting of chains of many linked nucleotides).

nucleus An organelle of eukaryote cells containing the genetic information (DNA) and is found in virtually all living cells (exceptions include mature sieve tube elements and mature mammalian red blood cells). DNA replication and transcription of RNA occur exclusively in the nucleus. It is the largest organelle, typically spherical and bounded by a double membrane, the *nuclear envelope* or *nuclear membrane*, which is perforated by many pores (*nuclear pores*) that allow exchange of materials with the cytoplasm. The outer nuclear membrane is an extension of the endoplasmic reticulum. In the nondividing (interphase) nucleus the genetic material is irregularly dispersed as chromatin; during nuclear division (mitosis or meiosis) this condenses into densely staining chromosomes, and the nuclear envelope disappears as do the nucleoli that are normally present. *Compare* nucleoid. *See* chromatin; chromosome; nucleolus.

Nurse, (Sir) Paul M. (1949–) British cell biologist who was awarded the 2001 Nobel Prize for physiology or medicine jointly with R. T. Hunt and L. H. Hartwell for discoveries of the key regulators of the cell cycle.

Nüsslein-Volhard, Christiane (1942–) German biochemist who was jointly awarded the 1995 Nobel Prize for physiology or medicine with E. B. Lewis and E. F. Wieschaus for work on the genetic control of early embryonic development.

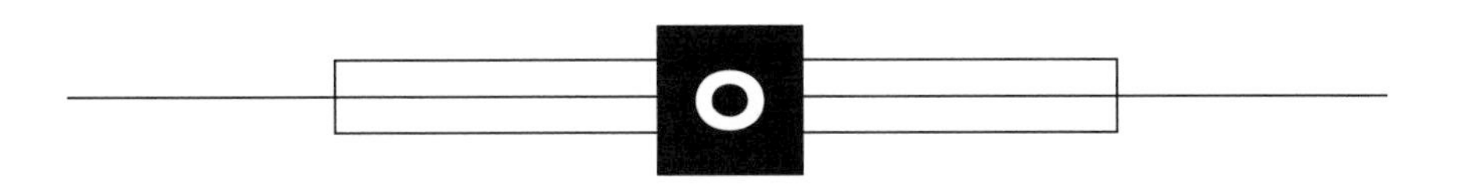

OAA *See* oxaloacetic acid.

Ochoa, Severo (1905–93) Spanish-born American biochemist. He was awarded the Nobel Prize for physiology or medicine in 1959 jointly with A. Kornberg for their discovery of the mechanisms in the biological synthesis of ribonucleic acid and deoxyribonucleic acid.

octadecanoic acid *See* stearic acid.

octadecenoic acid *See* oleic acid.

Okazaki fragment *See* replication.

oleate A salt or ester of oleic acid; i.e. an octadecenoate.

olefin *See* alkene.

oleic acid An unsaturated fatty acid occurring as the glyceride in oils and fats. Oleic acid occurs naturally in larger quantities than any other fatty acid. In many organisms oleic acid can be synthesized directly from stearic acid and further enzymatic paths exist for conversion to linoleic acid and linolenic acid. This pathway does not occur in humans and the higher animals so plant sources are an essential dietary element. The systematic name is *cis*-9-octadecenoic acid.

olfactory organ The organ involved in the detection of smells, which consists of a group of sensory receptors that respond to air- or water-borne chemicals. Vertebrates possess a pair of olfactory organs in the mucous membrane lining the upper part of the nose, which opens to the exterior via the external nares (nostrils). Chemicals from the environment are dissolved in the mucus secreted by the nasal epithelium and information is transmitted to the brain by the receptors via the olfactory nerve. Olfactory organs are found on the antennae in insects and in various positions in other invertebrates.

oligotrophic Describing lakes that are deficient in nutrients and consequently low in productivity. *Compare* eutrophic.

oncogene A gene that is capable of transforming a normal cell into a cancerous cell. Many retroviruses carry oncogenes that are derived from normal cellular genes of their hosts; these cellular counterparts of oncogenes are termed *proto-oncogenes*. It is believed that the cellular gene is picked up by the retrovirus and that this process changes it into an oncogene by altering its regulation or the regulation of its protein product. This can happen in various ways: the gene may suffer a deletion or other type of mutation, resulting in a protein product with abnormal and excessive activity; or its removal to a new setting may affect the regulation of its expression, causing it to be overexpressed. Any mutation (not necessarily by the action of a virus) that has one of these effects will cause a cellular protooncogene to become an oncogene. Protooncogenes normally perform various functions in the growth and differentiation of cells, for example their products may act as growth factors, be involved in signal transduction for growth factors (e.g. *ras*, *raf*, and *src*) or as regulators of gene expression. Oncogenes are therefore well placed to disrupt the normal activities of cells and cause the changes that lead to cancer, and their study is a major area of cancer research.

oncogenic Causing the production of a tumor, especially a malignant tumor (i.e. cancer). The term is usually applied to viruses known to be implicated in causing cancer; such viruses include some retroviruses, papovaviruses, adenoviruses, and herpesviruses.

one gene–one enzyme hypothesis The theory that each gene controls the synthesis of one enzyme, which was advanced following studies of nutritional mutants of fungi. Thus by regulating the production of enzymes, genes control the biosynthetic reactions catalyzed by enzymes and ultimately the character of the organism. Genes also code for proteins, or polypeptides that form proteins, other than enzymes, so the idea is perhaps more accurately expressed as the *one gene–one polypeptide hypothesis*.

onium ion An ion formed by addition of a proton (H^+) to a molecule. The hydronium ion (H_3O^+) and ammonium ion (NH_4^+) are examples.

ontogeny The course of development of an organism from fertilized egg to adult. Occasionally ontogeny is used to describe the development of an individual structure.

oocyte A reproductive cell in the ovary of an animal that gives rise to an ovum. The primary oocyte develops from an oogonium, which has undergone a period of multiplication and growth. It divides by meiosis and the first meiotic (or reduction) division produces a secondary oocyte, containing half the number of chromosomes, and a small polar body. The secondary oocyte undergoes the second meiotic division to form an ovum and a second polar body. In many species the second meiotic division is not completed until after fertilization.

OP Osmotic pressure. *See* osmosis.

operon A genetic unit found in prokaryotes and comprising a group of closely linked genes acting together and coding for the various enzymes of a particular biochemical pathway. At one end is an *operator*, which may under certain conditions be repressed by another gene outside the operon, the *regulator gene*. The regulator gene produces a protein (a *repressor*) that binds to the operator, renders it inoperative, and so prevents enzyme production. The presence of a suitable substrate prevents this binding, and so enzyme production can commence. Another site in the operon, the *promoter*, initiates the formation of the messenger RNA that carries the code for the synthesis of the enzymes determined by all the *structural genes* of the operon. The promoter and operator form the *control region* of an operon. *See* promoter; repressor molecule.

opium *See* morphine.

opsin The protein component of the retinal pigment RHODOPSIN, which is localized in the rod cells of the retina. Opsin is released from rhodopsin when light strikes the retina.

opsonin Any of various blood proteins that bind to microorganisms or other foreign material and enhance their susceptibility to engulfment by phagocytic cells. The main types of opsonins are immunoglobulin G antibodies and certain complement proteins. *See also* antibody.

opsonization The process by which a host coats the surface of an invading cell with opsonins to render it susceptible to phagocytosis.

optical activity The ability of certain compounds to rotate the plane of polarization of plane-polarized light when the light is passed through them. Optical activity can be observed in crystals, gases, liquids, and solutions. The amount of rotation depends on the concentration of the active compound.

Optical activity is caused by the interaction of the varying electric field of the light with the electrons in the molecule. It occurs when the molecules are asymmetric – i.e. they have no plane of symmetry. Such molecules have a mirror image that cannot be superimposed on the original molecule. In organic compounds this usually means that they contain a carbon atom attached to four different groups, forming a chiral center. The two mirror-image forms of an asymmetric molecule are optical isomers. One isomer will rotate the polarized light in one sense and the other by the same amount in the opposite sense. Such isomers are described as *dextrorotatory* or *levorotatory*, according to whether they rotate the plane to the 'right' or 'left' respectively (rotation to the left is clockwise to an observer viewing the light coming toward the observer). Dextrorotatory compounds are given the symbol *d* or (+) and levorotatory compounds *l* or (–). A mixture of the two isomers in equal amounts does not show optical activity. Such a mixture is sometimes called the (±) or *dl*-form, a *racemate*, or a *racemic mixture*

Optical isomers have identical physical properties (apart from optical activity) and cannot be separated by fractional crystallization or distillation. Their general chemical behavior is also the same, although they do differ in reactions involving other optical isomers. Many naturally occurring substances are optically active (only one optical isomer exists naturally) and biochemical reactions occur only with the natural isomer. For instance, the natural form of glucose is *d*-glucose and living organisms cannot metabolize the *l*-form.

The terms 'dextrorotatory' and 'levorotatory' refer to the effect on polarized light. A more common method of distinguishing two optical isomers is by their *D-form* (*dextro-form*) or *L-form* (*levo-form*). This

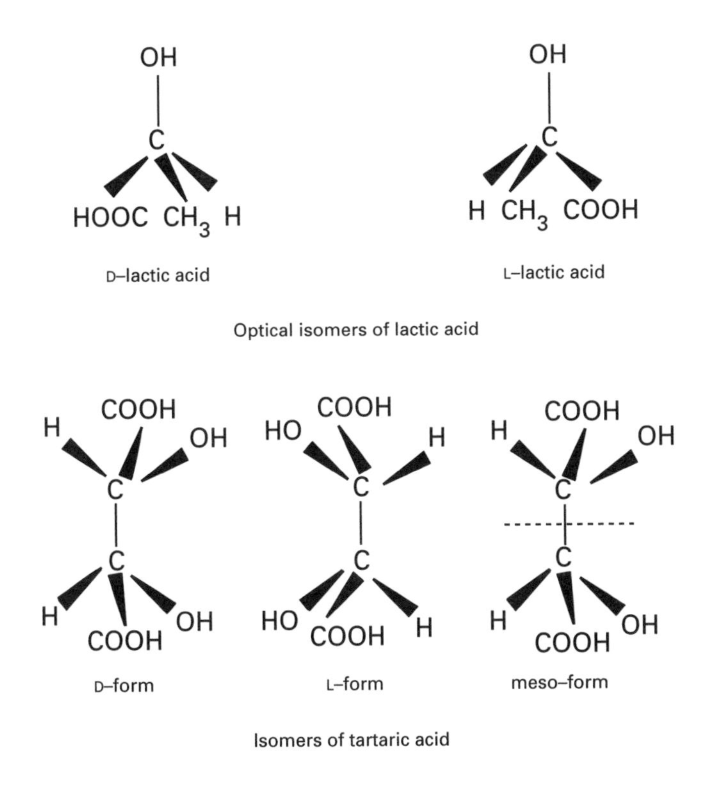

Optical activity

convention refers to the absolute structure of the isomer according to specific rules. Sugars are related to a particular configuration of glyceraldehyde (2,3-dihydroxypropanal). For alpha amino acids the 'corn rule' is used: the structure of the acid $RC(NH_2)(COOH)$ H is drawn with H at the top; viewed from the top the groups spell CORN in a clockwise direction for all D-amino acids (i.e. the clockwise order is $-COOH,R,NH_2$). The opposite is true for L-amino acids. Note that this convention does not refer to optical activity: D-alanine is dextrorotatory but D-cystine is levorotatory.

An alternative is the R/S system for showing configuration. There is an order of priority of attached groups based on the proton number of the attached atom:

I, Br, Cl, SO_3H, $OCOCH_3$, OCH_3, OH, NO_2, NH_2, $COOCH_3$, $CONH_2$, $COCH_3$, CHO, CH_2OH, C_6H_5, C_2H_5, CH_3, H

Hydrogen has the lowest priority. The chiral carbon is viewed such that the group of lowest priority is hidden behind it. If the other three groups are in descending priority in a clockwise direction, the compound is R-. If descending priority is anticlockwise it is S-.

The existence of a carbon atom bound to four different groups is not the strict condition for optical activity. The essential point is that the molecule should be asymmetric. Inorganic octahedral complexes, for example, can show optical isomerism. It is also possible for a molecule to contain asymmetric carbon atoms and still have a plane of symmetry. One structure of tar-

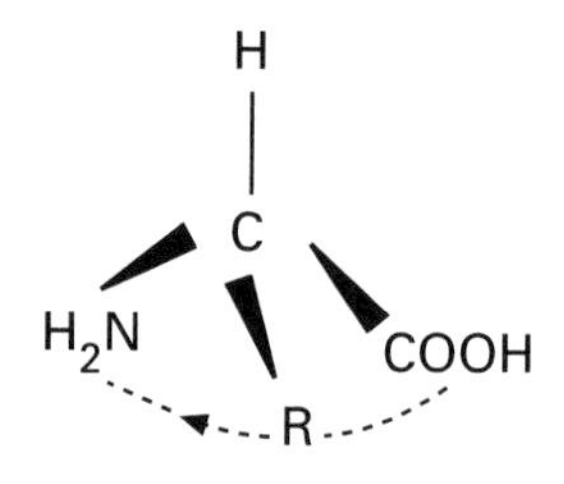

The corn rule for absolute configuration of alpha amino acids

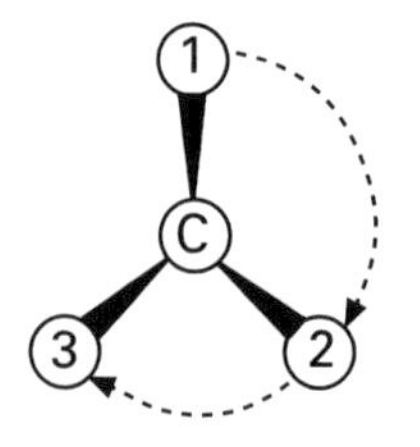

"R"–configuration

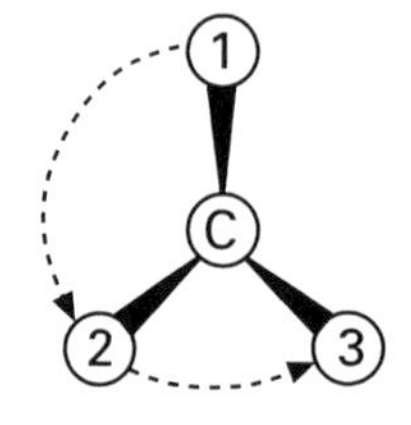

"S"–configuration

The R/S system

Optical activity

taric acid has two parts of the molecule that are mirror images, thus having a plane of symmetry. This (called the *meso-form*) is not optically active.

optical isomerism *See* isomerism; optical activity.

optical rotary dispersion (ORD) The phenomenon in which the amount of rotation of plane-polarized light by an optically active substance depends on the wavelength of the light. Plots of rotation against wavelength can be used to give information about the molecular structure of optically active compounds.

optical rotation Rotation of the plane of polarization of plane-polarized light by an optically active substance.

ORD *See* optical rotary dispersion.

order The sum of the indices of the concentration terms in the expression that determines the rate of a chemical reaction. For example, in the expression

$$\text{rate} = k[A]^x[B]^y$$

x is called the order with respect to A, y the order with respect to B, and $(x + y)$ the order overall. The values of x and y are not necessarily equal to the coefficients of A and B in the molecular equation. Order is an experimentally determined quantity derived without reference to any equation or mechanism. Fractional orders do occur. For example, in the reaction

$$CH_3CHO \rightarrow CH_4 + CO$$

the rate is proportional to $[CH_3CHO]^{1.5}$ i.e. it is of order 1.5.

ore A mineral source of a chemical element.

organic acid An organic compound that can release hydrogen ions (H^+) to a base, such as a carboxylic acid or a phenol. *See* acid, carboxylic acid, phenol.

organic base An organic compound that can function as a base. Organic bases are typically amines that gain H^+ ions. *See* amine, base.

organic chemistry The chemistry of compounds of carbon. Originally the term *organic chemical* referred to chemical compounds present in living matter, but now it covers any carbon compound with the exception of certain simple ones, such as the carbon oxides, carbonates, cyanides, and cyanates. These are generally studied in inorganic chemistry. The vast numbers of synthetic and natural organic compounds exist because of the ability of carbon to form chains of atoms (catenation). Other elements are involved in organic compounds; principally hydrogen and oxygen but also nitrogen, halogens, sulfur, and phosphorus.

organometallic compound An organic compound containing a carbon–metal bond. Tetraethyl lead, $(C_2H_5)_4Pb$, is an example of an organometallic compound, used as an additive in petrol.

organelle A discrete subcellular structure with a particular function. The largest organelle is the nucleus; other examples are chloroplasts, mitochondria, Golgi apparatus, vacuoles, and ribosomes. Organelles allow division of labor within the cell. Prokaryotic cells have very few organelles compared with eukaryotic cells.

organizer A part of an embryo whose presence causes neighboring tissue to develop in a particular way. Examples are the eye-cup of vertebrates, which causes lens, and later cornea, to be produced; the gut of snail embryos, which organizes shell gland and mantle; the dorsal lip of the blastopore of frogs, which becomes notochord and organizes all the axial structures of the embryo; and the dermal papilla of a hair, feather, or tooth, which organizes local epidermis and dermis to form the follicle and appendages. The term *primary organizer* is restricted to the first or most important initiator at gastrulation; for example, the dorsal lip of the amphibian blastopore or Hensen's node in mammals and birds. Determinants of major systems (e.g. notochord) are called *secondary organizers*; local centers of developmental activity (e.g. dermal papillae) are *tertiary organizers*.

origin of life Geological evidence strongly suggests that life originated on earth about 4000 million years ago. The basic components of organic matter – water, methane, ammonia, and related compounds – were abundant in the atmosphere. Energy from the sun (cosmic rays) and lightning storms caused these to recombine into increasingly complex organic molecules. Particular combinations of such complex substances eventually showed the characteristics of living organisms.

ornithine *See* amino acids.

ornithine cycle (**urea cycle**) The sequence of enzyme-controlled reactions by which urea is formed as a breakdown product of amino acids. It occurs in cells of the liver. The amino acid ornithine is combined with ammonia (from amino acids) and carbon dioxide, forming another amino acid, arginine, which is then split into urea (which is excreted) and ornithine.

ortho- Designating a benzene compound with substituents in the 1,2 positions. The position next to a substituent is the ortho position on the benzene ring. This was used in the systematic naming of benzene derivatives. For example, orthodinitrobenzene (or *o*-dinitrobenzene) is 1,2-dinitrobenzene.

osazones Distinctly shaped crystals produced by heating monosaccharides with phenylhydrazine hydrochloride and sodium acetate. The osazones are examined microscopically and used for identifying individual monosaccharides; fructose, mannose, and glucose however give identical osazones, but may be distinguished by other tests.

osmiophilic globules (**osmiophilic droplets**) *See* plastoglobuli.

osmium tetroxide A stain used in electron microscopy because it contains the heavy metal osmium. It also acts as a fixative, i.e. preserves material in a lifelike condition, often being used in conjunction with the fixative glutaraldehyde. It stains lipids, and therefore membranes, particularly intensely.

osmometer An instrument that is used to measure osmotic pressure.

osmoregulation The process by which animals regulate their internal osmotic pressure by controlling the amount of water and the concentration of salts in their bodies, thus counteracting the tendency of water to pass in or out by osmosis. In freshwater animals, water tends to enter the body and various methods have been developed to remove the excess, such as the contractile vacuole of protoctists, nephridia and Malpighian tubules in invertebrates, and kidneys with well-developed glomeruli in freshwater fish. Marine vertebrates prevent excess water loss and excrete excess salts by having kidneys with few glomeruli and short tubules. Terrestrial vertebrates avoid desiccation by having kidneys with long convoluted tubules, which increase the reabsorption of water and salts.

osmosis The movement of solvent from a dilute solution to a more concentrated solution through a membrane. For example, if a concentrated sugar solution (in water) is separated from a dilute sugar solution by a membrane, water molecules can pass through from the dilute solution to the concentrated one. A membrane of this type (which allows the passage of some kinds of molecule and not others) is called a *semipermeable membrane*. Membranes involved in living systems are not perfectly semipermeable, and are often called *differentially permeable membranes*

Osmosis between two solutions will continue until they have the same concentration. If a certain solution is separated from pure water by a membrane, osmosis also occurs. The pressure necessary to stop this osmosis is called the *osmotic pressure* (*OP*) of the solution. The more concentrated a solution, the higher its osmotic pressure. Osmosis is a very important feature of both plant and animal biology. Cell walls act as differentially permeable membranes and osmosis can occur into or out of the cell. It is necessary for an animal to have a mechanism of osmoregulation to stop the cells bursting or shrinking. In the case of plants, the cell walls are slightly 'elastic' – the concentration in the cell can be higher than that of the surroundings, and osmosis is prevented by the pressure exerted by the cell walls.

Osmosis is a phenomenon involving diffusion through the membrane; water diffuses from regions of high water concentration to low water concentration. Physiologists now describe the tendency for water to move in and out of cells in terms of water potential. *See* water potential.

osmotic potential *See* water potential.

osmotic pressure *See* osmosis.

ossification (**osteogenesis**) The transformation of embryonic or adult connective tissue (*intramembranous ossification*) or cartilage (*endochondral ossification*) into bone. Bone is produced by the action of special cells (*see* osteoblast), which deposit a network of collagen fibers impregnated with calcium salts; they eventually become enclosed in the bone matrix as bone cells (*see* osteocyte).

osteoblast Any of the cells that form layers of BONE in the early stages of OSSIFICATION. They are at first on the outside of the embryonic cartilage or membrane, but after it has been eroded by osteoclasts they accompany the ingrowing blood vessels and form temporary trabeculae of bone. Later the osteoblasts lay down the permanent structure of bone, and those that become trapped between the lamellae are called *osteocytes*. *See also* osteoclast; osteocyte.

osteoclast Any of the cells that attack and erode the calcified cartilage or membrane formed in the early stages of ossification of BONE. Blood vessels, preceded by osteoclasts, invade the tissue, and then osteoblasts lay down the permanent structure of bone. *See also* osteoblast.

osteocyte Any of the cells that secrete the hard matrix of BONE. They are found in small spaces (lacunae) between the concentric lamellae of bone that form the Haversian systems. Each osteocyte has many fine cytoplasmic processes that pass, in fine canaliculae, through the matrix and connect with each other and with blood vessels to maintain supplies of food and oxygen to the living cells. *See also* ossification; osteoblast.

Ostwald's dilution law *See* dissociation.

otolith One of several granules of calcium carbonate that are contained in a gelatinous mass and attached to hairlike processes of sensory cells within the utriculus and sacculus of the vertebrate inner ear. They respond to changes in the position of the head and so stimulate the sensory cells.

ovary 1. (*Botany*) The swollen base of the carpel in the gynoecium of plants, containing at least one ovule. The gynoecium of angiosperms may consist of more than one carpel that fuses in certain species forming a complex ovary. After fertilization, the ovary wall becomes the pericarp of the fruit enclosing seeds in its central hollow.
2. (*Zoology*) The female reproductive organ in animals, which produces egg cells (ova). There is usually a pair of ovaries in vertebrates (in birds, only the left is functional); they also produce sex hormones. In humans, the ovaries are cream-colored oval structures, about 4 cm long, which are attached to the posterior wall of the abdominal cavity, below the kidneys.

ovule Part of the female reproductive organs in seed plants. It consists of the nucellus, which contains the embryo sac, surrounded by the integuments. After fertilization the ovule develops into the seed. In angiosperms the ovule is contained within an ovary and may be orientated in different ways being upright, inverted, or sometimes horizontal. In gymnosperms ovules are larger but are not contained within an ovary. Gymnosperm seeds are thus naked while angiosperm seeds are contained within a fruit, which develops from the ovary wall.

ovum (**egg cell**) The immotile female reproductive cell (gamete) produced in the ovary of an animal. It consists of a central haploid nucleus surrounded by cytoplasm, containing a variable amount of yolk and a vitelline membrane. Size varies between species; in humans, it is about 0.15 mm in diameter. In chickens it is about 30 mm in diameter and further enlarged by a layer of albumen, more membranes, and a shell to become a true egg. A single ovum is released from the ovary at regular intervals; in humans, about once every 28 days. If fertilized by a spermatozoon it develops into a new individual of the same species. Sometimes fertilization occurs before the ovum is fully developed, i.e. at the oocyte stage.

oxalate A salt or ester of ethanedioic acid (oxalic acid).

oxalic acid (**ethanedioic acid**) A dicarboxylic acid, $(COOH)_2$, which occurs in rhubarb leaves, wood sorrel and the garden oxalis (hence the name).

oxaloacetic acid (**OAA**) A water-soluble carboxylic acid, structurally related to fumaric acid and maleic acid. Oxaloacetic acid forms part of the Krebs cycle, it is produced from L-malate in an NAD-requiring reaction and itself is a step towards the formation of citric acid in a reaction involving pyruvate ion and coenzyme A.

oxidation An atom, an ion, or a molecule is said to undergo oxidation or to be oxidized when it loses electrons. The process may be effected chemically, i.e. by reaction with an *oxidizing agent*, or electrically, in which case oxidation occurs at the anode. For example,

$$2Na + Cl_2 \rightarrow 2Na^+ + 2Cl^-$$

where chlorine is the oxidizing agent and sodium is oxidized, and

$$4CN^- + 2Cu^{2+} \rightarrow C_2N_2 + 2CuCN$$

where Cu^{2+} is the oxidizing agent and CN^- is oxidized.

The *oxidation state* of an atom is indicated by the number of electrons lost or effectively lost by the neutral atom, i.e. the oxidation number. The oxidation number of a negative ion is negative. The process of oxidation is the converse of reduction.

oxidative phosphorylation The production of ATP from phosphate and ADP as electrons are transferred along the electron-transport chain from NADH or $FADH_2$ to oxygen. Most of the NADH and $FADH_2$ is formed in the mitochondrial matrix by the Krebs cycle and fatty acid oxidation. Oxidative phosphorylation occurs in mitochondria and is the main source of ATP in aerobes. *See* electron-transport chain.

2-oxopropanoic acid *See* pyruvic acid.

oxyacid An acid in which the replaceable hydrogen atom is part of a hydroxyl group, including carboxylic acids, phenols and inorganic acids such as phosphoric(V) acid and sulfuric(VI) acid. *See* acid.

oxygen An element essential to living organisms both as a constituent of carbohydrates, fats, proteins, and their derivatives, and in aerobic respiration. It enters plants both as carbon dioxide and water, the oxygen from water being released in gaseous form as a by-product of photosynthesis. Plants are the main if not the only source of gaseous oxygen and as such are essential in maintaining oxygen levels in the air for aerobic organisms.

oxygen debt A physiological state that occurs when a normally aerobic animal is forced to respire anaerobically during a temporary shortage of oxygen (anoxia), e.g. due to violent muscular exertion. Pyruvate, a product of the first stage of cellular respiration, is converted anaerobically to lactic acid, which is toxic and requires oxygen for its breakdown, thereby building up an oxygen debt. The debt is repaid when oxygen is made available and allows oxidation of the lactic acid in the liver. *See also* glycolysis.

oxygen quotient The rate of oxygen consumption of an organism or tissue. It is usually expressed in microliters of oxygen per milligram of dry weight per hour. Small organisms tend to have higher oxygen quotients than larger ones.

oxyhemoglobin *See* hemoglobin.

oxytocin A peptide hormone, produced by the hypothalamus and posterior pituitary gland, that acts on smooth muscle. At the end of pregnancy the uterus becomes very sensitive to oxytocin, which promotes labor and also the release of milk from the mammary gland. It is used to induce labor artificially. *Compare* vasopressin.

pachytene In MEIOSIS, the stage in mid-prophase I that is characterized by the contraction of paired homologous chromosomes. At this point each chromosome consists of a pair of chromatids and the two associated chromosomes are termed a tetrad.

PAGE (**polyacrylamide gel electrophoresis**) *See* electrophoresis.

pairing *See* synapsis.

Palade, George Emil (1912–) Romanian-born American cell biologist. He was awarded the Nobel Prize for physiology or medicine in 1974 jointly with A. Claude and C. R. M. J. de Duve for their discoveries concerning the structural and functional organization of the cell.

palisade mesophyll *See* mesophyll.

pallium *See* cerebral cortex.

palmitic acid (**hexadecanoic acid**) A saturated fatty acid, $CH_3(CH_2)_{14}$, occurring widely in fats and oils of animal and vegetable origin. *See also* carboxylic acid; oleic acid.

pancreas A gland lying between the spleen and the duodenum, and having a duct that enters the duodenum. It secretes pancreatic juice which contains various enzymes, including: trypsin for breaking down proteins to amino acids, amylase for converting starch to maltose, and lipase for changing emulsified oils to glycerin and fatty acids. The gland also contains endocrine tissues and produces insulin and glucagon. *See* islets of Langerhans.

pantothenic acid (**vitamin B_5**) One of the water-soluble B-group of vitamins. Sources of the vitamin include egg yolk, kidney, liver, and yeast. As a constituent of coenzyme A, pantothenic acid is essential for several fundamental reactions in metabolism. A deficiency results in symptoms affecting a wide range of tissues; the overall effects include fatigue, poor motor coordination, and muscle cramps.

paper chromatography A chromatographic method using absorbent paper. A paper strip with a drop of test material at the bottom is dipped into the carrier liquid (solvent) and removed when the solvent front almost reaches the top of the strip. Two-dimensional chromatograms can be produced using square paper and two different solvents. The paper is removed from the first carrier liquid, turned at right angles and dipped into the second. This gives a two-dimensional 'map' of the constituents of the test drop. The identity of the constituents may be found by measuring the R_f values.

para- Designating a benzene compound with substituents in the 1,4 positions. The position on a benzene ring directly opposite a substituent is the para position. This was used in the systematic naming of benzene compounds. For example, para-dinitrobenzene (or *p*-dinitrobenzene) is 1,4-dinitrobenzene.

parabiosis The experimental or natural union of two similar animals (each a *parabiont*) so that their blood circulations are continuous. Experimental parabiosis is often performed on insects to study what effects chemicals in one individual have when passed to the other. Natural parabiosis occurs in Siamese twins.

parasympathetic nervous system (craniosacral nervous system) One of the two divisions of the autonomic nervous system, which supplies motor nerves to the smooth muscles of the internal organs and to cardiac muscles. Parasympathetic fibers emerge from the central nervous system via cranial nerves, especially the vagus nerve, and a few spinal nerves in the sacral region. Their endings release acetylcholine, which slows heart rate, lowers blood pressure, and promotes digestion, thereby antagonizing the effects of the sympathetic nervous system. *See also* autonomic nervous system; sympathetic nervous system.

parathyroid glands Four small oval-shaped structures embedded in the thyroid gland. The glands are composed of columns of cells with vascular channels between the columns. They produce *parathyroid hormone* and calcitonin, which control the blood calcium level. Secretion of the hormones is controlled by the blood calcium level, through a feedback mechanism. *See* calcitonin.

partial dominance *See* co-dominance.

parturition The process of giving birth to the fetus at the termination of pregnancy in viviparous animals. Parturition is triggered by hormonal changes initiated by the fetus. In humans, the fetal membranes and endometrium secrete prostaglandins, which stimulate rhythmic contractions of the uterine wall. Stretching of the cervix by the fetus's head triggers the mother's pituitary to secrete oxytocin, which reinforces the muscle contractions of labor, leading to delivery of the fetus.

Pasteur, Louis (1822–95) French chemist, microbiologist, and immunologist. Pasteur is remembered for his recognition of the role of microorganisms in fermentation and for devising methods of immunization against certain diseases. He also did pioneering work in stereochemistry.

pasteurization The partial sterilization of foodstuffs by heating to a temperature below boiling. This kills harmful microorganisms but retains the flavor. It is named after the pioneer of the method, Louis Pasteur, who used it to prevent spoilage of wine and beer. Milk is pasteurized by heating at 65°C for 30 minutes.

pathogen Any organism that is capable of causing disease or a toxic response in another organism. Many bacteria, viruses, fungi, and other microorganisms are pathogenic.

Pauling, Linus Carl (1901–94) American chemist noted for his early work on the nature of the chemical bond. He also studied a number of complex biochemical systems. He was awarded the Nobel Prize for chemistry in 1954. Pauling was also awarded the Nobel Peace Prize in 1962.

Pavlov, Ivan Petrovich (1849–1936) Russian physiologist and experimental psychologist. He was awarded the Nobel Prize for physiology or medicine in 1904 for his work on the physiology of the salivary glands.

PCR *See* polymerase chain reaction.

pectic substances Polysaccharides that, together with hemicelluloses, form the matrix of plant cell walls. They serve to cement the cellulose fibers together. Fruits are a rich source.

They are principally made from the group of sugar acids known as uronic acids. *Pectic acids*, the basis of the other pectic substances, are soluble unbranched chains of α-1,4 linked galacturonic acid units (derived from the sugar galactose). The acid is precipitated as insoluble calcium or magnesium pectate in the middle lamella of plant cells. *Pectinic acids* are slightly modified pectic acids. Under suitable conditions pectinic acids and *pectins* form gels with sugar and acid. Pectins are used commercially as gelling agents, e.g. in jams. Insoluble pectic substances are termed *protopectin* and this is the most important group in normal cell walls. Protopectin is hydrolyzed to soluble pectin by pectinase in ripening fruits, changing the fruit consistency.

pectin *See* pectic substances.

penetrance The appearance in the phenotype of characteristics that are genetically determined. Certain characters may fail to 'penetrate' even though the individual carries dominant or homozygous recessive genes in their genotype.

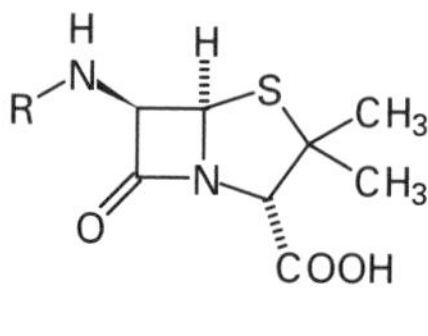

Penicillin

penicillin An antibiotic formed by the mold *Penicillium notatum*. This form is known as penicillin G. A class of similar compounds are active against bacteria.

pentanoic acid (**valeric acid**) A colorless liquid carboxylic acid, $CH_3(CH_2)_3$-COOH, used in making perfumes.

pentose A SUGAR that has five carbon atoms in its molecules.

pentose phosphate pathway (**hexose monophosphate shunt**) A pathway of glucose breakdown in which pentoses are produced, in addition to reducing power (NADPH) for many synthetic reactions. It is an alternative to glycolysis and is much more active in adipose tissue (where large amounts of NADPH are consumed) than in skeletal muscle.

PEP carboxylase *See* C_4 plant.

pepsin An enzyme that catalyzes the partial hydrolysis of proteins to polypeptides. It is secreted by the gastric glands in an inactive form, pepsinogen, and is activated by hydrogen ions. At pH values of 4.6 and less pepsin activates pepsinogen, i.e. it is autocatalytic. Pepsin initiates the digestion of proteins, splitting them into smaller fragments. The extent of this action is proportional to the length of time the protein is in contact with the enzyme.

peptidase *See* protease.

peptide A type of compound formed of a number of amino acid molecules linked together. Peptides can be regarded as formed by a reaction in which the carbonyl group of one amino acid reacts with the amino group of another amino acid with the elimination of water. This link between amino acids is called a *peptide bond*. According to the number of amino acids linked together they are called di-, tri-, oligo-, or polypeptides. In general, peptides have an amino group at one end of the chain (the amino, or N, terminus) and a carbonyl group at the other (the carboxy, or C, terminus). Peptides can be produced by the partial hydrolysis of proteins.

perennation A vegetative means of surviving unfavorable seasons, seen in biennial and perennial plants. The metabolic activities are reduced to a minimum, usually by die-back of aerial parts, and food for the next growing season is stored in swollen underground organs. Seeds may also be regarded as perennating organs.

perforin *See* natural killer cell.

periosteum The connective tissue membrane that surrounds a BONE. It is tough and fibrous, with many interlacing bundles of white collagen fibers. It contains OSTEOBLASTS, important in the formation of bone.

peripheral nervous system The system of nerves and their ganglia that run from the CENTRAL NERVOUS SYSTEM to the organs and peripheral regions of the body. It constitutes all parts of the nervous system not included in the central nervous system. In vertebrates it comprises the cranial and spinal nerves with their many branches. These convey impulses from sense organs for processing by the central nervous system and transmit the consequent motor impulses to muscles, glands, etc.

peroxide 1. An oxide containing the $^{-}O{-}O^{-}$ ion.
2. A compound containing the –O–O– group.

Perutz, Max Ferdinand (1914–) Austrian-born British molecular biologist who worked on hemoglobin and myoglobin. He was awarded the Nobel Prize for chemistry in 1962, the prize being shared with J. C. Kendrew for their studies on the structures of globular proteins.

peta- Symbol: P A prefix denoting 10^{15}.

petri dish A shallow circular glass or plastic container, fitted with a lid, that is used for tissue culture or for growing such microorganisms as bacteria, molds, etc., on nutrient agar or some other medium. It is named after the German bacteriologist J. R. Petri.

PG *See* prostaglandin.

pH A measure of the acidity or alkalinity of a solution on a scale 0–14. A neutral solution has a pH of 7. Acid solutions have a pH below 7; alkaline solutions have a pH above 7. The pH is given by $\log_{10}(1/[H^+])$, where $[H^+]$ is the hydrogen ion concentration in moles per liter.

phaeophytin A yellow-gray colored pigment of chlorophyll appearing in organic solvent extracts of chlorophyll and often seen during paper chromatography of such extracts.

phage (**bacteriophage**) A virus that infects bacteria. Phages usually have complex capsids composed of a polyhedral head, containing the nucleic acid (DNA or RNA), and a helical tail, through which nucleic acid is injected into the host. After reproduction of the viral nucleic acid the host cell usually undergoes lysis. Some phage are *lysogenic*, i.e. after infection viral DNA is integrated into the bacterial genome and replicated along with the bacterial DNA, producing infected daughter cells. In this state the phage is known as a *prophage*. The infected bacteria can be activated to lyse and release mature phage. In genetic engineering, nonviral DNA can be inserted into a phage, which is then used as a cloning vector.

phagocyte A cell that is capable of engulfing particles from its surroundings by a process termed phagocytosis. Examples are the neutrophils and macrophages in vertebrates, which play an important role in protecting the organism against infection. Many other cells are capable of phagocytosis, e.g. intestinal epithelial cells and certain protoctists (*see* protozoa).

phagocytosis *See* endocytosis; phagocyte.

phenetic Describing or relating to the observable similarities and differences between organisms. Phenetic classification systems are based on such characteristics, rather than evolutionary relationships between groups. *Compare* phyletic.

phenocopy A change in the appearance of an organism caused by the environment, but which is similar in effect to a change caused by gene mutation. Such changes, which are not inherited, are generally caused by environmental factors (e.g. malnutrition or radiation) affecting the organism at an early stage of development.

phenol 1. (carbolic acid, hydroxybenzene, C_6H_5OH) A white crystalline solid used to make a variety of other organic compounds.
2. A type of organic compound in which at least one hydroxyl group is bound directly to one of the carbon atoms of an aromatic ring. Phenols do not show the behavior typical of alcohols. In particular they are more acidic because of the electron-withdrawing effect of the aromatic ring.

phenolphthalein An acid–alkali indicator that is colorless in acids and red in alkalis. It has a pH range from 8.4–10.0 and is a frequently used indicator for the detection of pH change in acid–alkali titrations. Phenolphthalein, together with borax, is used to test for saccharide derivatives, e.g. glycerol. Such derivatives turn the solution from red to colorless but on boiling the red color returns.

phenotype The observable characteristics of an organism, which are determined by the interaction of the genotype with the environment. Many genes present in the genotype do not show their effects in the phenotype because they are masked by dominant alleles. Genotypically identical organisms may have very different phenotypes in different environments, an effect particularly noticeable in plants grown in various habitats.

phenylalanine *See* amino acids.

phenyl group The group C_6H_5–, derived from benzene.

phenylhydrazine A colorless liquid that reacts with aldehydes and ketones to give *phenylhydrazones*. These are white solids with definite melting points that can be used to identify the respective aldehydes and ketones. With monosaccharides, phenylhydrazine forms osazones, which are yellow crystalline compounds that are distinctive for most monosaccharides. *See* osazones.

phenylketonuria A genetic disorder resulting in the inability to metabolize phenylalanine to tyrosine and causing severe mental retardation if untreated. Phenylketonuria is caused by a defective recessive gene and therefore both parents must be carriers for the child to be affected; it may be diagnosed by the presence of phenylpyruvic acid (a precursor of phenylalanine) in the urine. If detected soon after birth, a diet low in phenylalanine will enable the infant to develop normally.

3-phenylpropenoic acid (cinnamic acid) A white pleasant-smelling crystalline carboxylic acid, $C_6H_5CH{:}CHCOOH$. It occurs in amber but can be synthesized and is used in perfumes and flavorings.

pheromone A substance that is excreted by an animal and causes a response in other animals of the same species (e.g. sexual attraction, development). *Compare* kairomone.

phloem Plant vascular tissue that transports food from areas where it is made to where it is needed or stored. It consists of sieve tubes, which are columns of living cells with perforated end walls, that allow passage of substances from one cell to the next.

phloroglucinol *See* staining.

phosphagen Creatine phosphate. *See* creatine.

phosphatidate *See* phospholipid.

phosphatide A glycerophospholipid. *See* lipid.

phosphatidylcholine (lecithin) One of a group of phospholipids that contain glycerol, fatty acid, phosphoric acid, and choline and are found widely in higher plants and animals, particularly as a component of cell membranes.

phosphocreatine *See* creatine.

phospholipid A lipid with a phosphate group attached by an ester linkage. They are the major class of lipid in all biological membranes and, together with glycolipids and cholesterol, are the main structural components. All membrane phosphlipids, except sphingosine, are derived from glycerol and are called *glycerophospholipids*. They consist of a glycerol backbone with two fatty acid chains esterified to carbons 1 and 2 and a phosphorylated alcohol esterified at carbon 3. The simplest glycerophospholipid is diacylglycerol 3-phosphate or *phosphatidate*, which has no alcohol part. Only small amounts exist naturally, but it is a key intermediate in the biosynthesis of other glycerophospholipids with the phosphate being esterified to one of several alcohols (serine, ethanolamine, choline, or glycerol) to form the major membrane glycerophospholipids: phosphatidylserine, phosphatidylethanolamine, phosphatidylcholine, and phosphatidylglycerol. Sphingomylein (like glycolipids) is derived from sphingosine and has a phosphorylcholine esterified to the primary hydroxyl group of sphingosine.

Another important glycerophospholipid is phosphatidylinositol, which is phosphorylated by specific kinases to phosphatidylinositol 4,5-bisphosphate. This is a key molecule in signal transduction that mediates the action of several hormones and other effectors that control important cellular functions, e.g. glycogenolysis, insulin secretion, the aggregation of platelets, and smooth muscle contraction.

phosphoprotein A conjugated protein formed by the combination of protein with phosphate groups. Casein is an example.

phosphorescence 1. The absorption of energy by atoms followed by emission of electromagnetic radiation. Phosphorescence is a type of luminescence, and is distinguished from fluorescence by the fact that the emitted radiation continues for some time after the source of excitation has been removed. In phosphorescence the excited atoms have relatively long lifetimes before they make transitions to lower energy states. However, there is no defined time distinguishing phosphorescence from fluorescence.
2. In general usage the term is applied to the emission of 'cold light' – light produced without a high temperature. The name comes from the fact that white phosphorus glows slightly in the dark as a result of a chemical reaction with oxygen. The light comes from excited atoms produced directly in the reaction – not from the heat produced. It is thus an example of *chemiluminescence*. There are also a number of biochemical examples termed bioluminescence; for example, phosphorescence is sometimes seen in the sea from marine organisms, or on rotting wood from certain fungi (known as 'fox fire').

phosphorus One of the essential elements in living organisms where it occurs as a phosphate. In vertebrates, calcium phosphate is the main constituent of the skeleton. Phospholipids are important in cell membrane structure, and phosphates are necessary for the formation of the sugar–phosphate backbone of nucleic acids. Phosphates are also necessary for the formation of high energy bonds in compounds such as ATP. Phosphate compounds are important in providing energy for muscle contraction in vertebrates (creatine phosphate) and invertebrates (arginine phosphate). Phosphate has many other important roles in living tissues, being a component of certain coenzymes and involved in enzyme regulation. The free phosphate ion, PO_4^{3-}, is an important buffer in cell solutions.

phosphorylation The transfer of a phosphoryl group $-PO(OH)_2$ from ATP to a protein by a protein kinase. Many metabolic enzymes are regulated by phosphorylation as are several signal pathways involved in cell growth. The removal of the phosphoryl group (dephosphorylation) is brought about by enzymes known as phosphatases.

photoautotrophism *See* autotrophism; phototrophism.

photoheterotrophism *See* heterotrophism; phototrophism.

photolysis Chemical breakdown caused by light. In photosynthesis the process is important in providing hydrogen donors by the splitting of water, as follows:

$$4H_2O \rightarrow 4[H] + 4[OH]$$
$$4[OH] \rightarrow 2H_2O + O_2$$
$$4[H] + CO_2 \rightarrow CH_2O + H_2O$$

photomicrograph *See* micrograph.

photoperiodism The response of an organism to changes in day length (*photoperiod*). In plants, leaf fall and flowering are common responses to seasonal changes in day length, as are migration, reproduction, molting, and winter-coat development in animals. Many animals, particularly birds, breed in response to an increasing spring photoperiod, a long-day response. Some animals (e.g. sheep, goats, and deer) breed in autumn in response to short days so that offspring are born the following spring. The substance MELATONIN, produced by the pineal gland, is thought to play a role in regulating such changes.

Plants are classified as short-day plants (SDPs) (e.g. cocklebur and chrysanthemum) or long-day plants (LDPs) (e.g. cucumber and barley) according to whether they flower in response to short or long days. Day-neutral plants (e.g. pea and tomato) have no photoperiodic requirement. The length of the dark period is also a critical factor since flowering of SDPs is inhibited by even a brief flash of red light in the dark period (a PHYTOCHROME response), and an artificial cycle of long days and long nights inhibits flowering in LDPs. Thus, it is the interaction between light and dark periods that in some way affects flowering through the mediation of phytochrome. The P_{FR} form of phytochrome inhibits flowering in SDPs (P_{FR} slowly disappears during long nights) and promotes flowering in LDPs (P_{FR} remains at high levels in short nights). The light stimulus is perceived by the leaves and in some unknown way transmitted to the floral apices.

photophosphorylation (**photosynthetic phosphorylation**) The conversion of ADP to ATP using light energy. *See* photosynthesis.

photoreceptor Any light-sensitive organ or organelle. The eyes of vertebrates and the ocelli and compound eyes of insects are photoreceptors, as are the organelles of such protoctists as *Euglena*.

photorespiration A light-dependent metabolic process of most green plants that resembles true respiration only in that it uses oxygen and produces carbon dioxide. It wastes carbon dioxide and energy, using more ATP than it produces. It is a means of recovering some of the carbon from the excess glycolate produced in C_3 plants as a result of the preference of the main carbon-fixing enzyme ribulose-bisphosphate carboxylase for oxygen over carbon dioxide. It is estimated that in C_3 plants 40% of the potential yield of photosynthesis is lost through photorespiration. C_4 plants are more efficient at photosynthesis as their method of carbon dioxide fixation results in less glycolate being produced. *See also* C_4 plant.

photosynthesis The synthesis of organic compounds using light energy absorbed by chlorophyll. With the exception of a small group of bacteria, organisms photosynthesize from inorganic materials. All green plants photosynthesize as well as certain prokaryotes (some bacteria). In green plants, photosynthesis takes place in chloroplasts, mainly in leaves. Directly or indirectly, photosynthesis is the source of carbon and energy for all except chemoautotrophic organisms. The mechanism is complex and involves two sets of stages: *light reactions* followed by *dark reactions*. The overall reaction in green plants can be summarized by the equation:

$$CO_2 + 4H_2O \rightarrow [CH_2O] + 3H_2O + O_2$$

In the light reactions, light energy is absorbed by chlorophyll (and other pigments), setting off a chain of chemical reactions in which water is split and gaseous oxygen evolved. The hydrogen from the water is attached to other molecules, and used to reduce carbon dioxide to carbohydrates in the later dark reactions.

The light reaction involves the conversion of ADP to ATP in a process known as photophosphorylation. At the same time, the hydrogen carrier NADP is reduced to $NADPH_2$. The enzymes and other components for this are located in the thylakoid membrane of the chloroplast, where they constitute an electron-transport chain similar to that involved in aerobic respiration in mitochondria. Photons of light are trapped by clusters of chlorophyll molecules; each cluster (*antenna complex*) channels the light energy to a single chlorophyll *a* molecule, the *reaction center*. This transfers excited electrons to its associated protein complex called a *photosystem*; there are two photosystems in the electron-transfer chain, labeled I and II. Additional

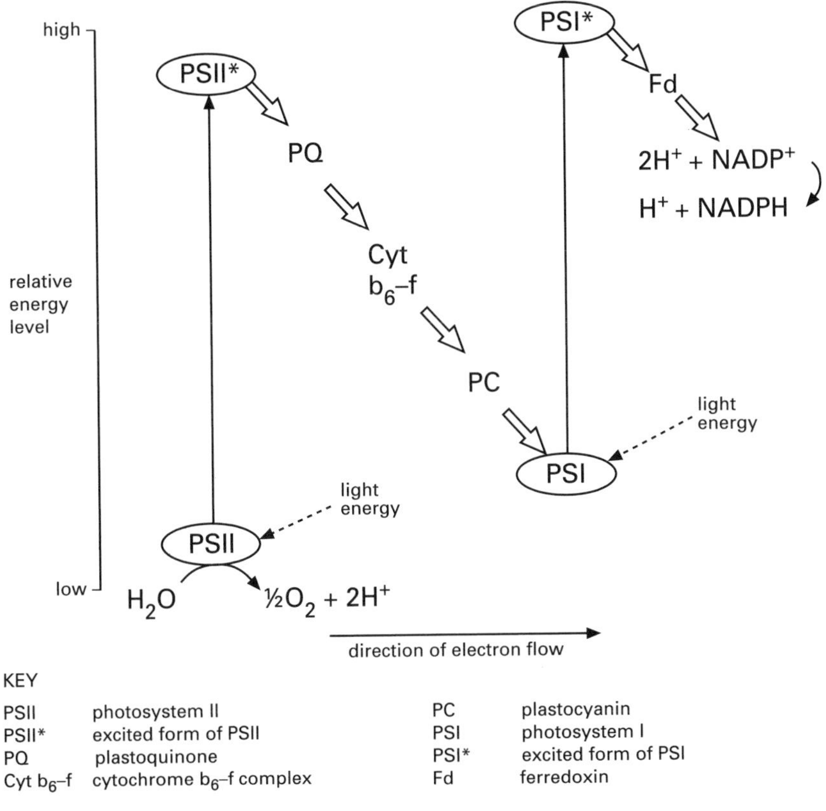

Photosynthesis

components are a group of cytochrome proteins, called the *b_6-f complex*, and certain electron-carrier molecules (plastoquinone, plastocyanin, and ferredoxin). According to the chemiosmotic theory, as electrons flow along the chain they cause hydrogen (H^+) ions to be pumped into the thylakoid space, creating a chemiosmotic gradient. As these H^+ ions diffuse back through the thylakoid membrane into the stroma of the chloroplast they drive ATP production by the enzyme ATP synthase, also located in the thylakoid membrane. Meanwhile, at the end of the electron-transport chain, electrons are transferred to $NADP^+$ in the formation of NADPH by the enzyme NADP reductase.

There are two patterns of electron flow. *Noncyclic electron flow* involves all components of the electron-transport chain, there is photolysis of water yielding oxygen, and both ATP and NADPH are generated. *Cyclic electron flow* involves only the b_6-f complex, plastocyanin, photosystem I, and ferredoxin. It produces the extra ATP needed for the dark reactions, but not NADPH, and occurs when there is insufficient $NADP^+$ to accept electrons.

During the dark reactions ATP and NADPH from the light reactions are used to reduce carbon dioxide to carbohydrate. The reactions take place in solution; in eukaryotes in the chloroplast stroma. They are called the dark reactions because they do not need light but they usually occur in the light. In most plants (C_3 plants) carbon dioxide is first fixed by combination with the 5-carbon sugar ribulose bisphosphate (RUBP) by the enzyme RUBP carboxylase, which is the most abundant protein in nature, to form two molecules of phosphoglyceric acid (PGA), the first product of photosynthesis. PGA is then reduced to phosphoglyceraldehyde (triose phosphate) using the NADPH and some of the ATP. Some of the triose phosphate and the rest of the ATP is used to regenerate the carbon dioxide acceptor RUBP in a complex cycle involving 3-, 4-, 5-, 6-, and 7-carbon sugar phosphates. Details of this cycle were elucidated by Benson, Bassham, and Calvin working with the green alga *Chlorella* and using the radioactive isotope ^{14}C and paper chromatography to identify the intermediates; it is now usually called the *Calvin cycle* or the reductive pentose-phosphate cycle. The rest of the triose phosphate can be used in synthesis of carbohydrates, fats, proteins, etc. *See* C_4 plant; electron-transport chain; photosynthetic pigments; phototrophism.

photosynthetic bacteria A group of bacteria able to photosynthesize through possession of a green pigment, *bacteriochlorophyll*, which is slightly different from the chlorophyll of plants. They do not use water as a hydrogen source, as do plants, and thus do not produce oxygen as a product of photosynthesis, but some oxidized by-product instead. Photosynthetic bacteria include the green sulfur bacteria, purple sulfur bacteria, and purple nonsulfur bacteria. *See also* Cyanobacteria.

photosynthetic pigments Pigments that absorb the light energy required in photosynthesis. They are located in the chloroplasts of plants and algae, whereas in most photosynthetic bacteria they are located in thylakoid membranes, typically distributed around the cell periphery. All photosynthetic organisms contain chlorophylls and carotenoids; some also contain phycobilins. Chlorophyll *a* is the *primary pigment* since energy absorbed by this is used directly to drive the light reactions of photosynthesis. The chlorophyll *a* that forms the reaction center of photosystem II has an absorption peak at 680 nm and that of photosystem I at 700 nm. The other pigments (chlorophylls *b*, *c*, and *d*, and the carotenoids and phycobilins) are *accessory pigments* that pass the energy they absorb on to chlorophyll *a*. They broaden the spectrum of light used in photosynthesis. *See* absorption spectrum.

photosystem *See* photosynthesis.

phototaxis (**phototactic movement**) A taxis in response to light. Many motile algae are positively phototactic, e.g. *Volvox*, while cockroaches are examples of negatively phototactic organisms.

phototrophism A type of nutrition in which the source of energy for synthesis of organic requirements is light. Most phototrophic organisms are autotrophic (i.e. show *photoautotrophism*); these comprise the green plants, Cyanobacteria, and some photosynthetic bacteria (the purple and green sulfur-bacteria). A few are heterotrophic (i.e. show *photoheterotrophism*); these are a group of photosynthetic bacteria (e.g. the purple nonsulfur bacteria) and a few algae. *Compare* chemotrophism. *See* autotrophism; heterotrophism. *See also* photosynthesis.

phototropism (**heliotropism; phototropic movement**) A directional growth movement of part of a plant in response to light. The phenomenon is clearly shown by the growth of shoots and coleoptiles towards light (positive phototropism). According to one model, the stimulus is perceived in the region just behind the shoot tip. If light falls on only one side of the apex then auxins produced in the apex tend to diffuse towards the shaded side. Thus more auxin diffuses down the stem from the shaded side of the tip. This results in greater elongation of cells on the shaded side thus causing the stem to bend towards the light source. However, in some cases curvature of the shoot tip can occur without apparent differential transport of auxins; moreover, another group of plant hormones, abscisins, may alo be involved. Most roots are light-insensitive but some (e.g. the adventitious roots of climbers such as ivy) are negatively phototropic. *See also* tropism.

phycobilins A group of accessory photosynthetic pigments found in Cyanobacteria and red algae. Chemically they are linear tetrapyrroles in contrast to chlorophyll, which is a cyclic tetrapyrrole. They absorb light in the middle of the spectrum not absorbed by chlorophyll, an important function in algae living under water where blue and red light are absorbed in the surface layers. They comprise the blue *phycocyanins*, which absorb extra orange and red light, and the red *phycoerythrins*, which absorb green light, enabling red algae to grow at depth in the sea. *See also* absorption spectrum; photosynthetic pigments.

phycocyanin A photosynthetic pigment. *See* phycobilins.

phycoerythrin A photosynthetic pigment. *See* phycobilins.

phyletic (**phylogenetic**) Relating to or reflecting the evolutionary history of an organism. Some developmental structures or processes, such as gill pouches in mammal embryos, are regarded as phyletic. Phyletic classifications are based on the assumed evolutionary relationships between organisms rather than their observable characteristics. *Compare* phenetic.

phylloquinone *See* vitamin K.

phylogenetic *See* phyletic.

phylogeny The evolutionary history of a group of organisms.

physiological saline A solution of sodium chloride and various other salts in which animal tissues are bathed *in vitro* to keep them alive during experiments. It must be isotonic with, and of the same pH as, body fluids. One of the most commonly used is *Ringer's solution*, which contains (in addition to sodium chloride) calcium, magnesium, and potassium chlorides. Other solutions may also contain a food supply, e.g. glucose. *See also* tissue culture.

physiology The way in which organisms or parts of organisms function. *Compare* morphology.

phytoalexin A nonspecific antibiotic produced by a plant, usually in response to infection by a fungus or to injury.

phytochrome A proteinaceous pigment found in low concentrations in most plant organs, particularly meristems and dark-grown seedlings. It exists in two interconvertible forms. P_R (or P_{660}) has an absorption peak at 660 nm (red light) and P_{FR} (or P_{730}) at 730 nm (far-red light). Natural white light favors formation of P_{FR}, the physiologically active form. Light intensities required for conversion are very low and it occurs within seconds.

Phytochrome plays a vital role as a photoreceptor in a wide range of light-induced physiological processes: e.g. photoperiodic responses; photomorphogenesis, including leaf expansion, leaf unrolling in grasses and cereals, and greening; and germination of light-sensitive seeds such as lettuce. P_{FR} is thought to induce changes in membrane permeability and the subsequent events often involve growth substances, particularly gibberellins, cytokinins, and possibly florigen. *See* photoperiodism.

phytohormone *See* plant hormone.

pico- Symbol: p A prefix denoting one million-millionth, or 10^{-12}. For example, 1 picogram (pg) = 10^{-12} gram.

pineal gland A gland that arises from an outgrowth on the dorsal surface of the forebrain of the vertebrate brain and secretes the hormone MELATONIN. It is thought to have given rise to the third eye in fossil vertebrates.

pinocytosis *See* endocytosis.

pipette A device used to transfer a known volume of solution from one container to another; in general, several samples of equal volume are transferred for individual analysis from one stock solution. Pipettes are of two types, bulb pipettes, which transfer a known and fixed volume, and graduated pipettes, which can transfer variable volumes. Pipettes were at one time universally mouth-operated but safety pipettes using a plunger or rubber bulb are now preferred.

pituitary gland An endocrine gland in the vertebrate brain situated beneath the thalamencephalon behind the optic chiasma. It is regarded as the 'master' endocrine gland because many of its hormones control the secretions of other endocrine glands. The more important pituitary hormones include:
1. Growth hormone (somatotropin), which affects protein metabolism. Excess production leads to gigantism and deficiency results in dwarfism.
2. Vasopressin (antidiuretic hormone), which stimulates reabsorption of water from the kidneys.
3. Adrenocorticotropic hormone (ACTH), which stimulates the secretions of the adrenal gland.
4. Gonadotropic hormones (e.g. follicle-stimulating hormone, luteinizing hormone), which stimulate gonad development.
5. Oxytocin, which stimulates the uterine walls to contract during birth.
6. Prolactin, which stimulates milk production by the mammary glands.
7. Thyrotropin (thyroid-stimulating hormone), which stimulates the secretion of the thyroid glands.

placenta 1. (*Zoology*) A disk-shaped organ that develops within the womb (uterus) of a pregnant mammal and establishes a close association between embryonic and maternal tissues for the exchange of materials. It is composed of both embryonic and maternal tissues; embryonic membranes develop numerous finger-like projections (villi) that grow into the highly vascular uterus wall. Into these villi extend embryonic capillaries from the umbilical arteries and vein. This brings the embryonic and maternal circulation into close contact and the fetus is able to obtain oxygen, nutrients, etc., and have waste metabolic products, such as carbon dioxide and nitrogenous compounds, removed. The fetal and maternal blood are never in direct contact. The placenta is discharged soon after the birth of the young.

2. (*Botany*) The region of tissue occurring on the inner surface of the ovary wall of the carpels of flowering plants where the ovules develop.

plane polarization A type of polarization of electromagnetic radiation in which the vibrations take place entirely in one plane.

plant An organism that can make its own food by taking in simple inorganic substances and building these into complex molecules by a process termed photosynthesis. This process uses light energy, absorbed by a green pigment called chlorophyll, which is found in all plants but no animals. One major characteristic that distinguishes plants from other plantlike organisms, such as algae or fungi, is the possession of an embryo that is retained and nourished by maternal tissue. Fungi and algae lack embryos and develop from spores. Plants are also characterized by having cellulose cell walls, not found in animals, and by the inability to move around freely except for some mobile microscopic plants.

plant hormone (**phytohormone**) One of a group of essential organic substances produced in plants. They are effective in very low concentrations and control growth and development by their interactions. Examples are auxins, gibberellins, cytokinins, abscisic acid, and ethylene.

plasma *See* blood plasma.

plasma cell A mature B-lymphocyte (white blood cell) that is programed to secrete just one particular type of antibody. A single plasma cell can secrete up to 2000 antibody molecules per second. *See also* antibody; B-cell; lymphocyte.

plasmagel (**ectoplasm**) The gel-like region of cytoplasm located in a thin layer just beneath the plasma membrane of cells that move in an ameboid fashion, such as amebas and macrophages. It consists of a three-dimensional network of actin microfilaments cross-linked by molecules of another protein, filamin, and its semisolid state gives shape to the cell and transmits tension to the substrate. It is thought that the movement of ameboid cells by extension and retraction of cytoplasmic projections (pseudopods) involves reversible changes between plasmagel and the fluid plasmasol in the cell's interior. The gel–sol transition is brought about by dismantling and assembly of the microfilament network through the action of various other proteins, possibly regulated by calcium ion concentration. *See* plasmasol.

plasmagene A gene contained in a self-replicating cytoplasmic particle. Inheritance of the characters controlled by such genes is not Mendelian because appreciable amounts of cytoplasm are passed on only with the female gametes. Mitochondria and plastids contain plasmagenes.

plasmalemma *See* plasma membrane.

plasma membrane (**cell membrane; plasmalemma**) The membrane that surrounds all living cells. *See* membrane.

plasmasol (**endoplasm**) The sol-like form of cytoplasm, located inside the plasmagel. It is free-flowing and contains the cell organelles. Ameboid movement involves sol-gel conversions, i.e. the conversion of plasmasol to plasmagel and *vice versa*. *Compare* plasmagel.

plasmid An extrachromosomal genetic element found within bacterial cells that replicates independently of the chromosomal DNA. Plasmids typically consist of circular double-stranded DNA molecules of molecular weight 10^6–10^8. They carry a variety of genes, including those for antibiotic resistance, toxin production, and enzyme formation, and may be advantageous to the cell. Plasmids are widely used as cloning vectors in genetic engineering.

plasmin (**fibrinolysin**) A proteolytic enzyme that breaks down fibrin in blood clots, restoring the fluidity of the blood. It exists in the blood as an inactive precursor, *plasminogen*, which can be converted to the active form by a variety of factors, including urokinase, trypsin, and leukocyte protease. Plasmin can also lyse other proteins, such as Factor VIII and immunoglobulin.

plasmolysis Loss of water from a walled cell (e.g. of a plant or bacterium) to the point at which the protoplast shrinks away from the cell wall. The point at which this is about to happen is called *incipient plasmolysis*. Here the cell wall is not being stretched; i.e. the cell has lost its turgidity or become *flaccid* (wall pressure is zero). Wilting of herbaceous plants occurs here. As plasmolysis proceeds parts of the protoplast may remain attached to the cell wall, giving an appearance characteristic of the species. Plasmolysis occurs when a cell is surrounded by a more concentrated solution. More concentrated solutions have lower water potential and this normally occurs only under experimental conditions. *See* water potential.

plastid An organelle enclosed by two membranes (the envelope) that is found in plants and certain protoctists (e.g. algae), and develops from a proplastid. Various types exist, but all contain DNA and ribosomes. *See* chloroplast; chromoplast.

plastocyanin An electron carrier in photosynthesis.

plastoglobuli (**osmiophilic globules**) Spherical lipid-rich droplets found in varying numbers inside chloroplasts. They stain intensely with osmium tetroxide and so appear black and circular with the electron microscope. *See* chloroplast.

plastoquinone A lipid-soluble compound used as an electron carrier in photosynthesis.

platelet (**thrombocyte**) A tiny particle found in blood plasma. Platelets are 2–3 μm in diameter and there are about 250 000 per cubic millimeter of blood. They are made in red bone marrow, being derived from large cells (called megakaryocytes) from which fragments of cytoplasm are pinched off. When they come into contact with a rough surface, such as a damaged tissue, platelets aggregate to form a plug. They also start the chain of reactions leading to the formation of a blood clot. Platelets release serotonin, which causes constriction of blood vessels, so reducing capillary bleeding; and platelet-derived growth factor, which stimulates tissue cells to grow and repair the wound. *See also* blood clotting.

poikilothermy The condition of having a body temperature that varies approximately with that of the environment. Most animals other than birds and mammals are poikilothermic ('cold-blooded'). *Compare* homoiothermy.

polar Describing a compound with molecules that have a permanent dipole moment. Hydrogen chloride and water are examples of polar compounds.

polar bond A covalent bond in which the bonding electrons are not shared equally between the two atoms. A bond between two atoms of different electronegativity is said to be polarized in the direction of the more electronegative atom, i.e. the electrons are drawn preferentially towards the atom. This leads to a small separation of charge and the development of a bond dipole moment as in, for example, hydrogen fluoride, represented as H→F or as $H^{\delta+}$–$F^{\delta-}$ (F is more electronegative).

The charge separation is much smaller than in ionic compounds; molecules in which bonds are strongly polar are said to display partial ionic character.

polarized light Light in which the electric and magnetic fields are restricted to single planes. Light is a transverse wave motion; it is composed of electric and magnetic fields vibrating at right angles to the direction of propagation. In 'normal' light the fields vibrate in all directions perpendicular to the propagation direction. Polarized light is produced, for example, by reflection or passage through Polaroid.

polar solvent *See* solvent.

pollution Any damaging or unpleasant change in the environment that results from the physical, chemical, or biological side-effects of human industrial or social activities. Pollution can affect the atmosphere, rivers, seas, and the soil.

Air pollution is caused by the domestic and industrial burning of carbonaceous fuels, by industrial processes, and by vehicle exhausts. Among recent problems are industrial emissions of sulfur dioxide causing acid rain, and the release into the atmosphere of chlorofluorocarbons, used in refrigeration, aerosols, etc., has been linked to the depletion of ozone in the stratosphere. Carbon dioxide, produced by burning fuel and by motor vehicle exhausts, is slowly building up in the atmosphere, which could result in an overall increase in the temperature of the atmosphere (*greenhouse effect*). Vehicle exhausts also contain carbon monoxide and other hazardous substances, such as fine particulate dusts. Lead was formerly a major vehicle pollutant, but the widespread introduction of lead-free petrol has eliminated this problem in most countries. Photochemical smog, caused by the action of sunlight on hydrocarbons and nitrogen oxides from vehicle exhausts, is a problem in many major cities.

Water pollutants include those that are biodegradable, such as sewage effluent, which cause no permanent harm if adequately treated and dispersed, as well as those which are nonbiodegradable, such as certain chlorinated hydrocarbon pesticides (e.g. DDT) and heavy metals, such as lead, copper, and zinc in some industrial effluents. The latter accumulate in the environment and can become very concentrated in food chains. The pesticides DDT, aldrin, and dieldrin are now banned. Water supplies can become polluted by leaching of nitrates from agricultural land, or of a wide range of potentially toxic substances from domestic and industrial waste tips. The discharge of waste heat can cause thermal pollution of the environment, but this is reduced by the use of cooling towers. In the sea, oil spillage from tankers and the inadequate discharge of sewage effluent are the main problems.

Other forms of pollution are noise from airplanes, traffic, and industry and the disposal of radioactive waste.

polyamine An aliphatic compound which has two or more amino and/or imino groups. Polyamines are often found associated with DNA and RNA in bacteria and viruses. This may stabilize the nucleic acid molecule in a way analogous to the action of histones on DNA in eukaryote cells. Examples of polyamines include spermine, spermidine, cadaverine, and putrescine.

polygene A gene with an individually small effect on the phenotype that interacts with other polygenes controlling the same character to produce the continuous quantitative variation typical of such traits as height, weight, and skin color.

polymer A compound in which there are very large molecules made up of repeating molecular units (monomers). Polymers do not usually have a definite relative molecular mass, as there are variations in the lengths of different chains. They may be natural substances (e.g. polysaccharides or proteins) or synthetic materials (e.g. nylon or polyethene). The two major classes of synthetic polymers are thermosetting (e.g. Bakelite) and thermoplastic (e.g. polyethene). The former are infusible, and heat may only make them harder, whereas the latter soften on heating.

polymerase An enzyme that regulates the synthesis of a polymer. Examples include *RNA polymerases* and *DNA polymerases*. There is only one type of RNA polymerase in prokaryotes, but in eukaryotes there are three different types: type I makes ribosomal RNA, type II makes messenger RNA precursors, and type III makes transfer RNA and 5S ribosomal RNA. They are all large enzymes made of many subunits and require the presence of a number of additional proteins to function. *See* transcription.

DNA polymerases are involved either in the synthesis of double-stranded DNA from single-stranded DNA or in the repair of DNA by scanning the DNA molecule and removing damaged nucleotides. *See* replication; reverse transcriptase.

polymerase chain reaction (**PCR**) A technique for amplifying small samples of DNA rapidly and conveniently. Developed in 1983, it is now used widely in research and forensic science, e.g. to produce a suitable quantity of DNA for genetic fingerprinting from the minute amounts present in traces of blood or other tissue. To amplify a particular segment of DNA it is necessary first to know the sequence of bases flanking it at either end. This enables the construction of short single DNA strands (primers) that are complementary to and will bind with these flanking regions. Then the sample is incubated with the primers, nucleotides, and enzymes, especially a DNA polymerase that is stable at high temperatures, in a water bath. By varying the temperature precisely and rapidly, amplification proceeds in cycles of DNA denaturation, annealing of primers, and replication of new DNA strands, each lasting about 20 seconds. After 30 cycles, some 10^9 copies of the original DNA are produced. The temperature cycles are controlled automatically. RNA can also be amplified, but it must first be converted to DNA by a reverse transcriptase.

polymorph (**polymorphonuclear leukocyte**) A white blood cell (leukocyte) with a lobed nucleus and granules in the cytoplasm. The term can be used for any GRANULOCYTE.

polynucleotide *See* nucleotide.

polypeptide A compound that contains many amino acids linked together by peptide bonds. *See* peptide.

polyploid The condition in which a cell or organism contains three or more times the haploid number of chromosomes. Polyploidy is far more common in plants than in animals and very high chromosome numbers may be found; for example in octaploids and decaploids (containing eight and ten times the haploid chromosome number). Polyploids are often larger and more vigorous than their diploid counterparts and the phenomenon is therefore exploited in plant breeding, in which the chemical colchicine can be used to induce polyploidy. Polyploids may contain multiples of the chromosomes of one species (autopolyploids) or combine the chromosomes of two or more species (allopolyploids). Polyploidy is rare in animals because the sex-determining mechanism is disturbed. For example a tetraploid XXXX would be sterile.

polyribosome *See* ribosome.

polysaccharide A polymer of monosaccharides joined by glycosidic links (*see* glycoside). They contain many repeated units in their molecular structures and are of high molecular weight. They can be broken down to smaller polysaccharides, disaccharides, and monosaccharides by hydrolysis or by the appropriate enzyme. Important polysaccharides are inulin (hydrolyzed to fructose), starch (hydrolyzed to glucose), glycogen (also known as animal starch) and hydrolyzed to glucose, and cellulose (hydrolyzed to glucose but not metabolized by humans). *See also* carbohydrates; glycan; sugar.

polysome *See* ribosome.

polytene Describing the chromosome condition caused by chromatids not separating after duplication. It leads to the formation of *giant chromosomes* consisting of numerous identical chromatids lying parallel to each other. Giant chromosomes have characteristic bands, caused by the degree of coiling of the DNA-histone fiber – highly coiled regions stain darker when prepared for microscopy. They are used to study gene activity and make chromosome maps. Polytene chromosomes are common in the salivary gland cells of dipterous insects, e.g. *Drosophila*.

porphyrins Cyclic organic structures that have the important characteristic property of forming complexes with metal ions. Examples of such *metalloporphyrins* are the iron porphyrins (e.g. heme in hemoglobin) and the magnesium porphyrin, chlorophyll, the photosynthetic pigment in plants. In nature, the majority of metalloporphyrins are conjugated to proteins to form a number of very important molecules, e.g. hemoglobin, myoglobin, and the cytochromes.

Porter, Rodney Robert (1917–85) British biochemist and immunochemist noted particularly for his isolation of the constituent polypeptide chains of antibody molecules. He was awarded the Nobel Prize for physiology or medicine in 1972 jointly with G. M. Edelman.

post-translational modification Any enzyme-catalysed change made to a protein after synthesis, e.g. proteolytic cleavage, phosphorylation.

potassium One of the essential elements in plants and animals. It is absorbed by plant roots as the potassium ion, K^+, and in plants is the most abundant cation in the cell sap. Potassium ions are required in high concentrations in the cell for efficient protein synthesis, and for glycolysis in which they are an essential cofactor for the enzyme pyruvate kinase. In animals the gradient of potassium and sodium ions across the cell membrane is responsible for the potential difference across the membrane, which is important for the transmission of nerve impulses. *See also* sodium.

P_R (P660) *See* phytochrome.

precipitin An antibody that combines with and precipitates soluble antigen, used in the *precipitin reaction* for identifying antigens.

precursor A substance from which another substance is formed in a chemical reaction.

Prelog, Vladimir (1906–98) Bosnian-born Swiss organic chemist. He was awarded the Nobel Prize for chemistry in 1975 for his research into the stereochemistry of organic molecules and reactions. The prize was shared with J. W. Cornforth.

pressure potential *See* water potential.

prion An infectious protein particle that causes various nervous diseases in humans and other animals, including bovine spongiform encephalopathy (BSE or 'mad cow disease') in cattle, and a form of Creutzfeldt-Jakob disease (CJD) in humans. Prions are apparently unique in that unlike viruses, virions, and all other infectious agents they lack any form of genetic material (i.e. DNA or RNA). It is thought that the infectious prion is a variant of a membrane protein normally produced in cells, which induces changes in the folding of the normal proteins, causing them to form rod-shaped aggregates in the central nervous system, which are responsible for the symptoms of disease.

probiotic Any compound produced by a microorganism that promotes growth in other microorganisms. *Compare* antibiotic.

procaryote *See* prokaryote.

progesterone A steroid hormone secreted by the corpus luteum in the ovary after ovulation. It initiates the preparation of the uterus for implantation of the ovum, the development of the placenta, and the development of the mammary gland in preparation for lactation. *See also* estrogen.

progestogen Any hormone whose effects resemble those of progesterone. Synthetic progestogens are used in therapy and oral contraceptives. *See also* estrogen.

programmed cell death *See* apoptosis.

prohormone The inactive form of a hormone: the form in which it is stored. Activation usually involves enzymatic removal of some part of the prohormone; for example, removal of amino acids from the polypeptide prohormone, proinsulin, to form insulin.

prokaryote (**procaryote**) An organism whose genetic material (DNA) is not enclosed by membranes to form a nucleus but lies free in the cytoplasm. Organisms can be divided into prokaryotes and eukaryotes, the latter having a true nucleus. This is a fundamental division because it is associated with other major differences. Prokaryotes comprise Bacteria and Archaea constitute the kingdom Prokaryotae. All other organisms are eukaryotes. Prokaryote cells evolved first and gave rise to eukaryote cells. *See* cell. *Compare* eukaryote.

prolactin A hormone, formerly called luteotropic hormone, produced by the anterior pituitary gland. In mammals it stimulates and controls lactation after the mammary gland has been prepared for milk production by estrogens, progesterone, and other hormones. In birds prolactin stimulates secretion of crop milk from the crop glands. *See also* gonadotropin.

proline *See* amino acids.

promoter A specific DNA sequence at the start of a gene that initiates transcription by binding RNA polymerase. In *Escherichia coli* the RNA polymerase has a protein 'sigma factor' that recognizes the promoter; in the absence of this factor the enzyme binds to, and begins transcription at, random points on the DNA strand. In eukaryotic cells, binding of RNA polymerase to the promoter involves proteins called transcription factors. *See* transcription.

prophage *See* phage.

prophase The first stage of cell division in MEIOSIS and MITOSIS. During prophase the chromosomes become visible and the nuclear membrane dissolves. The events occurring differ in meiosis and mitosis, notably in that bivalents (pairs of homologous chromosomes) are formed in meiosis, whereas homologous chromosomes remain separate in mitosis. Prophase in meiosis may be divided into successive stages termed leptotene, zygotene, pachytene, diplotene, and diakinesis.

prostaglandin (PG) One of a group of fatty acid derivatives, originally identified in human prostate secretions but now known to be present in all tissues. They have many physiological effects, notably stimulating smooth muscle contraction in the uterus. Different prostaglandins often have opposing actions, e.g. PGE and PGA reduce blood pressure while PGF raises it. They have been shown to affect the secretion of hormones by their vasodilatory and vasoconstrictory actions on the blood vessels supplying the endocrine glands. They are also implicated in pain production, being released during inflammation. Most prostaglandins are synthesized locally and are rapidly metabolized by enzymes in the tissues.

prostate gland A gland in male mammals surrounding the urethra in the region where it leaves the bladder. It releases a fluid containing various substances, including enzymes and an antiagglutinating factor, that contribute to the production of semen. Its size and secretory function is under the control of hormones (androgens).

prosthetic group The nonprotein component of a conjugated protein. Thus the heme group in hemoglobin is an example of a prosthetic group, as are the coenzyme components of a wide range of enzymes.

protamine One of a group of polypeptides formed from a few amino acids. They are soluble in water, dilute acids, and bases. On heating they do not coagulate. When protamines are hydrolyzed they yield a large proportion of basic amino acids, particularly arginine, alanine, and serine. They occur in the sperm of vertebrates, packing the DNA into a condensed form.

protease (**peptidase; proteinase**) An enzyme that catalyzes the hydrolysis of peptide bonds in proteins to produce peptide chains and amino acids. Individual proteases are highly specific in the type of peptide bond they hydrolyze.

proteasome *See* proteolysis.

protein One of a large number of substances that are important in the structure and function of all living organisms. Proteins are polypeptides; i.e. they are made up of amino acid molecules joined together by peptide links. Their molecular weight may vary from a few thousand to several million. About 20 amino acids are present in proteins. Simple proteins contain only amino acids. In conjugated proteins, the amino acids are joined to other groups. Proteins may consist of one or several polypeptide chains.

The primary structure of a protein is the particular sequence of amino acids present. The secondary structure is the way in which this chain is arranged; for example, coiled in an alpha helix or held in beta pleated sheets. The secondary structure is held by hydrogen bonds. The tertiary structure of the protein is the way in which the polypeptide chain is folded into a three-dimensional structure. This may be held by cystine bonds and by attractive forces between atoms. A protein must have the correct tertiary structure in order to function properly and abnormally folded proteins can cause disease, e.g. prion diseases, Alzheimer's. Quaternary structure is the interaction between polypeptide chains (or subunits).

proteinase *See* protease.

protein kinase An enzyme that phosphorylates one or more hydroxyl or phenolic groups in proteins using ATP as a donor. Over 100 protein kinases have been identified so far and they are split into two classes: those that phosphorylate the hydroxyl groups of serine and threonine residues and those that phosphorylate the phenolic group of tyrosine residues. Both classes are important regulatory enzymes. The serine/threonine kinases include the enzymes that regulate metabolism and many of the enzymes involved in regulating the cell cycle are tyrosine kinases.

protein sequencing The determination of the primary structure of proteins, i.e. the type, number, and sequence of amino acids in the polypeptide chain. This is done by progressive hydrolysis of the protein using specific proteases to split the polypeptides into shorter peptide chains. Terminal amino acids are labeled, broken off by a specific enzyme, and identified by chromatography. The first protein to be sequenced was insulin, by Frederick Sanger at Cambridge University in 1954.

protein synthesis The process whereby proteins are synthesized on the ribosomes of cells. The sequence of bases in messenger RNA (mRNA), transcribed from DNA, determines the sequence of amino acids in the polypeptide chain: each codon in the mRNA specifies a particular amino acid. As the ribosomes move along the mRNA in the process of *translation*, each codon is 'read', and amino acids bound to different transfer RNA molecules are brought to their correct positions along the mRNA molecule. These amino acids are polymerized to form the growing polypeptide chain. *See also* messenger RNA; transfer RNA; translation. The polypeptide chain can then be further modified (post-translational modification) e.g. glycosylated, folded into its three-dimensional shape and transported to its destination.

proteoglycan (**mucoprotein**) A type of glycoprotein consisting of long branched heterogeneous chains of glycosaminoglycan molecules linked to a protein core of amino acids. Unlike more typical glycoproteins, they have a greater carbohydrate content, the protein core is rich in serine, and they have a higher molecular weight.

proteolysis The hydrolysis of proteins into their amino acids. Enzymes that catalyze this are *proteases* or *proteolytic enzymes*. Proteolysis to breakdown unwanted proteins in the cell occurs either in the lysosome (*see* lysosome) or in a large complex called the *proteasome*. Proteins are marked for degradation by the proteasome by the addition of a small protein, *ubiquitin*, which is abundant in both the nucleus and cytoplasm. The process of adding ubiquitin is tightly controlled as important regulatory proteins are degraded in the proteasome.

proteome The complete set of proteins expressed in an organism; their structure, function, expression patterns and interactions. Investigation of the proteome is called *proteomics* and involves a wide range of techniques. Now that the goal of the Human Genome Project and other genome sequencing projects has been largely achieved, attention is being switched to the proteome of humans and other organisms and a Human Proteome Project was set up in 2001 to coordinate international research efforts. The goal of the Project, the complete knowledge of all (human) proteins, has immense implications for medicine.

proteomics *See* proteome.

proteolytic enzyme *See* proteolysis.

prothrombin The inactive form of the enzyme thrombin in blood plasma. It is activated during blood clotting by another enzyme, thrombokinase, in the presence of calcium ions.

Protista In some classifications, a kingdom of simple organisms including the bacteria, algae, fungi, and protozoans. It was introduced to overcome the difficulties of assigning such organisms, which may show both animal and plantlike characteristics, to the kingdoms Animalia or Plantae. Today the grouping is considered artificial and many taxonomists support a system whereby the bacteria and fungi are both assigned to separate kingdoms, while algae and protozoans constitute various phyla of protoctists.

Protoctista A kingdom of simple eukaryotic organisms that includes the algae, slime molds, fungus-like oomycetes, and the organisms traditionally classified as protozoa, such as flagellates, ciliates, and sporozoans. Most are aerobic, some are capable of photosynthesis, and most possess undulipodia (flagella or cilia) at some stage of their life cycle. Protoctists are typically microscopic single-celled organisms, such as the amebas, but the group also has large multicellular members, for example the seaweeds and other conspicuous algae.

protoplasm The living contents of a cell, comprising the cytoplasm plus nucleus. *See* cytoplasm; nucleus.

protoplast The protoplasm and plasma membrane of a cell (e.g. of a plant, alga, or bacterium) after removal of the cell wall. This can be achieved by physical means or by enzymic digestion. Protoplasts can be grown in culture and make possible certain observational or experimental work such as study of new cell wall formation, pinocytosis, and fusion of cells. Fusion of protoplasts of different species is being investigated by plant breeders as a means of crossing otherwise incompatible plants. Under suitable culture conditions the hybrid cell can develop to form a mature fertile plant. The ability to regenerate mature plants from single transformed protoplasts is essential for certain genetic engineering techniques.

protozoa A group of single-celled heterotrophic often motile eukaryotic organisms, traditionally classified as animals and constituting a phylum, or subkingdom, Protozoa. In more recent classifications they are placed with other single-celled or simple multicelled eukaryotes in the kingdom Protoctista. They range from plant-like forms (e.g. *Euglena*, *Chlamydomonas*) to members that feed and behave like animals (e.g. *Amoeba*, *Paramecium*). There are over 30 000 species living universally in marine, freshwater, and damp terrestrial environments. Some form colonies (e.g. *Volvox*) and many are parasites (e.g. *Plasmodium*). Protozoa vary in body form but specialized organelles (e.g. cilia and flagella) are common. Reproduction is usually by binary fission although multiple fission and conjugation occur in some species. The main protozoan phyla are: Rhizopoda (rhizopods); Zoomastigina (flagellates); Apicomplexa (sporozoans); and Ciliophora (ciliates).

provirus A viral chromosome that is integrated in a host chromosome and multiplies with it. Proviruses do not leave the host chromosome and begin a normal cycle of viral replication unless triggered to do so (*see* virus). *See also* retrovirus.

proximal Denoting the part of an organ, limb, etc., that is nearest the origin or point of attachment. *Compare* distal.

Prusiner, Stanley B. (1942–) American biochemist who discovered prions. He was awarded the 1997 Nobel Prize for physiology or medicine.

pseudoallele A mutation in a gene that produces an effect identical to another mutation at a different site in the same gene locus. The two pseudoalleles thus act as a single gene but do not occupy the same position, as evidenced by the occasional rare recombinations between them that result in the *cis-trans* effect.

pseudogene A DNA sequence that is similar to a gene but cannot be transcribed. Although they have no immediate function, pseudogenes have the potential to form new genes.

pseudopodium A temporary finger-like projection or lobe on the body of an ameboid cell. It is formed by a flowing action of the cytoplasm and functions in locomotion and feeding.

psychrophilic Describing microorganisms that can live at temperatures below 20°C. *Compare* mesophilic; thermophilic.

pteroylglutamic acid *See* folic acid.

ptyalin (**salivary amylase**) An enzyme present in the saliva of humans and some animals. It belongs to the group of carbohydrate-hydrolyzing enzymes known as *amylases*. It catalyzes the conversion of starch to maltose. *See* amylase.

puff (**Balbiani ring**) A swelling that is seen in certain areas of the giant (polytene) chromosomes found in the salivary glands and other tissues of certain dipterous insects. Puffs originate in different regions of the chromosome in a certain sequence and their occurrence can be correlated with specific developmental events. Others occur only in certain tissues. The puffs are sites of active transcription of probably just a single gene, albeit present as numerous copies. *See* polytene.

punctuated equilibrium A theory of evolution proposing that there have been long periods of geological time, lasting for several million years, when there is little evolutionary change, punctuated by short periods of rapid speciation of less than 100 000 years. This is in contrast to the traditional theory (*see* neo-Darwinism) in which it is postulated that species have evolved gradually throughout geological time.

Purcell, Edward Mills (1912–97) American electrical engineer, physicist, and radioastronomer notable for his discovery (independently of F. Bloch) of the phenomenon of nuclear magnetic resonance. He was awarded the Nobel Prize for physics in 1952 jointly with F. Bloch.

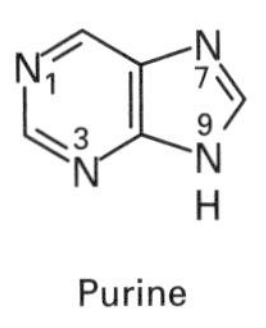

Purine

purine A simple nitrogenous organic molecule with a double ring structure. Members of the purine group include adenine and guanine, which are constituents of the nucleic acids, and certain plant alkaloids, e.g. caffeine and theobromine.

putrescine An amine, $H_2N[CH_2]_4NH_2$, produced from ornithine in decaying meat or fish.

pyranose A SUGAR that has a six-membered ring form (five carbon atoms and one oxygen atom).

pyrenoid A protein structure found in the chloroplasts of green algae and hornworts (*Anthoceros*). Pyrenoids are associated with the storage of starch.

pyridine (C_5H_5N) An organic liquid of formula C_5H_5N. The molecules have a hexagonal planar ring and are isoelectronic with benzene. Pyridine is an example of an aromatic heterocyclic compound, with the electrons in the carbon–carbon pi bonds and the lone pair of the nitrogen delocalized over the ring of atoms.

pyridoxine (**vitamin B_6**) One of the water-soluble B-group of vitamins. Good sources include yeast and certain seeds (e.g. wheat and corn), liver, and to a limited extent, milk, eggs, and leafy green vegetables. There is also some bacterial synthesis of the vitamin in the intestine. Pyridoxine gives rise to a coenzyme involved in various aspects of amino acid metabolism. *See also* vitamin B complex.

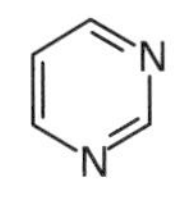

Pyrimidine

pyrimidine A simple nitrogenous organic molecule whose ring structure is contained in the pyrimidine bases cytosine, thymine, and uracil, which are constituents of the nucleic acids, and in thiamine (vitamin B_1).

pyrrhole A heterocyclic liquid aromatic compound with a five-membered ring containing four carbon atoms and one nitrogen atom, $(CH)_4NH$. It has important biochemical derivatives, including chlorophyll and heme.

pyruvate An intermediate in several metaoblic pathways, including glycolysis and gluconeogenesis, with the formula CH_3COCOO^-.

The last step in glycolysis is the virtually irreversible conversion of phosphoenolpyruvate into pyruvate by pyruvate kinase. The reactions that turn glucose into pyruvate (glycolysis) are common to all cells and do not depend on the presence of oxygen. The subsequent fate of pyruvate is dependent on the amount of oxygen present. If oxygen is available i.e. aerobic conditions pyruvate is converted to acetyl CoA by pyruvate dehydrogenase with the release of NADH and carbon dioxide. Acetyl CoA enters the Krebs cycle and generates more NADH and $FADH_2$, which enter the electron transport chain and transfer electrons to oxygen, leading to the synthesis of ATP. The oxidation of glucose in aerobic conditions leads to the formation of 36 or 38 molecules of ATP. Pyruvate is therefore the link between glycolysis and the Krebs cycle.

If limited amounts of oxygen are available i.e. anaerobic conditions, most microorganisms and the cells of higher organisms convert pyruvate to lactate. Some microorganisms e.g. brewer's yeast convert pyruvate to ethanol (fermentation) under anaerobic conditions. These anaerobic conversions do not generate NADH or FADH2 and therefore do not generate ATP. They regenerate NAD^+, which sustains glycolysis in anaerobic conditions. The only ATP synthesised is the two molecules of ATP made during glycolysis. In higher organisms lactate is made during intense muscle activity and cannot be metabolised further. It diffuses out of the muscles and enters the liver where it is oxidised to pyruvate, which is converted into glucose by the reactions of gluconeogensis. Glucose can then re-enter the blood and be taken up by muscle. See

Pyruvate is also the precursor for the synthesis of the amino acids alanine, valine, and leucine.

pyruvic acid A carboxylic acid, CH_3-COCOOH. The systematic name is 2-*oxopropanoic acid.*

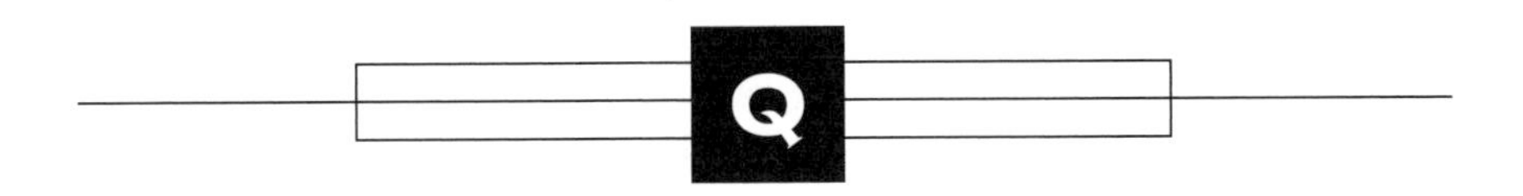

quantitative inheritance (**polygenic inheritance; multifactorial inheritance**) The pattern of inheritance shown by traits, such as height in humans or grain yield in wheat, that show continuous variation within a certain range of values. Such traits are typically controlled by many different genes (polygenes) distributed among the genome. The identification and manipulation of genetic loci determining quantitative traits (*quantitative trait loci*, QTLs) is of great importance in plant and animal breeding, where the goal is often to improve such traits.

quaternary ammonium compound A compound formed from an amine by addition of a proton to produce a positive ion. Quaternary compounds are salts, the simplest example being ammonium compounds formed from ammonia and an acid, for example:

$$NH_3 + HCl \rightarrow NH_4^+Cl^-$$

Other amines can also add protons to give analogous compounds. For instance, methylamine (CH_3NH_2) forms the compound

$$[CH_3NH_3]^+X^-$$

where X^- is an acid radical.

The formation of quarternary compounds occurs because the lone pair on the nitrogen atom can form a coordinate bond with a proton. This can also occur with heterogeneous nitrogen compounds, such as adenine, cytosine, thymine, and guanine. Such compounds are known as *nitrogenous bases.*

quinine A poisonous alkaloid present in the bark of the South American cinchona tree. It is used to treat malaria.

racemate *See* optical activity.

racemic mixture *See* optical activity.

racemization The conversion of an optical isomer into an equal mixture of isomers, which is not optically active.

radioactive dating Any method of dating that uses the decay rates of naturally occurring radioactive isotopes to assess the age of a specimen. Organic matter less than 7000 years old can be dated using radiocarbon dating. This uses the fact that the isotope carbon-14 is found in the atmosphere and taken in by plants when they photosynthesize, and subsequently assimilated by the animals that feed on them. When plants and animals die, no more carbon is taken in and the existing ^{14}C decays to the nonradioactive isotope carbon-12. If the proportion of ^{14}C to ^{12}C in the atmosphere and the decay rate of ^{14}C to ^{12}C are both known, as they are, then the sample may be dated by finding the present proportion of ^{14}C to ^{12}C. Specimens over 7000 years old can be dated by other radioisotope methods, e.g. potassium–argon dating.

radiochemistry The chemistry of radioactive isotopes of elements. Radiochemistry involves such topics as the preparation of radioactive compounds, the separation of isotopes by chemical reactions, the use of radioactive labels in studies of mechanisms, and experiments on the chemical reactions and compounds of transuranic elements.

radioimmunoassay (RIA) A type of immunoassay used for finding the concentration of a particular substance, for example a protein, in a biological sample. The substance acts as an antigen, binding to specific antibodies. Also required is a preparation of the substance labeled with a radioisotope. Two series of mixtures are prepared, each using the same known concentrations of antibody and labeled (antigenic) substance; in one series various standard solutions of the substance are added; in the second series, various dilutions of the sample are added. In each case, the unlabeled antigen competes with the labeled antigen for binding sites on the antibody. When the antigen–antibody complexes are separated from the reaction mixture, the ratio of labeled antigen/unlabeled antigen can be measured, and the concentration of substance in the sample found by comparison with the series of standard solutions. The technique is highly sensitive, and is widely used in medicine and research to make accurate determinations of a huge range of enzymes, hormones, drugs, and other substances.

radioisotope A radioactive isotope of an element. Tritium, for instance, is a radioisotope of hydrogen. Radioisotopes are extensively used in research as souces of radiation and as tracers in studies of chemical reactions. Thus, if an atom in a compound is replaced by a radioactive nuclide of the element (a *label*) it is possible to follow the course of the chemical reaction. Radioisotopes are also used in medicine or diagnosis and treatment.

radiolysis A chemical reaction produced by high-energy radiation (x-rays, gamma rays, or particles).

r.a.m. *See* relative atomic mass.

Ramón y Cajal, Santiago (1852–1934) Spanish anatomist and histologist. He was awarded the Nobel Prize for physiology or medicine in 1906 jointly with C. Golgi in recognition of their work on the structure of the nervous system.

raphides Bunches of needle-like crystals of calcium oxalate found in certain plant cells.

rate constant (**velocity constant; specific reaction rate**) Symbol: k The constant of proportionality in the rate expression for a chemical reaction. For example, in a reaction A + B → C, the rate may be proportional to the concentration of A multiplied by that of B; i.e.

$$\text{rate} = k[\text{A}][\text{B}]$$

where k is the rate constant for this particular reaction. The constant is independent of the concentrations of the reactants but depends on temperature; consequently the temperature at which k is recorded must be stated. The units of k vary depending on the number of terms in the rate expression, but are easily determined remembering that rate has the units s^{-1}.

rate-determining step (**limiting step**) The slowest step in a multistep reaction. Many chemical reactions are made up of a number of steps in which the one with the lowest rate is the one that determines the rate of the overall process.

rate of reaction A measure of the amount of reactant consumed in a chemical reaction in unit time. It is thus a measure of the number of effective collisions between reactant molecules. The rate at which a reaction proceeds can be measured by the rate the reactants disappear or by the rate at which the products are formed. The principal factors affecting the rate of reaction are temperature, pressure, concentration of reactants, light, and the action of a catalyst. The units usually used to measure the rate of a reaction are mol dm^{-3} s^{-1}.

reactant A compound taking part in a chemical reaction.

reading frame Any series of DNA bases that can be transcribed into a polypeptide chain. A reading frame in a DNA sequence is determined by the position of first base that is read because this determines the grouping of bases into triplets. For example, the sequence –ATCGTTGACA– could be grouped:

–ATC/GTT/GAC/A– or
–AT/CGT/TGA/CA– or
–A/TCG/TTG/ACA–.

A change in reading frame, which can be caused by the insertion or deletion of a single base, can lead to incorrect amino acids being incorporated into the polypeptide chain or transcription terminating in the wrong place.

reagent A compound that reacts with another (the substrate). The term is usually used for common laboratory chemicals – sodium hydroxide, hydrochloric acid, etc. – used for experiment and analysis.

receptor A cellular protein that specifically binds a ligand and causes a change in cell function. Cell surface receptors, e.g. the insulin receptor and cytokine receptors, span the cell membrane and bind their ligand on the extracellular side. Intracellular receptors, e.g. the steroid hormone receptors, bind their ligand within the cell. Receptors can also have an enzyme activity that is activated by ligand binding, e.g. several cytokine receptors are protein kinases.

recessive An allele that is only expressed in the phenotype when it is in the homozygous condition.

recombinant DNA **1.** Any DNA fragment or molecule that contains inserted foreign DNA, whether from another organism or artificially constructed. Recombinant DNA is fundamental to many aspects of genetic engineering, particularly the introduction of foreign genes to cells or organisms. There are now many techniques for creating recombinant DNA, depending on the nature of the host cell or organism receiving the foreign DNA. Particular genes or DNA sequences are cut from the parent molecule using specific

type II restriction endonucleases, or are assembled using a messenger RNA template and the enzyme reverse transcriptase. In gene cloning using cultures of bacterial or eukaryote tissue cells, the foreign gene is inserted into a vector, e.g. a bacterial plasmid or virus particle, which then infects the host cell. Inside the host cell the recombinant vector replicates and the foreign gene is expressed. Plasmids are also used to insert foreign DNA into plants. One of the most common is the Ti (tumor-inducing) plasmid of the bacterium *Agrobacterium tumefaciens*. This causes crown gall tumors in plants, and its plasmid has been used on a range of crop plants. Some cells, of for example mouse embryos, can be injected with DNA. The embryos are then implanted into receptive mothers, which give birth to transgenic offspring. The same principle is used with other mammals, including sheep. Another technique, used in transfecting certain plant cells or cell organelles, for example, is to shoot DNA-coated microprojectiles, such as tungsten or gold particles, at the host cell target. This is termed *biolistics*. *See* gene cloning; genetic engineering; restriction endonuclease; reverse transcriptase; vector.
2. DNA formed naturally by recombination, e.g. by crossing over in meiosis or by conjugation in bacteria. *See* recombination.

recombination The regrouping of genes that regularly occurs during meiosis as a result of the independent assortment of chromosomes into new sets, and the exchange of pieces of chromosomes (crossing over). Recombination results in offspring that differ both phenotypically and genotypically from both parents and is thus an important means of producing variation.

red blood cell (red blood corpuscle) *See* erythrocyte.

reduction The gain of electrons by such species as atoms, molecules, or ions. It often involves the loss of oxygen from a compound, or addition of hydrogen. Reduction can be effected chemically, i.e. by the use of *reducing agents* (electron donors), or electrically, in which case the reduction process occurs at the cathode. For example,

$$2Fe^{3+} + Cu \rightarrow 2Fe^{2+} + Cu^{2+}$$

where Cu is the reducing agent and Fe^{3+} is reduced, and

$$2H_2O + SO_2 + 2Cu^{2+} \rightarrow 4H^+ + SO_4^{2-} + 2Cu^+$$

where SO_2 is the reducing agent and Cu^{2+} is reduced.

reduction division The first division of MEIOSIS, including prophase, metaphase I, and anaphase I. It results in a haploid number of chromosomes gathering at each end of the nuclear spindle.

refractory period The period following passage of an impulse along a nerve when either no stimulus, however large, will evoke a further impulse (the absolute refractory period) or only an abnormally large stimulus will evoke further impulse (the relative refractory period). During this time the resting potential of the cell membrane is recovered by the active pumping of sodium ions out of the cell.

regulator gene *See* operon.

Reichstein, Tadeus (1897–1996) Polish-born Swiss organic chemist and botanist noted for his isolation and study of 29 different steroids from the cortex of the adrenal gland. He also invented an industrial method of synthesizing ascorbic acid from glucose. He was awarded the Nobel Prize for physiology or medicine in 1950 jointly with E. C. Kendall and P. S. Hench for their discoveries relating to the hormones of the adrenal cortex.

relative atomic mass (r.a.m.) Symbol: *Ar* The ratio of the average mass per atom of the naturally occurring element to 1/12 of the mass of an atom of nuclide 12C. It was formerly called *atomic weight*

relative molecular mass Symbol: M_r The ratio of the average mass per molecule of the naturally occurring form of an element or compound to 1/12 of the mass of an atom of nuclide ^{12}C. This was formerly called *molecular weight*. It does not have to be used only for compounds that have discrete molecules; for ionic compounds (e.g. NaCl) and giant-molecular structures (e.g. BN) the formula unit is used.

relaxin A hormone produced by the corpus luteum and responsible for the inhibition of uterine contraction. During birth, relaxin stimulates dilation of the cervix and relaxes the pubic symphysis enabling the pelvic girdle to widen.

rem (*r*adiation *e*quivalent *m*an) A unit for measuring the effects of radiation dose on the human body. One rem is equivalent to an average adult male absorbing one rad of radiation. The biological effects depend on the type of radiation as well as the energy deposited per kilogram.

renin A proteolytic enzyme produced in the kidney and released into the bloodstream that splits angiotensin from its precursor, angiotensinogen. *See* angiotensin.

rennin *See* chymosin.

repetitive DNA DNA that consists of multiple repeats of the same nucleotide sequences. Unlike prokaryotic cells, eukaryotic cells contain appreciable amounts of repetitive DNA; much of this is not transcribed and apparently serves no useful function. It constitutes part of the so-called 'junk DNA'. Satellite DNA is made up of about one million tandem (head-to-tail) repeats of the same sequence of some 200 nucleotides. Blocks of satellite DNA are found in the region of chromosome centromeres, and is described as *highly repetitive DNA*. Tandem repeats of fewer nucleotides is called minisatellite DNA. The number of repeats is very variable and gives the VNTRs that are used as genetic markers (*see* genetic fingerprinting). The genomes of higher eukaryotes also contains highly repetitive sequences that are not clustered together. They are associated with transposons. Some repetitive DNA is accounted for by multiple copies of particular genes, e.g. genes encoding histones and ribosomal RNA. *See also* satellite DNA; selfish DNA; transposons.

replica A thin detailed copy of a biological specimen, obtained by spraying the surface with a layer of plastic and carbon. Replicas are used in electron-microscope work.

replication The mechanism by which exact copies of the genetic material are formed. Replicas of DNA are made when the double helix unwinds and the separated strands serve as templates along which complementary nucleotides are assembled through the action of the enzyme DNA polymerase. The result is two new molecules of DNA each containing one strand of the original molecule, and the process is termed *semiconservative replication*. Because the two DNA strands are antiparallel and DNA polymerase only works in one direction, one DNA strand, the *leading strand*, is synthesized continuously and the other, the *lagging strand*, is synthesized in fragments. These fragments are called *Okazaki fragments* and are linked together by DNA ligase. Most prokaryotes have three types of DNA polymerase and eukaryotic cells have at least four. It is important for survival that DNA replication does not introduce incorrect bases that would cause mutation so DNA polymerase 'proof reads' the newly synthesized DNA and mistakes are corrected by the mismatch repair system. In RNA viruses an RNA polymerase is involved in the replication of the viral RNA.

reporter molecule A molecule having a characteristic property, e.g. fluorescence, ultraviolet absorbance, that is sensitive to polarity. The molecule is introduced into a protein so that changes in the property can be monitored in order to measure changes in the environment of the protein.

repressor molecule A protein molecule that prevents protein synthesis by binding to the operator sequence of the gene and preventing transcription. The molecule is produced by a regulatory gene and may act either on its own or in conjunction with a *corepressor*. In some cases another molecule, an *inducer*, may bind to the repressor, weakening its bonds with the operator and derepressing the gene, allowing transcription to proceed. *See* operon.

residual body *See* lysosome.

residue Any of the amino-acid parts of a peptide or protein.

resin One of a group of acidic substances occurring in many trees and shrubs (e.g. conifers and some poplars) either as sticky glassy solids or in solution in essential oils, i.e. as balsams, such as turpentine. Resins may be phenolic derivatives or oxidation products of terpenes, a group of substances which have branched-chain carbon skeletons consisting of 5-carbon units. They are usually secreted by special cells into long resin ducts or canals. Sometimes they are produced in response to injury or infection. They sometimes form a sticky covering to buds (e.g. horse chestnut), reducing transpiration and giving protection. Some are important commercially, for example oleoresin from pine tree bark is a source of rosin which, on distillation, yields turpentine.

resolution (of racemates) The separation of a racemate into the two optical isomers. This cannot be done by normal methods, such as crystallization or distillation, because the isomers have identical physical properties. The main methods are:
1. Mechanical separation. Certain optically active compounds form crystals with distinct left- and right-handed shapes. The crystals can be sorted by hand.
2. Chemical separation. The mixture is reacted with an optical isomer. The products are then not optical isomers of each other, and can be separated by physical means. For instance, a mixture of D- and L-forms of an acid, acting with a pure L-base, produces two salts that can be separated by fractional crystallization and then reconverted into the acids.
3. Biochemical separation. Certain organic compounds can be separated by using bacteria that feed on one form only, leaving the other.

resolving power (**resolution**) The ability of an optical system to form separate images of closely spaced objects. *See* microscope.

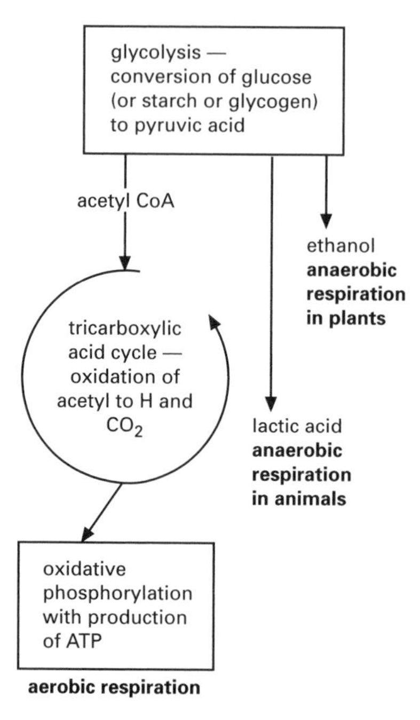

Respiration

respiration The oxidation of organic molecules to provide energy in plants and animals. In animals, food molecules are respired, but autotrophic plants respire molecules that they have themselves synthesized by photosynthesis. The energy from respiration is used to attach a high-energy phosphate group to ADP to form the short-term energy carrier ATP, which can then be used to power energy-requiring processes within the cell. The actual chemical reactions of respiration are known as *internal* (*cell* or *tissue*) *respiration* and they normally require oxygen from the environment (*aerobic respiration*). Some organisms are able to respire, at least for a short period, without the use of oxygen (*an-*

aerobic respiration), although this process produces far less energy than aerobic respiration, e.g. 38 molecules of ATP are generated for each molecule of glucose oxidized in aerobic respiration, compared with only 2 in anerobic respiration. Respiration usually involves an exchange of gases with the environment; this is known as *external respiration*. In small animals and all plants exchange by diffusion is adequate, but larger animals generally have special respiratory organs with large moist and ventilated surfaces (e.g. lungs, gills) and there is often a circulatory system to transport gases internally to and from the respiratory organs. *See also* electron-transport chain; glycolysis; Krebs cycle.

respiratory chain The electron-transport chain in aerobic respiration.

respiratory pigments Colored compounds that can combine reversibly with oxygen. HEMOGLOBIN is the blood pigment in all vertebrates and a wide range of invertebrates. Other blood pigments, such as haemoerythrin (containing iron) and hemocyanin (containing copper), are found in lower animals, and in many cases are dissolved in the plasma rather than present in cells. Their affinity for oxygen is comparable with hemoglobin, though oxygen capacity is generally lower.

respiratory quotient (**RQ**) The ratio of the volume of carbon dioxide expired by an organism compared to the volume of oxygen consumed during the same period of respiration. A theoretical RQ can be calculated for the various foodstuffs used in respiration, giving a value of 1 for carbohydrates, 0.7 for fats, and 0.8 for proteins. However, in practice, more than one foodstuff is respired at one time and other metabolic processes may produce carbon dioxide or use oxygen, so an RQ measurement for an organism gives unreliable information about the type of foodstuff respired.

response A change in an organism or in part of an organism that is produced as a reaction to a stimulus.

resting potential The potential difference that exists across the cell membrane of a nonconducting neurone. It is produced and maintained by the sodium pump, which actively expels sodium ions from the cell and thereby builds up a positive charge on the outside of the membrane. The sodium pump also pumps potassium ions into the cell, but in smaller numbers than sodium ions pumped out, so there is a net outflow of positive charge resulting in a potential of about +70 mV outside (relative to the inside). As the membrane is slightly permeable to sodium and potassium in its resting state, the sodium pump functions continuously, at a slow rate, using metabolic energy to maintain the resting potential. *See also* nerve impulse.

restriction endonuclease (**restriction enzyme**) A type of enzyme, found mainly in bacteria, that can cleave and fragment DNA internally (*see* endonuclease). They are so named as they restrict their activity to foreign DNA, such as the DNA of an invading virus; thus their function is protection of the cell. Some restriction endonucleases cleave DNA at random, but a particular group of enzymes, known as *class II restriction endonucleases*, cleave DNA at specific sites. Most recognize a sequence of six nucleotides, but some five or four. The specific sites on the cell's own DNA are protected from this enzyme activity by methylation, which is controlled by another type of site-specific enzyme.

The resulting fragments of DNA may be blunt-ended, i.e. the two strands finish at the same point, or may have a cohesive or 'sticky' end with a single strand extending. Both types of fragment can be inserted into a cloning vector using a DNA ligase enzyme. The discovery of these enzymes formed the basis for the development of GENETIC ENGINEERING, since they enable the isolation of particular gene sequences and the DNA fragments can be easily replicated. About 2500 type II restriction endonucleases have been discovered, with some 200 different cleavage site specificities. They have been named according to the organism in which they occur. For example, *Eco*R1 is obtained from *Esche-*

richia coli, strain R, and was the first enzyme to be isolated in this bacterium.

restriction fragment length polymorphism (RFLP) Variation among the members of a population in the sites at which class II restriction enzymes cleave the DNA, and hence in the size of the resulting DNA fragments. It results from differences between individuals in nucleotide sequences at the cleavage sites (restriction sites). The presence or absence of particular restriction sites can be ascertained using DNA probes in the technique called Southern blotting. Restriction sites vary enormously, and this variation is exploited in analyzing and comparing the genomes of different individuals, e.g. to establish how closely related they may be. Restriction sites are also invaluable as genetic markers in chromosome mapping, and can be used to track particular genes and identify mutations, e.g. for cystic fibrosis. *See* chromosome map; Southern blotting.

restriction map A map of a segment of DNA showing the cleavage sites of restriction endonucleases and their physical distance apart, usually measured in base pairs. It can be used to reveal variations in restriction sites between individuals of the same species or between different species (*See* restriction fragment length polymorphism). This variation serves as a key to the organism's genes, since the restriction sites can be used as markers to identify closely linked genetic loci and allow investigation of deletions, insertions, or other mutations. They are an essential tool in chromosome mapping.

The first step in constructing a restriction map is to label the ends of the DNA with a radioisotope. Then the DNA is subjected to a series of total and partial digests with one or more restriction enzymes. The fragments resulting from each digest are separated according to size by gel electrophoresis, and the order of fragments, and hence restriction sites, deduced from the various fragment sizes and the labeled ends. *See* restriction endonuclease. *See also* genetic fingerprinting.

reticulocyte An immature red blood cell. It develops in the red marrow in bones from a *proerythroblast* (red cell precursor) that gradually accumulates hemoglobin until it is a fully formed reticulocyte. In mammals it then loses its nucleus and is released into the blood as an erythrocyte.

reticuloendothelial system The system of macrophage cells, which are scattered throughout the body and are capable of engulfing foreign particles. The reticuloendothelial system is important in defending the body against disease and in destroying worn-out erythrocytes. *See* macrophage.

reticulum The second region of the specialized stomach of ruminants (e.g. the cow). The cud passes back into the reticulum after it has been regurgitated and chewed. It is lined with tough cornified stratified squamous epithelium and is formed in sectional folds like a honeycomb.

retina The innermost light-sensitive layer of the vertebrate eye. It consists of two types of photosensitive cells (rods and cones), adjacent to, and at right angles to, the choroid. The rods and cones are connected by synapses to bipolar and ganglion nerve cells. From the ganglion cells, nerve fibers pass over the inner surface of the retina to the optic nerve. Light entering the eye through the pupil has to pass through all the layers of the retina before it reaches the sensitive ends of the rods and cones, except at the fovea.

retinal (**retinene**) An aldehyde derivative of retinol (vitamin A). Retinal is a constituent of the light-sensitive conjugated protein, rhodopsin, which occurs in the rod cells of the retina. *See* rhodopsin.

retinene *See* retinal.

retrovirus An RNA-containing virus whose genome is transcribed into double-stranded DNA by the (viral) enzyme reverse transcriptase and becomes integrated into the host DNA and then replicates with it. One group of retroviruses can cause cancerous changes in their host cells (i.e. they are oncogenic) by means of the activity of one or more of their genes (*see* oncogene), e.g. the Rous sarcoma virus (RSV), which was discovered in 1911 and causes cancer in chickens. HIV, the virus causing AIDS, is a retrovirus.

reverse transcriptase An enzyme that catalyzes the synthesis of DNA from RNA (i.e. the reverse of transcription, in which mRNA is synthesized from a DNA template). The enzyme occurs in certain RNA viruses (*see* retrovirus) and enables the viral RNA to be 'transcribed' into DNA, which is then integrated into the host DNA and replicates with it. Reverse transcriptase does not have the proof-reading ability of DNA polymerase and causes a high mutation rate in the viral genome. Retroviruses therefore mutate rapidly, even within a single infection. As eukaryotic cells do not naturally contain reverse transcriptase, many antiretroviral drugs, e.g. AZT, are designed to inhibit it. It is also used in genetic engineering to make complementary DNA (cDNA) from an RNA template.

reversible reaction A chemical reaction that can proceed in either direction. For example, the reaction:

$$N_2 + 3H_2 \leftrightarrow 2NH_3$$

is reversible. In general, there will be an equilibrium mixture of reactants and products.

RFLP *See* restriction fragment length polymorphism.

R_f value In paper chromatography, the distance traveled by the solute divided by the distance traveled by the solvent front, the latter always being taken as one. The R_f (relative front) value is thus always between zero and one and is characteristic of a particular molecule. In this way various amino acids, chlorophylls, etc., may be identified. *See* paper chromatography.

rhesus factor (Rh factor) An antigen attached to human red blood cells, so named because it is also present in the rhesus monkey. The antigen is present in most people, who are therefore described as rhesus-positive (Rh-positive), but absent in others (Rh-negative).

Normally, neither type of blood contains the anti-Rh antibody. However, it may be present in the blood of Rh-negative women who have borne Rh-positive children: during pregnancy, the red cells of the fetus, carrying the Rh factor, can diffuse across the placenta and stimulate anti-Rh antibody production in the mother's blood. If the woman subsequently becomes pregnant with another Rh-positive fetus, the anti-Rh antibody could diffuse into the circulation of the fetus, producing a serious condition called hemolytic disease of the newborn, in which the red cells of the fetus are destroyed. To prevent this, Rh-negative mothers are injected soon after delivery of the first child with a concentrated immunoglobulin that destroys any Rh-positive fetal red cells in her circulation, thus preventing the production of anti-Rh antibodies.

Rhizobium A spherical or rod-shaped bacterium that can live either freely in the soil or symbiotically in the root nodules of leguminous plants and a few other species, such as alder. The bacteria can move slowly through the soil by means of flagella and are attracted to and infiltrate the root hairs of leguminous plants. They produce infection threads that penetrate the cells of the root cortex, which are stimulated to divide rapidly and form a swollen mass of tissue, the root nodule. The central region of the nodule consists of enlarged cells containing large numbers of bacteria. The outer region of the nodule contains vascular strands linking with the vascular bundles of the root. The bacteria in the nodules use atmospheric nitrogen and the resulting nitrates can be passed to the plant. In return, the bacterium is supplied with carbohydrates, such as sugars. *See also* nitrogen cycle; nitrogen fixation.

rhodopsin (visual purple) A light-sensitive pigment in the retina. It has a protein component, opsin, linked to a nonprotein molecule, retinal, which is a derivative of vitamin A. It is localized in the rod cells. When light strikes the retina, rhodopsin is split into its separate components, opsin and retinal, from which it is subsequently regenerated. The biochemical mechanism for cone vision is analogous to rod vision, retinal being used as the chromophore, but the protein component being different. *See* rod.

RIA See radioimmunoassay.

riboflavin (**vitamin B_2**) One of the water-soluble B-group of vitamins. It is found in cereal grains, peas, beans, liver, kidney, and milk. Riboflavin is a constituent of several enzyme systems (flavoproteins), acting as a coenzyme for hydrogen transfer in the reactions catalyzed by these enzymes. Two forms of phosphorylated riboflavin are known to exist in various enzyme systems: FMN (flavin mononucleotide) and FAD (flavin adenine dinucleotide). *See also* vitamin B complex.

ribonuclease (**RNase**) Any nuclease enzyme that cleaves the phosphodiester bonds between adjacent nucleotides in RNA. *Exoribonucleases* cleave nucleotides from one or both ends of the RNA molecule, while *endoribonucleases* cleave bonds within the molecule. There are many types of both classes of ribonuclease, each having a specific action. For example ribonuclease T1 degrades RNA to mono- and oligonucleotides terminating in a 3′-guanine nucleotide, while those produced by ribonuclease T2 terminate in a 3′-adenine nucleotide.

ribonucleic acid *See* RNA.

ribose A monosaccharide, $C_5H_{10}O_5$; a component of RNA.

ribosomal RNA (**rRNA**) The RNA found in the ribosomes. *See* ribosome.

ribosome A small organelle found in large numbers in all cells that acts as a site for protein synthesis. Ribosomes are often bound to endoplasmic reticulum but may occur free in the cytoplasm. In most species they are composed of roughly equal amounts of protein and ribosomal RNA. The ribosome consists of two unequally sized rounded subunits arranged on top of each other like a cottage loaf. Eukaryotic cells have larger ribosomes than prokaryotic cells but the ribosomes in mitochondria and chloroplasts are about the same size as prokaryotic ribosomes.

During translation the ribosome moves along the messenger RNA (mRNA), enabling the peptide linkage of amino acids delivered to the site by transfer RNA molecules according to the code in mRNA. Several ribosomes may be actively engaged in protein synthesis along the same mRNA molecule, forming a polyribosome, or polysome.

ribozyme Any RNA molecule that acts as an enzyme. It is now known that some introns (noncoding messenger RNA sequences) catalyze their own removal from the primary messenger RNA transcript and the splicing together of the cleaved ends, in a process called *self-splicing*. Following its removal from the primary mRNA transcript, the intron RNA sequence may catalyze further reactions, including splitting of RNA molecules and even peptide bond formation. Such ribozymes have remarkable similarities with viroids, the minute plant pathogens consisting simply of RNA circles, and it has been proposed that viroids are escaped introns.

ribulose bisphosphate (**RUBP**) A 5-carbon compound that accepts carbon dioxide during photosynthesis. Each molecule is then converted to two molecules of 3-carbon phosphoroglyceric acid. The ribulose bisphosphate is regenerated in the Calvin cycle when the carbon dioxide is converted into carbohydrate. *See* photosynthesis.

ring A closed loop of atoms in a molecule, as in benzene or cyclohexane. A *fused ring* is one joined to another ring in such a way that they share two atoms. Naphthalene is an example of a fused-ring compound.

ring closure A reaction in which one part of an open chain in a molecule reacts with another part, so that a ring of atoms is formed.

Ringer, Sydney (1835–1910) British physician and physiologist who devised Ringer's solution.

Ringer's solution *See* physiological saline.

r.m.m. *See* relative molecular mass.

RNA (ribonucleic acid) A nucleic acid found mainly in the cytoplasm and involved in protein synthesis. It is a single polynucleotide chain similar in composition to a single strand of DNA except that the sugar ribose replaces deoxyribose and the pyrimidine base uracil replaces thymine. RNA is synthesized on DNA in the nucleus and exists in three forms (*see* messenger RNA; transfer RNA; ribosome). In certain viruses, RNA is the genetic material.

RNAi *See* RNA interference.

RNA interference (RNAi) The selective post-transcriptional silencing of specific genes by short double-stranded RNA molecules. The double-stranded RNA forms a complex with various nuclease enzymes in the cytoplasm and selectively targets the mRNA of a particular gene by base pairing of the antisense strand with mRNA. This reaction is highly sequence-specific. RNAi is used by plants as a method of resisting viral attacks. Mammalian cells have recently been shown to be susceptible to RNAi, which has great therapeutic potential.

RNA polymerase *See* polymerase.

RNA splicing The enzymatic removal of introns from hnRNA and the subsequent joining of the cut ends of the exons to form a continuous mRNA. Some mitochondrial RNA catalyzes its own splicing (*see* ribozyme). hnRNA transcribed in the nucleus is spliced in the nucleus by a complex assembly of RNA and protein called a *spliceosome*. Alternative splicing can produce different mRNAs by splicing different exons together.

Roberts, Richard J. (1943–) British molecular biologist. He was awarded the Nobel Prize for physiology or medicine in 1993 jointly with P. A. Sharp for their discovery of split genes.

Robinson, (Sir) Robert (1886–1975) British organic chemist who worked on plant products, especially the alkaloids. He was awarded the Nobel Prize for chemistry in 1947.

rod One of the two types of light-sensitive cells in the retina of the vertebrate eye. Rods are concerned with vision in dim light; they are found chiefly in the periphery of the retina and are absent from the fovea. They contain a pigment, visual purple (rhodopsin), that is bleached by light energy. This photochemical reaction breaks down rhodopsin into a protein (opsin) and retinal (a derivative of vitamin A) and causes nerve impulses to pass from the rod cells to the brain. The rhodopsin is continually reformed from retinal, using energy from ATP in the mitochondria of the rod. In bright light, the reformation does not keep pace with the destruction, so that rods can only function in dim light. Several rods connect with the same bipolar cell (retinal convergence) so that sharp images are not seen. They have low visual acuity, but high sensitivity due to summation of the impulses from several rods starting an impulse from the bipolar cell. *See also* cone; retina.

Rodbell, Martin (1925–98) American biochemist. He was awarded the Nobel Prize for physiology or medicine in 1994 jointly with A. G. Gilman for their discovery of G-proteins and the role of these proteins in signal transduction in cells.

root nodule *See* nitrogen fixation.

RQ See respiratory quotient.

rRNA See ribosomal RNA.

rubber A natural or synthetic polymeric elastic material. Natural rubber is a polymer of methylbuta-1,3-diene (isoprene). Various synthetic rubbers are made by polymerization; for example chloroprene rubber (from 2-chlorobuta-1,3-diene) and silicone rubbers.

RUBP See ribulose bisphosphate.

RUBP carboxylase See C_4 plant.

rumen The first region of the specialized stomach of ruminants (e.g. the cow). It is sometimes called the *paunch*. Here the food is degraded and fermented by millions of symbiotic bacteria and protoctists, some of which produce the enzyme cellulase. Mammals are unable to produce this enzyme themselves, and since plant food consists mainly of cellulose, they cannot digest it without the help of these bacteria. After some time in the rumen the partly digested food, now called the *cud*, is regurgitated back to the mouth and chewed again, before being swallowed and passed into the reticulum.

Ružička, Leopold (1887–1976) Croatian-born Swiss chemist who worked on terpenes. He was awarded the Nobel Prize for chemistry in 1939. The prize was shared with A. F. J. Butenandt.

saccharide *See* sugar.

Saccharomyces **(yeasts)** A genus of unicellular ascomycete fungi that can live in both aerobic and anaerobic conditions. They are important in the brewing and baking industries, respectively, for the alcohol and carbon dioxide they produce by anaerobic respiration. Reproduction is generally asexual by budding although in adverse conditions sexual spores may be formed. These yeasts are also used in gene cloning as convenient eukaryotic hosts to express inserted foreign genes. They are also used to study eukaryotic genetics – the cell cycle control genes were first identified in yeast.

safranin *See* staining.

Sakmann, Bert (1942–) German physiologist. He was awarded the Nobel Prize for physiology or medicine in 1991 jointly with E. Neher for their work on single ion channels in cells.

salicylate **(hydroxybenzoate)** A salt or ester of salicylic acid.

salicylic acid **(2-hydroxybenzoic acid)** A crystalline aromatic carboxylic acid, $C_6H_4(OH)COOH$. It is used in medicines, as an antiseptic, and in the manufacture of azo dyes. Its ethanoyl (acetyl) ester is aspirin.

saliva A secretion produced by the salivary glands of animals, consisting mainly of mucus. It is used to moisten and lubricate the food and in some animals contains enzymes. For example, in humans and certain insects, the enzyme amylase is present and starts off the process of starch digestion. In some insects (e.g. mosquito) the saliva contains an anticoagulant.

salt A compound with an acidic and a basic radical, or a compound formed by total or partial replacement of the hydrogen in an acid by a metal. In general terms a salt is a material that has identifiable cationic and anionic components.

saltatory conduction The mode of transmission of a nerve impulse along a myelinated nerve fiber whereby the impulse leaps between the nodes of Ranvier, considerably speeding up its passage. The fastest nerve impulses known, traveling up to 120 m/s, occur in vertebrate myelinated fibers.

The myelin sheath insulates against loss of local currents between the nodes; they are therefore transmitted along the fiber axis to a node, where in the absence of myelin they generate an action potential. *See also* nerve impulse.

Samuelsson, Bengt Ingemar (1934–) Swedish biochemist. He was awarded the Nobel Prize for physiology or medicine in 1982 jointly with S. K. Bergström and J. R. Vane for their discoveries concerning prostaglandins and related biologically active substances.

Sanger, Frederick (1918–) British biochemist. Sanger was awarded the Nobel Prize for chemistry in 1958 for his work on the structure of proteins, especially his work on the structure of insulin. He also shared the Nobel Prize for chemistry in 1980 with W. Gilbert for their work on the determination of base sequences in nucleic acids. The 1980 prize was shared with P. Berg.

saprophyte An organism that derives its nourishment by absorbing the products or remains of other organisms. Many fungi and bacteria are saprophytes and are important in food chains in returning nutrients to the soil by putrefaction and decay.

saprozoic Describing an organism that feeds on organic material in solution, rather than on solid organic material.

sarcomere The contractile element in a striated muscle fibril (myofibril). Each sarcomere is joined to the next one by *Krause's membrane* (Z line). Thick filaments of the protein myosin form a dark central *A band.* On either side of this is a light area, the *I band*, in which are thin filaments of another protein, actin. The two types of filament overlap in the dark band except in the center, leaving a slightly lighter *H band* (Hensen's disk). According to the *sliding filament theory* of muscle contraction, projecting parts of myosin molecules form crossbridges to connect with binding sites on adjacent actin molecules. ATP from mitochondria in the myofibril provides energy for the bridges to oscillate and pull the actin filaments along in a ratchet action, thus making the whole sarcomere shorter. The shortening of all the sarcomeres of all the myofibrils results in the contraction of a muscle fiber when it is stimulated by nerve impulses. *See also* myofibril; skeletal muscle.

sarcoplasm The cytoplasm of the fibers of striated muscle, excluding the myofibrils.

sarcoplasmic reticulum A modified form of smooth endoplasmic reticulum found in striated and cardiac muscle. Muscle contraction requires calcium ions and the network of sarcoplasmic reticulum releases them quickly to all parts of the muscle fiber in response to nervous impulses. Calcium activates an ATPase, which catalyzes breakdown of ATP, thus releasing the energy required for contraction. After contraction, the sarcoplasmic reticulum reabsorbs the calcium.

satellite DNA A type of DNA that can be separated by centrifugation from the main DNA fraction. It is highly repetitive and includes DNA from the chromosomal region adjacent to the centromere, has a base composition unlike that of most DNA, and DNA from mitochondria, chloroplasts, and ribosomes.

saturated compound An organic compound that does not contain any double or triple bonds in its structure. A saturated compound will undergo substitution reactions but not addition reactions since each atom in the structure will already have formed its maximum possible number of single bonds. *Compare* unsaturated compound.

scanning electron microscope *See* microscope.

Schally, Andrew Victor (1926–) Polish-born American biochemist and physiologist. He was awarded the Nobel Prize for physiology or medicine in 1977 jointly with R. Guillemin for their discoveries concerning the peptide hormone production of the brain. The prize was shared with R. Yalow.

Schultze's solution A solution of zinc chloride, potassium iodide, and iodine used mainly for testing for cellulose and hemicellulose. Both materials stain a blue color with the reagent, that of hemicellulose being weaker.

Schwann cell A cell that makes a section of the myelin sheath of a medullated nerve fiber. During development, the cell becomes spirally wrapped around the nerve fiber. *See* myelin sheath.

scleroprotein One of a group of proteins obtained from the exoskeletal structures of animals. They are insoluble in water, salt solutions, dilute acids, and alkalis. This group exhibits a wide range of both physical and chemical properties. Typical examples of scleroproteins are keratin (hair), elastin (elastic tissue), and collagen (connective tissue).

SCP *See* single-cell protein.

secondary messenger (second messenger) *See* signal transduction.

secretin A polypeptide hormone secreted by the mucosa of the duodenum and jejunum when the stomach empties its contents into the intestine. It stimulates alkaline pancreatic secretions to neutralize the acidic chyme from the stomach. Its secretion is also stimulated by the release of bile.

secretion Substances or fluids produced in cells and released to the surrounding medium. The secretion may be a fluid (e.g. sweat) or molecules (e.g. enzymes, hormones). The term is also used for the process of producing the secretion.

sedimentation The settling of a suspension, either under gravity or in a centrifuge. The speed of sedimentation can be used to estimate the average size of the particles. This technique is used with an ultracentrifuge to find the relative molecular masses of macromolecules.

seed The structure that develops from the ovule following fertilization in angiosperms or gymnosperms. In flowering plants one or more seeds are contained within a fruit developed from the ovary wall. The individual seeds are composed of an embryo and, in those seeds in which food is not stored in the embryo cotyledons, a nutritive endosperm tissue. This difference enables seeds to be classified as nonendospermic or endospermic. The whole is surrounded by a testa developed from the integuments of the ovule. In gymnosperms the seeds do not develop within a fruit but are shed 'naked' from the plant. Following dispersal from the parent plant, seeds may germinate immediately to form a seedling or may remain in a relatively inactive dormant state until conditions are favorable for germination. In annual plants, seeds provide the only mechanism for surviving the cold or dry seasons. Seeds may be formed asexually in certain plants by apomixis, e.g. in dandelion.

segregation The separation of the two alleles of a gene into different gametes, brought about by the separation of homologous chromosomes during meiosis. *See* Mendel's laws.

selfish DNA Repetitive DNA sequences that can move around within the genome of an organism or insert copies of itself at various sites without serving any apparent useful function and sometimes causing harm to the host by giving rise to genetic diseases. The prime examples of selfish DNA are the mobile genetic elements called transposons. Some biologists also regard introns as selfish DNA. *Compare* junk DNA. *See* transposon.

Seliwanoff's test A standard test for the presence of fructose in solution. A few drops of Seliwanoff's reagent, resorcinol in hydrochloric acid, are heated with the test solution. A red color or red precipitate indicates fructose.

semipermeable membrane *See* osmosis.

senescence **1.** The advanced phase of the ageing process of an organism or part of an organism, prior to natural death. It is usually characterized by a reduction in capacity for self-maintenance and repair of cells, and hence deterioration. The degree of senescence varies between groups and its mechanism remains largely obscure: some believe it is a genetically controlled event, others suggest it is an accumulation of metabolic disorders. It is believed to involve lysosomal activity.
2. The deterioration of the leaves of a deciduous plant towards the end of the growing season, culminating in abscission (leaf fall).

sensitization The increase in the reaction of an organism or cell to an antigen to which it has been previously exposed. It may occur naturally or be artificially induced, e.g. following vaccination.

sequestration The formation of a complex with an ion in solution, so that the ion does not have its normal activity. Sequestering agents are often chelating agents, such as edta.

serine *See* amino acids.

serology The *in vitro* study of reactions between antigens and antibodies in the blood serum. Various serological tests involving specific types of reaction enable the identification of blood groups, pathogens, diseases, etc. *See also* agglutination.

serotonin (hydroxytryptamine) A substance that serves as a neurohormone that acts on muscles and nerves, and a neurotransmitter found in both the central and peripheral nervous systems. It controls dilation and constriction of blood vessels and affects peristalsis and gastrointestinal tract motility. Within the brain it plays a role in mood behavior. Many hallucinogenic compounds (e.g. LSD) antagonize the effects of serotonin in the brain.

serous membrane The tissue that lines cavities, in vertebrates, that do not open to the exterior, e.g. the pleural and peritoneal cavities. It consists of mesothelium and underlying connective tissue.

Sertoli cells Large pillar-like cells in the germinal epithelium of the vertebrate testis, which protect and nourish developing spermatozoa. They also secrete the hormone inhibin, which inhibits secretion of follicle-stimulating hormone by the pituitary.

serum *See* blood serum.

sesqui- Prefix indicating a 2/3 ratio. A sesquioxide, for example, would have the formula M_2O_3. A sesquicarbonate is a mixed carbonate and hydrogencarbonate of the type $Na_2CO_3.NaHCO_3.2H_2O$, which contains 2 CO_3 units and 3 sodium ions.

sex chromosomes The chromosomes that determine sex in most animals. There are two types: in mammals these are called the *X chromosome* and the *Y chromosome.* In the heterogametic sex (XY) they can usually be distinguished from the other chromosomes, because the Y chromosome is much shorter than the X chromosome with which it is paired (unlike the remaining chromosomes, which are in similar homologous pairs). *See* sex determination; sex linkage.

sex determination In species having almost equal numbers of males and females sex determination is genetic. Very occasionally a single pair of alleles determine sex but usually whole chromosomes, the sex chromosomes, are responsible. The 1:1 ratio of males to females is obtained by crossing of the homogametic sex (XX) with the heterogametic sex (XY). In most animals, including humans, the female is XX and the male XY, but in birds, butterflies and some fishes this situation is reversed. In some species sex is determined more by the number of X chromosomes than by the presence of the Y chromosome, but in humans the Y chromosome is important in determining maleness. Many genes are involved in determining all aspects of maleness or femaleness, but it is thought that in humans one particular gene on the Y chromosome acts as sex switch to initiate male development. In the absence of this male switch gene (i.e. in XX individuals) the fetus develops as a female. Rarely, sex is subject to environmental control, in which case unequal numbers of males and females develop. In bees and some other members of the Hymenoptera, females develop from fertilized eggs and are diploid while males develop from unfertilized eggs and are haploid, the numbers of each sex being controlled by the queen bee.

sex hormone Any of several steroid hormones responsible for the development and functioning of the reproductive organs. They are also involved in the development of secondary sex characteristics. They are secreted mainly by the gonads and include androgens in males and estrogens and progesterone in females.

sex linkage The coupling of certain genes (and therefore the characters they control) to the sex of an organism because they happen to occur on the X sex chromosome. The heterogametic sex (XY), which in humans is the male, has only one X chromosome and thus any recessive genes carried on it are not masked by their dominant alleles (as they would be in the homogametic sex). Thus in humans recessive forms of the sex-linked genes appear in the male phenotype far more frequently than in the female (in which they would have to be double recessives). Color blindness and hemophilia are sex linked. Heterogametic organisms that have one copy of the recessive allele but do not display the phenotype are called carriers.

sexual reproduction The formation of new individuals by fusion of two nuclei or sex cells (gametes) to form a zygote. In unicellular organisms whole individuals may unite but in most multicellular organisms only the gametes combine. In organisms showing sexuality, the gametes are of two types: male and female (in animals, spermatozoa and ova). They are produced in special organs (carpel and anther in plants; ovary and testis in animals), which, with associated structures, form a reproductive system and aid in the reproductive process. Individuals containing both systems are termed monoecious or hermaphrodite.

Generally meiosis occurs before gamete formation, resulting in the gametes being haploid (having half the normal number of chromosomes). At fertilization, when the haploid gametes fuse, the diploid number of chromosomes is restored. In this way sexual reproduction permits genetic recombination, which results in greater variety in offspring and so provides a mechanism for evolution by natural selection.

Apomixis and parthenogenesis are usually regarded as modified forms of sexual reproduction.

shadowing A method of preparation of material for electron microscopy enabling surface features to be studied. It can be used for small entire structures, subcellular organelles, or even large molecules (e.g. DNA). The specimen is supported on a plastic or carbon film on a small grid and sprayed with vaporized metal atoms from one side while under vacuum. The coated specimen appears blacker (more electron-opaque) where metal accumulates, and the lengths and shapes of 'shadows' cast (regions behind the objects not coated with metal) give structural information. It is often used in association with freeze fracturing. *See* freeze fracturing.

Sharp, Phillip Allen (1944–) American molecular biologist. He was awarded the Nobel Prize for physiology or medicine in 1993 jointly with R. J. Roberts for their discovery of split genes.

shikimic acid pathway The main metabolic pathway in the biosynthesis of the aromatic amino acids tyrosine, phenylalanine, and tryptophan. The first step is erythrose-4-phosphate, which comes from the pentose phosphate pathway, condensing with phosphenolpyruvate, from glycolysis. This product cyclizes to shikimate, which is then converted by various pathways into the amino acids.

shotgun cloning *See* gene cloning.

side chain In an organic compound, an aliphatic group or radical attached to a longer straight chain of atoms in an acyclic compound or to one of the atoms in the ring of a cyclic compound. *See* chain.

side reaction A chemical reaction that takes place to a limited extent at the same time as a main reaction. Thus the main product of a reaction may contain small amounts of other compounds.

sievert Symbol: Sv The SI unit of dose equivalent. It is the dose equivalent when the absorbed dose produced by ionizing radiation multiplied by certain dimensionless factors is 1 joule per kilogram (1 J kg^{-1}). The dimensionless factors are used to modify the absorbed dose to take account of the fact that different types of radiation cause different biological effects.

signal transduction The transfer of a signal from the outside of a cell to the inside of a cell without the signaling molecule crossing the membrane. The signaling molecule, e.g. a growth factor, binds to a receptor that leads to the synthesis of a *secondary messenger* within the cell. There are a number of secondary messenger system, e.g. cyclic AMP, phosphatidyl inositol, protein phosphorylation, and the opening of ion channels. The secondary messenger starts a cascade of reactions, which culminates in the response of the cell to the signal, e.g. protein synthesis.

silicon A trace element found in many animals and plants, although not essential for growth in most organisms. It is found in large quantities in the cell walls of certain algae (e.g. desmids, diatoms) and horsetails, and in smaller amounts in the cell walls of many higher plants. It forms the skeleton of certain marine animals, e.g. the siliceous sponges. Silicon is also found in connective tissue.

single-cell protein (**SCP**) Protein produced from microorganisms, such as bacteria, yeasts, mycelial fungi, and unicellular algae, used as food for man and other animals.

SI units (**Système International d'Unités**) The internationally adopted system of units used for scientific purposes. It has seven base units (the meter, kilogram, second, kelvin, ampere, mole, and candela) and two supplementary units (the radian and steradian). Derived units are formed by multiplication and/or division of base units; a number have special names. Standard prefixes are used for multiples and submultiples of SI units.

skeletal muscle (**striated muscle; striped muscle; voluntary muscle**) Muscle that moves the bones of the skeleton. Each muscle is made up of many microscopic *muscle fibers*, bound together with connective tissue and surrounded by a sheath (*epimysium*). Skeletal muscle has a typically striped appearance. The muscle fibers are long and narrow with tapering ends. Each has an outer membrane (*sarcolemma*) inside which are many oval nuclei. The cytoplasm (*sarcoplasm*) contains many large mitochondria and longitudinal myofibrils, which contain the contractile elements – the sarcomeres – giving the striated appearance. The epimysium is continuous with the nonelastic fibers of the tendons attached to the tapering ends of the muscle. The tendons penetrate the tissues of the bones to which the muscle is joined, the *origin* of the muscle being on the stationary bone and the *insertion* on the movable one. When the muscle contracts it becomes shorter and fatter and the tendons pull on the bones, bringing about movement at the joint. All skeletal muscles are under the voluntary control of the central nervous system. *See also* myofibril; sarcomere.

skeleton A hard structure that supports and maintains the shape of an animal. It may be external to the body (exoskeleton) or within the body (endoskeleton).

skin The outer layer of the body of an animal. In vertebrates it protects the animal from excessive loss of water, from the entry of disease-causing organisms, from damage by ultraviolet radiation, and from mechanical injury. It contains numerous nerve endings and therefore also acts as a peripheral sense organ. In warm-blooded animals it plays a part in the regulation of body temperature. It consists of two layers, the inner dermis and the outer epidermis. The former originates from the mesoderm and the latter from the ectoderm.

Skou, Jens C. (1918–) Danish biochemist noted for his discovery of Na+/K+ ATPase. He was awarded the 1997 Nobel Prize for chemistry. The prize was shared with P. D. Boyer and J. E. Walker.

Smith, Hamilton Othanel (1931–) American molecular biologist. He was awarded the Nobel Prize for physiology or medicine in 1978 jointly with W. Arber and D. Nathans for the discovery of restriction enzymes and their application to problems of molecular genetics.

BASE AND SUPPLEMENTARY SI UNITS

Physical quantity	*Name of SI unit*	*Symbol for SI unit*
length	meter	m
mass	kilogram	kg
time	second	s
electric current	ampere	A
thermodynamic temperature	kelvin	K
luminous intensity	candela	cd
amount of substance	mole	mol
*plane angle	radian	rad
*solid angle	steradian	sr

*supplementary units

DERIVED SI UNITS WITH SPECIAL NAMES

Physical quantity	*Name of SI unit*	*Symbol for SI unit*
frequency	hertz	Hz
energy	joule	J
force	newton	N
power	watt	W
pressure	pascal	Pa
electric charge	coulomb	C
electric potential difference	volt	V
electric resistance	ohm	Ω
electric conductance	siemens	S
electric capacitance	farad	F
magnetic flux	weber	Wb
inductance	henry	H
magnetic flux density	tesla	T
luminous flux	lumen	lm
illuminance (illumination)	lux	lx
absorbed dose	gray	Gy
activity	becquerel	Bq
dose equivalent	sievert	Sv

DECIMAL MULTIPLES AND SUBMULTIPLES USED WITH SI UNITS

Submultiple	*Prefix*	*Symbol*	*Multiple*	*Prefix*	*Symbol*
10^{-1}	deci-	d	10^{1}	deca-	da
10^{-2}	centi-	c	10^{2}	hecto-	h
10^{-3}	milli-	m	10^{3}	kilo-	k
10^{-6}	micro-	μ	10^{6}	mega-	M
10^{-9}	nano-	n	10^{9}	giga-	G
10^{-12}	pico-	p	10^{12}	tera-	T
10^{-15}	femto-	f	10^{15}	peta-	P
10^{-18}	atto-	a	10^{18}	exa-	E
10^{-21}	zepto-	z	10^{21}	zetta-	Z
10^{-24}	yocto-	y	10^{24}	yotta-	Y

Smith, Michael (1932–) British-born Canadian biochemist. He was awarded the Nobel Prize for chemistry in 1993 for his fundamental contributions to the establishment of oligonucleotide-based, site-directed mutagenesis and its development for protein studies. The prize was shared with K. B. Mullis.

smooth muscle (**involuntary muscle**) The muscle of all internal organs (viscera) and blood vessels (except the heart). Usually it is in the form of tubes or sheets, which may be up to several layers in thickness. The cells are long, narrow, and tapering, with a long nucleus and cytoplasm containing fine longitudinal filaments of contractile protein. It is not under voluntary control, being supplied by the autonomic nervous system. It contracts when stretched, may have spontaneous rhythmic contractions, and can remain in a state of continuous contraction (tonus) for long periods without fatigue. All invertebrates except arthropods have only this type of muscle.

Snell, George Davis (1903–96) American geneticist. He was awarded the Nobel Prize for physiology or medicine jointly with B. Benacerraf and Jean Dausset in 1980 for their discoveries concerning genetically determined structures on the cell surface that regulate immunological reactions.

sodium An element essential in animal tissues, and often found in plants although it is believed not to be essential in the latter. It is found in bones, and is the most abundant ion in the blood and cell fluids, being extremely important in maintaining the osmotic balance of animal tissues.

The resting potential difference across the membranes of animal cells is due to the ratio of Na^+ and K^+ ions across the cell membrane. In order to keep this ratio constant, Na^+ ions must be continually pumped out of the cell against a concentration gradient. This requires energy and is achieved by a number of transporter proteins, the main one being the sodium/potassium ATPase.

solute potential *See* water potential.

solvation The attraction of a solute species (e.g. an ion) for molecules of solvent. In water, for example, a positive ion will be surrounded by water molecules, which tend to associate around the ion because of attraction between the positive charge of the ion, and the negative part of the polar water molecule. The energy of this solvation (hydration in the case of water) is the 'force' needed to overcome the attraction between positive and negative ions when an ionic solid dissolves. The attraction of the dissolved ion for solvent molecules may extend for several layers. In the case of transition metal elements, ions may also form complexes by coordination to the nearest layer of molecules.

solvent A liquid capable of dissolving other materials (solids, liquids, or gases) to form a solution. The solvent is generally the major component of the solution. Solvents can be divided into classes, the most important being:

Polar. A solvent in which the molecules possess a moderate to high dipole moment and in which polar and ionic compounds are easily soluble. Polar solvents are usually poor solvents for non-polar compounds. For example, water is a good solvent for many ionic species, such as sodium chloride or potassium nitrate, and polar molecules, such as the sugars, but does not dissolve paraffin wax.

Non-polar. A solvent in which the molecules do not possess a permanent dipole moment and consequently will solvate non-polar species in preference to polar species. For example, benzene and tetrachloromethane are good solvents for iodine and paraffin wax, but do not dissolve sodium chloride.

Amphiprotic. A solvent which undergoes self-ionization and can act both as a proton donator and as an acceptor. Water is a good example and ionizes according to:

$$2H_2O = H_3O^+ + OH^-$$

Aprotic. A solvent which can neither accept nor yield protons. An aprotic solvent is therefore the opposite to an amphiprotic solvent.

somatic Describing the cells of an organism other than germ cells. Somatic cells divide by mitosis producing daughter cells identical to the parent cell. A somatic mutation is a mutation in any cell not destined to become a germ cell; such mutations are therefore not heritable.

somatomedin *See* insulin-like growth factor.

somatostatin A peptide hormone produced by the hypothalamus that inhibits the release of growth hormone from the pituitary, and also the release of insulin and glucagon from the pancreas in fasted animals. It is found in various types of tissues.

somatotropin *See* growth hormone.

sorbitol A hexahydric alcohol that occurs in rose hips and rowan berries, $HOCH_2(CHOH)_4CH_2OH$. It can be synthesized by the reduction of glucose. Sorbitol is used to make vitamin C (ascorbic acid) and surfactants. It is also used in medicines and as a sweetener (particularly in foods for diabetics).

Southern blotting A technique for transferring DNA fragments from an electrophoretic gel to a nitrocellulose filter or nylon membrane, where they can be fixed in position and probed using DNA probes. Named after its inventor, E. M. Southern (1938–), it is widely used in genetic analysis. The DNA is first digested with restriction enzymes and the resulting mixture of fragments separated according to size by electrophoresis on agarose gel. The double-stranded DNA is then denatured to single-stranded DNA using sodium hydroxide, and a nitrocellulose filter pressed against the gel. This transfers, or blots, the single-stranded DNA fragments onto the nitrocellulose, where they are permanently bound by heating. The DNA probe can then be applied to locate the specific DNA fragment of interest, while preserving the electrophoretic separation pattern. *See* DNA probe. *Compare* Western blotting.

SP (suction pressure) *See* osmosis.

specific Denoting a physical quantity per unit mass. For example, volume (V) per unit mass (m) is called specific volume:

$$V = Vm$$

In certain physical quantities the term does not have this meaning: for example, *specific gravity* is more properly called relative density.

specific reaction rate *See* rate constant.

specific rotatory power Symbol: αm The rotation of plane-polarized light in degrees produced by a 10 cm length of solution containing 1 g of a given substance per milliliter of stated solvent. The specific rotatory power is a measure of the optical activity of substances in solution. It is measured at 20°C using the D-line of sodium.

spectral line A particular wavelength of light emitted or absorbed by an atom, ion, or molecule.

spectrograph An instrument for producing a photographic record of a spectrum.

spectrographic analysis A method of analysis in which the sample is excited electrically (by an arc or spark) and emits radiation characteristic of its component atoms. This radiation is passed through a slit, dispersed by a prism or a grating, and recorded as a spectrum, either photographically or photoelectrically. The photographic method was widely used for qualitative and semiquantitative work but photoelectric detection also allows wide quantitative application.

spectrometer **1.** An instrument for examining the different wavelengths present in electromagnetic radiation. Typically, spectrometers have a source of radiation, which is collimated by a system of lenses and/or slits. The radiation is dispersed by a prism or grating, and recorded photographically or by a photocell. There are

many types for producing and investigating spectra over the whole range of the electromagnetic spectrum. Often spectrometers are called *spectroscopes*.
2. Any of various other instruments for analyzing the energies, masses, etc., of particles.

spectrophotometer A form of spectrometer able to measure the intensity of radiation at different wavelengths in a spectrum, usually in the visible, infrared, or ultraviolet regions.

spectroscope An instrument for examining the different wavelengths present in electromagnetic radiation. *See also* spectrometer.

spectroscopy **1.** The production and analysis of spectra. There are many spectroscopic techniques designed for investigating the electromagnetic radiation emitted or absorbed by substances. Spectroscopy, in various forms, is used for analysis of mixtures, for identifying and determining the structures of chemical compounds, and for investigating energy levels in atoms, ions, and molecules. In the visible and longer wavelength ultraviolet, transitions correspond to electronic energy levels in atoms and molecules. The shorter wavelength ultraviolet corresponds to transitions in ions. In the x-ray region, transitions in the inner shells of atoms or ions are involved. The infrared region corresponds to vibrational changes in molecules, with rotational changes at longer wavelengths.
2. Any of various techniques for analysing the energy spectra of beams of particles or for determining mass spectra.

spectrum (*pl.* **spectra**) **1.** A range of electromagnetic radiation emitted or absorbed by a substance under particular circumstances. In an *emission spectrum*, light or other radiation emitted by the body is analyzed to determine the particular wavelengths produced. The emission of radiation may be induced by a variety of methods; for example, by high temperature, bombardment by electrons, absorption of higher-frequency radiation, etc. In an *absorption spectrum* a continuous flow of radiation is passed through the sample. The radiation is then analyzed to determine which wavelengths are absorbed.
2. In general, any distribution of a property. For instance, a beam of particles may have a spectrum of energies. A beam of ions may have a mass spectrum (the distribution of masses of ions).

sperm *See* spermatozoon.

spermatozoon (**sperm**) The small motile mature male reproductive cell (gamete) formed in the testis. It differs in form and size between species; in man it is about 52–62 μm long and comprises a head region containing a haploid nucleus, a middle region containing mitochondria, and a long tail region containing an undulipodium (flagellum). It is covered by a small amount of cytoplasm and a plasma membrane. They remain inactive until they pass from the testis during coitus, when secretions from the prostate gland and seminal vesicles stimulate undulating movements to pass along the tail and effect locomotion. About 200–300 million spermatozoa may be released in a single ejaculation, although only one may fertilize each ovum.

S phase *See* cell cycle.

spindle The structure formed during MITOSIS and MEIOSIS that is responsible for moving the chromatids and chromosomes to opposite poles of the cell. The spindle consists of a longitudinally orientated system of protein microtubules whose synthesis starts late in interphase under the control of a microtubule-organizing center. In plants and animals this is the centrosome. A special region of the centromeres of each pair of sister chromatids, the *kinetochore*, becomes attached to one or a bundle of spindle microtubules. During anaphase, the kinetochore itself acts as the motor, dissassembling the attached microtubules and hauling the chromatid towards the spindle pole. Later in anaphase, the unattached interpolar microtubules actively

COMMON STAINS FOR LIGHT MICROSCOPY

Stains	*Final Color*	*Suitable for*
aniline (cotton) blue	blue	fungal hyphae and spores
aniline sulphate or hydrochloride	yellow	lignin
borax carmine	pink	nuclei; particularly for whole amounts (large pieces) of animal material
eosin	pink	cytoplasm; *see* hematoxylin
	red	cellulose
Feulgen's stain	red/purple	DNA; particularly to show chromosomes during cell division
hematoxylin	blue	nuclei; mainly used for sections of animal tissue with eosin as counterstain for cytoplasm; also for smears
iodine	blue–black	starch; therefore for plant storage organs
Leishman's stain	red–pink	blood cells
	blue	white blood cell nuclei
light green or fast green	green	cytoplasm and cellulose; *see* safranin
methylene blue	blue	nuclei; suitable as a vital stain
phloroglucinol	red	lignin
safranin	red	nuclei. Lignin and suberin. Mainly used for sections of plant tissue with light green as counterstain for cytoplasm

slide past each other, elongating the entire spindle.

spleen A lymphoid organ situated just beneath the stomach in vertebrates. It produces LYMPHOCYTES and destroys and stores red blood cells. The spleen consists of loose connective tissue containing lymphoid tissue (Malpighian bodies), which surrounds a network of sinuses. The circulation of blood through the spleen is slow and blood can leak out of the sinuses into the lymphoid tissue. Thus, there is ample opportunity for phagocytosis of red blood cells and bacteria.

staggered conformation *See* conformation.

staining A procedure that is designed to heighten contrast between different structures. Normally biological material is lacking in contrast, protoplasm being transparent, and therefore staining is essential for an understanding of structure at the microscopic level. *Vital stains* are used to stain and examine living material. Most stains require dead or nonliving material. Staining is done after fixation and either during or after dehydration. *Double staining* involves the use of two stains; the second is called the *counterstain. Acidic stains* have a colored anion, *basic stains* have a colored cation. Some stains are neutral. Materials can be described as *acidophilic* or *basophilic* depending on whether they are stained by acidic or basic dyes respectively. Basic stains are suitable for nuclei, staining DNA. Stains for light microscopy are colored dyes; those for electron microscopy contain heavy metals, e.g. uranyl acetate, lead citrate, and osmium tetroxide.

Stanley, Wendell Meredith (1904–71) American biochemist. He was awarded the Nobel Prize for chemistry in 1946 jointly with J. H. Northrop for their preparation of enzymes and virus proteins in a pure form. The prize was shared with J. B. Sumner.

Staphylococcus A genus of Gram-positive spherical nonmotile bacteria. They are facultative anaerobes and do not form spores. Many species are parasites or pathogens of animals and some cause wound infections, abscesses, and a type of food poisoning. They are killed by pasteurization and many common disinfectants.

starch A polysaccharide that occurs exclusively in plants. Starches are extracted commercially from maize, wheat, barley, rice, potatoes, and sorghum. They exist in the plant cells as granules dispersed in the cytoplasm. The starches are storage reservoirs for plants; they can be broken down by enzymes to simple sugars and then metabolized to supply energy needs. Starch is a dietary component of animals. In humans it is digested by salivary and pancreatic amylase then further degraded by maltase to yield glucose, which may be stored as glycogen (animal starch). Excess starch, i.e. above the maximum liver and muscle storage capacity, is converted to lipids and stored as fat. Starch is composed of amylose (water-soluble, blue color with iodine) and amylopectin (not water-soluble, violet color with iodine). The composition is amylose 10–20%, amylopectin 80–90%.

steam distillation A method of isolating or purifying substances by lowering the boiling point of the mixture. When two immiscible liquids are distilled, the boiling point will be lower than that of the more volatile component and consequently will be below 100°C if one component is water. The method is particularly useful for recovering materials from tarry mixtures.

stearic acid (**octadecanoic acid**) A saturated carboxylic acid, $(CH_2)_{16}COOH$, which is widely distributed in nature as the glyceride ester. It is present in most fats and oils of animal and vegetable origin, particularly the so-called hard fats, i.e. those of higher melting point.

Stein, William Howard (1911–80) American protein chemist. He was awarded the Nobel Prize for chemistry jointly with S. Moore in 1972 for their contribution to the understanding of the connection between chemical structure and catalytic activity of the active center of the ribonuclease molecule. The prize was shared with C. B. Anfinsen.

stem cell Any cell that is not terminally differentiated and capable of unlimited division in order to provide new cells for growth or replacement of tissues. After division it may produce more stem cells or cells that differentiate into specialized cells.

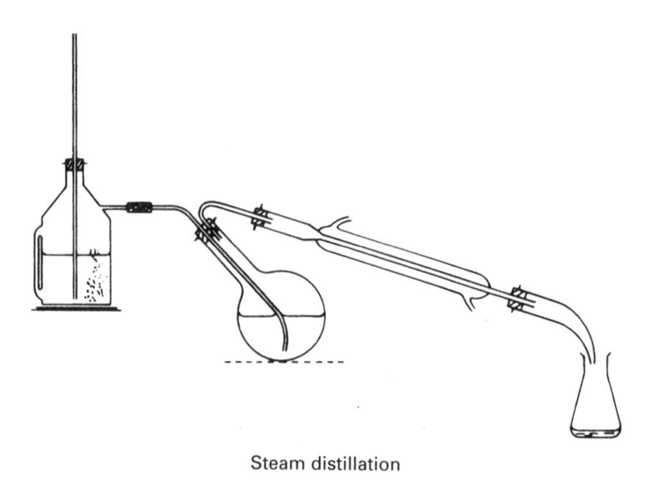

Steam distillation

Stem cells are found in adults and embryos but have different properties. Adult stem cells are found in the bone marrow and epithelial cells of the skin. They are partially differentiated and have limited potential for further differentiation. Embryonic stem cells are found in early stage embryos. They are undifferentiated and can produce cells that can differentiate into any other type of cell. Embryonic stem cells therefore have much greater potential but their use in research and medicine is highly controversial. *See* totipotency.

stenohaline Describing organisms that are unable to tolerate wide variations of salt concentrations in the environment. *Compare* euryhaline.

stereochemistry The branch of chemistry concerned with the shapes of molecules and the way these affect the chemical properties.

stereoisomerism *See* isomerism.

stereospecific Describing a chemical reaction involving an asymmetric atom (usually carbon) that results in only one geometric isomer. *See* isomerism; optical activity.

steric effect An effect in which the shape of a molecule influences its reactions. A particular example occurs in molecules containing large groups, which hinder the approach of a reactant (*steric hindrance*).

steric hindrance *See* steric effect.

steroid Any member of a group of compounds having a complex basic ring structure. Examples are corticosteroid hormones (produced by the adrenal gland), sex hormones (progesterone, androgens, and estrogens), bile acids, and sterols (such as cholesterol). *See also* anabolic steroid; sterol.

sterol A steroid with long aliphatic side chains (8–10 carbons) and at least one hydroxyl group. They are lipid-soluble and often occur in membranes (e.g. cholesterol and ergosterol).

stimulus A change in the external or internal environment of an organism that elicits a response in the organism. The stimulus does not provide the energy for the response.

stoichiometry The proportions in which elements form compounds. A *stoichiometric compound* is one in which the atoms have combined in small whole numbers.

stomach The part of the alimentary canal in vertebrates that lies between the esophagus and the duodenum. It acts as a storage organ so that food can be eaten at intervals instead of continuously. It is a large sac with thick muscular walls and is closed at each end by a ring of muscle (sphincter). It expands to hold a meal for several hours, during which time the food is churned by muscular contractions, mixed with hydrochloric acid, and the protein in it is partly digested by the enzymes of gastric juice. When the food has been reduced to semiliquid chyme, it is passed, a little at a time, through the pylorus to the duodenum.

The stomach is lined with mucous membrane containing simple tubular glands (gastric pits). These contain oxyntic cells, which secrete hydrochloric acid; peptic or chief cells, which secrete the enzymes pepsin and rennin; and goblet cells, which secrete mucus.

In birds, the posterior part of the stomach is the gizzard. In some herbivores, there are several compartments.

Strecker synthesis A method of synthesizing amino acids in the laboratory. Hydrogen cyanide forms an addition product (cyanohydrin) with aldehydes:

$$RCHO + HCN \rightarrow RCH(CN)OH$$

With ammonia further substitution occurs:

$$RCH(CN)OH + NH_3 \rightarrow RCH(CN)(NH_2) + H_2O$$

Acid hydrolysis of the cyanide group then produces the amino acid

$RCH(CN)(NH_2) \rightarrow RCH(COOH)(NH_2)$
The method is a general synthesis for α-amino acids (with the $-NH_2$ and $-COOH$ groups on the same carbon atom).

Streptococcus A genus of spherical Gram-positive bacteria that usually occur in pairs or chains; most strains are non-motile. Most species are parasites or pathogens of animals, often occurring in the respiratory or alimentary tracts. Some species are hemolytic (i.e. they destroy red blood cells) and cause such diseases as scarlet fever and rheumatic fever. Streptococci are killed by pasteurization and common disinfectants; penicillin, tetracycline, and other antibiotics are effective against hemolytic strains.

striated muscle *See* skeletal muscle.

stroma **1.** (*Botany*) The space between the inner membrane and the thylakoids (*see* granum) inside a chloroplast. It contains a solution of the enzymes and other components that catalyze the fixation of carbon dioxide and formation of starch (*see* photosynthesis).
2. (*Botany*) A mass of fungal hyphae, sometimes including host tissue, in which fruiting bodies may be produced. An example is the compact black fruiting body of the ergot fungus, *Claviceps purpurea*.
3. (*Zoology*) A tissue that acts as a framework; for example, the connective tissue framework of the ovary or testis that surrounds the cells concerned with gamete production.

structural formula The formula of a compound showing the numbers and types of atoms present, together with the way in which these are arranged in the molecule. Often this can be done by grouping the atoms, as in the structural formula of ethanoic acid $CH_3.CO.OH$.

structural isomerism *See* isomerism.

strychnine A poisonous alkaloid found in certain plants.

suberin A mixture of substances produced in the walls of cork tissue. It is similar in properties and functions to cutin. The Casparian band in roots and some stems contains suberin, lignin, or similar substances.

substituent An atom or group substituted for another in a compound. Often the term is used for groups that have replaced hydrogen in organic compounds; for example, in benzene derivatives.

substitution reaction A reaction in which an atom or group of atoms in an organic molecule is replaced by another atom or group. The substitution of a hydrogen atom in an alkane by a chlorine atom is an example.

The term 'substitution' is very general and several reactions that can be considered as substitutions are more normally given special names (e.g. esterification, hydrolysis, and nitration).

substrate **1.** The substance upon which an enzyme acts.
2. The nonliving material upon which an organism lives or grows.

succinic acid A dicarboxylic acid formed by fermentation of sugars. It occurs in algae, lichens, sugars, and other plant substances. Succinate ions have an important role in the Krebs cycle.

succus entericus The viscous alkaline secretion produced by Brunner's glands in the wall of the duodenum. It consists of water, hydrogencarbonate ions, and mucoproteins, and serves to protect the walls of the small intestine from the corrosive effects of the acidic and proteolytic chyme entering from the stomach. It was formerly thought to contain digestive enzymes, but all duodenal enzymes are now known to be localized on or within the epithelial cells.

sucrose (**cane sugar**) A sugar that occurs in many plants. It is extracted commercially from sugar cane and sugar beet. Sucrose is a disaccharide formed from a glucose unit and a fructose unit. It is hydrolyzed to a mixture of fructose and glucose by the enzyme invertase. Since this mixture has a different optical rotation (levorotatory) from the original sucrose, the mixture is called *invert sugar*.

suction pressure (SP) *See* osmosis.

sugar (saccharide) One of a class of sweet-tasting carbohydrates that are soluble in water. Sugar molecules consist of linked carbon atoms with –OH groups attached, and either an aldehyde or ketone group. The simplest sugars are the *monosaccharides*, such as glucose and fructose, which cannot be hydrolyzed to sugars with fewer carbon atoms. They can exist in a chain form or in a ring formed by reaction of the ketone or aldehyde group with an –OH group on one of the carbons at the other end of the chain. It is possible to have a six-membered (*pyranose*) ring or a five-membered (*furanose*) ring. Monosaccharides are classified according to the number of carbon atoms: a *pentose* has five carbon atoms and a *hexose* six. Monosaccharides with aldehyde groups are *aldoses*; those with ketone groups are *ketoses*. Thus, an *aldohexose* is a hexose with an aldehyde

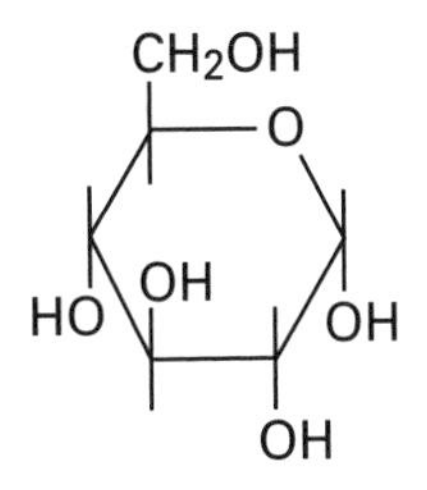

glucose – a pyranose ring

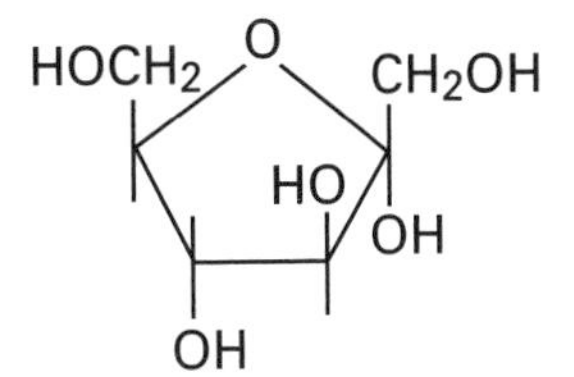

fructose – in a furanose ring form

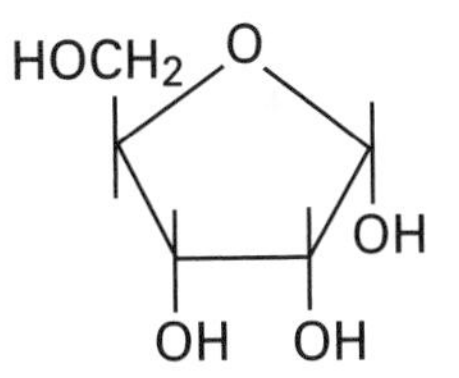

ribose – a pyranose ring

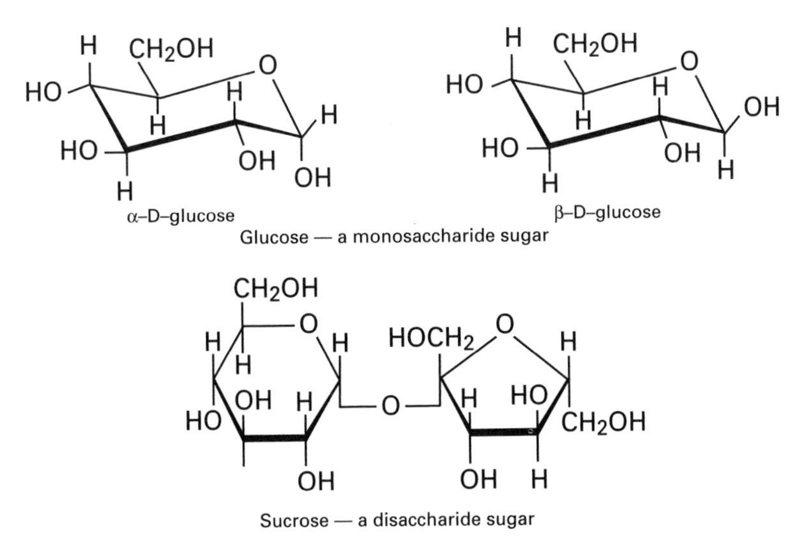

Glucose — a monosaccharide sugar

Sucrose — a disaccharide sugar

Sugar

group; a *ketopentose* is a pentose with a ketone group, etc.

Two or more monosaccharide units can be linked in *disaccharides* (e.g. sucrose), *trisaccharides*, etc. *See also* fructose; glucose; polysaccharide; sucrose.

sugar acid An acid formed from a monosaccharide by oxidation. Oxidation of the aldehyde group (CHO) of the aldose monosaccharides to a carboxyl group (COOH) gives an *aldonic acid*; oxidation of the primary alcohol group (CH_2OH) to COOH yields *uronic acid*; oxidation of both the primary alcohol and carboxyl groups gives an *aldaric acid*. The uronic acids are biologically important, being components of many polysaccharides, for example glucuronic acid (from glucose) is a major component of gums and cell walls, while galacturonic acid (from galactose) makes up pectin. Ascorbic acid or vitamin C is an important sugar acid found universally in plant tissues, particularly in citrus fruits.

sugar alcohol (**alditol**) An alcohol derived from a monosaccharide by reduction of its carbonyl group (CO) so that each carbon atom of the sugar has an alcohol group (OH). For example, glucose yields sorbitol, common in fruits, and mannose yields mannitol.

sulfonamide One of a group of bacteriostatic drugs having a sulfonamide group (SO_2NH_2). The bacteriostatic action is believed to be due to the similarity in chemical structure of sulfonamides to para-aminobenzoic acid, an essential growth substance in some bacteria. The bacteria are unable to distinguish the two and take up the sulfonamide if it is present in higher concentration, which prevents development and reproduction. Examples of sulfonamides are sulfanilamide, sulfafurazole, and sulfamerazine.

sulfonation A reaction introducing the $-SO_2OH$ (sulfonic acid) group into an organic compound. Sulfonation of aromatic compounds is usually accomplished by refluxing with concentrated sulfuric acid for several hours. The attacking species is SO_3 (sulfur(VI) oxide) and the reaction is an example of electrophilic substitution.

sulfonic acid A type of organic compound containing the $-SO_2.OH$ group. The simplest example is benzenesulfonic acid ($C_6H_5SO_2OH$). Sulfonic acids are strong acids. Electrophilic substitution can introduce other groups onto the benzene ring; the $-SO_2.OH$ group directs substituents into the 3-position.

sulfonium compound An organic compound of general formula R_3SX, where R is an organic radical and X is an electronegative radical or element; it contains the ion R_3S^+. An example is diethylmethylsulfonium chloride, $(C_2H_5)_2.CH_3.S^+Cl^-$, made by reacting diethyl sulfide with chloromethane.

sulfoxide An organic compound of general formula RSOR′, where R and R′ are organic radicals. An example is dimethyl sulfoxide, $(CH_3)_2SO$, commonly used as a solvent.

sulfur An essential element in living tissues, being contained in the amino acids cysteine and methionine and hence in nearly all proteins. Sulfur atoms are also found bound with iron in ferredoxin, one of the components of the electron transport chain in photosynthesis. Plants take up sulfur from the soil as the sulfate ion SO_4^{2-}. The sulfides released by decay of organic matter are oxidized to sulfur by sulfur bacteria of the genera *Chromatium* and *Chlorobium*, and further oxidized to sulfates by bacteria of the genus *Thiobacillus*. There is thus a cycling of sulfur in nature.

sulfur bacteria Filamentous autotrophic chemosynthetic bacteria that derive energy by oxidizing sulfides to elemental sulfur and build up carbohydrates from carbon dioxide. An example is *Beggiatoa*. *See also* photosynthetic bacteria.

summation 1. The additive effect of several impulses arriving at a synapse of a nerve and/or muscle cell, when individually

the impulse cannot evoke a response. The impulses either arrive simultaneously at different synapses at the same cell (*spatial summation*) or in succession at one synapse (*temporal summation*). Stimulation of the synapse elicits a graded postsynaptic potential and if the potential exceeds the threshold level, a postsynaptic impulse is triggered. Summation is one of the major mechanisms of integration in the nervous system. *Compare* facilitation.
2. The interaction of two substances with similar effects in a given system, such that the combined effect is greater than their separate effects.

Sumner, James Batcheller (1887–1995) American biochemist who discovered that enzymes can be crystallized. He was awarded the Nobel Prize for chemistry in 1946. The prize was shared with J. H. Northrop and W. M. Stanley.

supergene A collection of closely linked genes that tend to behave as a single unit because crossing over between them is very rare.

Sutherland, Earl Wilbur Jr (1915–74) American biochemist who worked on the mechanisms of hormones and discovered cyclic AMP. He was awarded the Nobel Prize for physiology or medicine in 1971.

Svedberg, Theodor (1884–1971) Swedish colloid chemist noted for his invention (1923) of the ultracentrifuge. He was awarded the Nobel Prize for chemistry in 1926.

sympathetic nervous system (**thoracolumbar nervous system**) One of the two divisions of the autonomic nervous system, which supplies motor nerves to the smooth muscles of internal organs and to heart muscle. Sympathetic nerve fibers arise via spinal nerves in the thoracic and lumbar regions. Their endings release mainly norepinephrine, which increases heart rate and breathing rate, raises blood pressure, and slows digestive processes, thereby preparing the body for 'fight or flight' and antagonizing the effects of the parasympathetic nervous system. The medulla of the adrenal gland is supplied only by sympathetic fibers, which trigger the release of epinephrine into the bloodstream, thus enhancing the effects of the sympathetic system. *See also* autonomic nervous system; parasympathetic nervous system.

synapse The junction between two neurones or between a neurone and a muscle cell across which nerve impulses can be transmitted. Synapses occur between the knoblike axon endings of one neurone and the dendrites or cell body of another. One neurone may have many synapses with other neurones. Each synapse consists of adjacent specialized regions in the cell membranes of both neurones, separated by a narrow gap (synaptic cleft).

A nerve impulse arriving at the axon ending of the presynaptic cell causes small vesicles to release a chemical (neurotransmitter), which diffuses across the cleft and combines with receptor sites in the cell membrane of the postsynaptic cell. Depending on the neurones involved, this may act either to start a nerve impulse in the postsynaptic cell (excitation) or to prevent impulses from other neurones being transmitted (inhibition). Most synapses will only transmit nerve impulses in one direction. *See* facilitation; neurotransmitter; summation.

synapsis (**pairing**) The association of homologous chromosomes during the prophase stage of meiosis that leads to the production of a haploid number of bivalents. Homologous chromosomes pair point to point so that corresponding regions lie in contact.

syncytium An area of animal cytoplasm containing many nuclei, the whole being bounded by a continuous cell membrane. This gives rise to a multinucleate condition. The term may be applied to an area of cytoplasm partially divided by membranes into discrete cells but with extensive cytoplasmic continuity. Such structures are to be found in striped and cardiac muscle, insect eggs, and some protoctists.

synergism **1.** The interaction of two substances, e.g. drugs or hormones, which have similar effects in a given system, such that the effect produced is greater than the sum of their separate effects.
2. The coordinated action of muscles to produce a particular movement.

syngamy *See* fertilization.

Synge, Richard Laurence Millington (1914–94) British biochemist and peptide chemist. He was awarded the Nobel Prize for chemistry in 1952 jointly with A. J. P. Martin for their invention of partition chromatography.

syngraft *See* graft.

synovial membrane The membrane that forms the capsule surrounding a joint. It consists of tough connective tissue with a high proportion of white collagen fibers and it secretes the viscous lubricating *synovial fluid*.

Szent-Györgyi, Albert (1893–1986) Hungarian-born American biochemist distinguished for his work in a variety of fields, including the isolation of ascorbic acid, the role of dicarboxylic acids in oxidative metabolism, and the biochemistry of muscle contraction. He was awarded the Nobel Prize for physiology or medicine in 1937.

tactic movement *See* taxis.

tannin One of a mixed group of substances that, as defined by industry, combine with hide to form leather. Tannins are also used in dyeing and ink manufacture. Many plants accumulate tannins, particularly in leaves, fruits, seed coats, bark, and heartwood. Their astringent taste may deter animals from eating the plant and they may discourage infection. Tannins precipitate proteins and hence inactivate enzymes; they are therefore segregated in cell vacuoles, organelles, or cell walls. Chemically they are polymers derived either from carbohydrates and phenolic acids by condensation reactions, or from flavonoids.

tartaric acid A crystalline hydroxy carboxylic acid with the formula:

$$HOOC(CHOH)_2COOH$$

Its systematic name is 2,3-dihydroxybutanedioic acid. *See also* optical activity.

tartrate A salt or ester of tartaric acid.

taste bud A small bulblike group of chemical receptor cells in vertebrates that is responsible for the sense of taste. In terrestrial vertebrates, taste buds are usually embedded in small projections (papillae) of the epithelium of the throat and mouth, especially the tongue. Chemical substances in solution stimulate the cells to send nerve impulses to the brain for interpretation as taste. Humans are considered to have four kinds of taste buds, which distinguish sweet, sour, salt, and bitter chemicals. Aquatic vertebrates may have taste buds anywhere on the body surface.

Tatum, Edward Lawrie (1909–75) American microbiologist. He was awarded the Nobel Prize for physiology or medicine in 1958 jointly with G. W. Beadle for their discovery that genes act by regulating definite chemical events. The prize was shared with J. Lederberg.

tautomerism Isomerism in which each isomer can convert into the other, so that the two isomers are in equilibrium. The isomers are called *tautomers*. Tautomerism often results from the migration of a hydrogen atom. *See* keto–enol tautomerism.

taxis (**tactic movement**) Movement of an entire cell or organism (i.e. locomotion) in response to an external stimulus, in which the direction of movement is dictated by the direction of the stimulus. Movement toward the stimulus is positive taxis and away from the stimulus is negative taxis. It is achieved by protoplasmic streaming, extrusion of cell substances, or by locomotory appendages, such as cilia and flagella. *See* aerotaxis; phototaxis. *See also* nastic movements; tropism.

taxonomy The area of systematics that covers the principles and procedures of classification.

TCA cycle (**tricarboxylic acid cycle**) *See* Krebs cycle.

T-cell (**T-lymphocyte**) A LYMPHOCYTE involved primarily in cell-mediated immunity. Like B-cells, T-cells originate in the bone marrow, but during embryological development they migrate to the thymus where they mature and differentiate (B-cells are believed to mature in the bone marrow). There are several types of T-cell, which act collaboratively in identifying and destroying virus-infected body cells, and possibly cancerous cells. All of them possess on their surface special receptors, T-cell receptors, that are structurally related to the antibody molecules secreted by B-cells, and that bind to specific viral antigens on the surface of infected body cells. Unlike antibodies, they do not bind soluble antigens. A T-cell will generally only recognize its specific antigen if the antigen is bound to a marker protein (an MHC protein; *see* major histocompatibility complex) on the surface of a cell. Binding of the T-cell receptor to the MHC–antigen complex acts as the stimulus for the T-cell's rapid division, to form a clone with four types of T-cell.

Cytotoxic T-cells (T_C cells) recognize other body cells that display the same antigens on their surface, and lyse the target cells by releasing a protein called perforin. This makes a hole in the cell membrane of the infected cell, causing it to burst. Hence, viral replication by that cell ceases. *Helper cells* (T_H cells) release substances, called lymphokines, that stimulate the growth and maturation of both T_C cells and B-cells. In fact, T_H cells are essential for antibody production by B-cells, but will only release lymphokines if their receptor sites bind to an antigen-MHC complex on the surface of *antigen-presenting cells*. These are cells, such as macrophages and B-cells, that possess class II MHC proteins on their surface. T_H cells are destroyed by HIV to produce the symptoms of AIDS. A third type of T-cell, the *T-suppressor cell* (T_S cell) suppresses the activities of T_H cells and thus limits antibody production by B-cells. *See* AIDS; B-cell; immunity; killer cell.

telomere The structure found at the ends of linear chromosomes. It consists of simple tandemly repeated sequences with one strand being G rich and the other C rich. The repeated sequences are added by the enzyme *telomerase* and exist to maintain chromosome length and prevent the loss of coding sequences.

telophase The final stage in mitosis and meiosis before cells enter interphase. During this stage chromosomes uncoil and disperse, the nuclear spindle degenerates, and a new nuclear membrane forms. The cytoplasm may also divide during this phase.

Temin, Howard Martin (1934–94) American virologist. He was awarded the Nobel Prize for physiology or medicine in 1975 jointly with D. Baltimore and R. Dulbecco for their discoveries concerning the interaction between tumor viruses and the genetic material of the cell.

temperate phage A bacteriophage that becomes integrated into the bacterial DNA and multiplies with it, rather than replicating independently and causing lysis of the bacterium. *See* phage. *Compare* virulent phage.

tendon The tough nonelastic connective tissue that joins a muscle to a bone. It consists of a mass of parallel white collagen fibers, which are continuous with those of the muscle sheath (epimysium) and the periosteum of the bone. When the muscle contracts, the tendon pulls on the bone, causing movement at the joint.

tera- Symbol: T A prefix denoting 10^{12}. For example, 1 terawatt (TW) = 10^{12} watts (W).

teratogen Any environmental factor that causes physical defects (teratomas) in a fetus. Teratogens include various drugs (e.g. thalidomide), infections (e.g. German measles), and irradiation. Teratogens interfere with essential growth mechanisms, causing arrested or distorted growth; the fetus is particularly sensitive during the first two months, when rudimentary growth patterns are being established. In later life, when growth patterns are well established, teratogens have no effect.

teratology The study of plant and animal abnormalities (*teratomas*).

terminalization The movement of *chiasmata* to the end of the bivalent arms, a process that may occur during late prophase I of meiosis. The chiasmata can slip off the ends of the bivalents, and thus chiasma frequency may be reduced by terminalization.

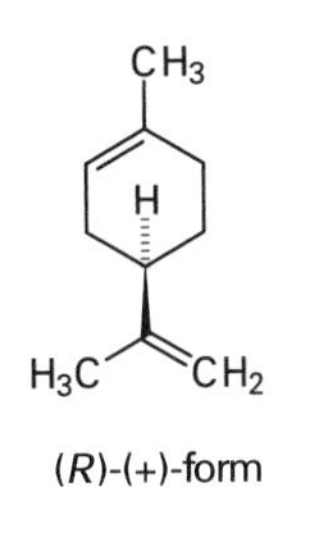

(*R*)-(+)-form

Limonene (a terpene)

terpene One of a complex group of lipids based on the hydrocarbon skeleton C_5H_8 (isoprene). *Monoterpenes* are built from two C_5 residues ($C_{10}H_{16}$), *diterpenes* from four, etc. The C_{10} to C_{20} terpenes are present in essential oils, giving the characteristic scent of some plants (e.g. mint). Some terpenoid substances are physiologically active, e.g. vitamin A.

testis (**testicle**) The male reproductive organ of animals, which produces spermatozoa. In vertebrates there is a pair of testes, which also produce sex hormones (androgens). Each is comprised of a fibrous capsule (the *tunica albuginea*), which surrounds a mass of seminiferous tubules separated into compartments by fibrous tissue. Between the tubules lie interstitial (or Leydig's) cells, which produce androgens. Spermatozoa are continuously produced within the seminiferous tubules from the onset of sexual maturity to the age of about 70. They migrate via efferent ducts to the epididymis for temporary storage.

testosterone A naturally occurring androgen secreted by the testis under the influence of luteinizing hormone. Its secretion during adult life is responsible for the development, function, and maintenance of secondary male sexual characteristics, male sex organs, and spermatogenesis. Testosterone is also secreted from the adrenal cortex and the ovaries. It is metabolized in the liver, its metabolites (e.g. androsterone) being excreted in the urine. *See* androgen.

tetrad **1.** A group of four spores formed as a result of meiosis in a spore mother cell. **2.** In MEIOSIS, the association of four homologous chromatids seen during the pachytene stage of prophase.

tetraploid A cell or organism containing four times the haploid number of chromosomes. Tetraploid organisms may arise by the fusion of two diploid gametes that have resulted from the NONDISJUNCTION of chromosomes at meiosis. Tetraploids may also arise through nondisjunction of the chromatids during the mitotic division of a zygote. *See also* polyploid.

tetrapyrrole A structure of four pyrrole molecules linked together, found in heme, chlorophyll, and other compounds. Usually a metal ion is coordinated to the four nitrogen atoms on the pyrrole rings. *See* porphyrin.

theobromine An alkaloid found in the cacao bean. Its action is similar to caffeine. The systematic name is *3,7-dimethylxanthine*.

theophylline An alkaloid similar in action to caffeine. Its systematic name is *1,3-dimethylxanthine*.

Theorell, Axel Hugo Theodor (1903–82) Swedish biochemist. He was awarded the Nobel Prize for physiology or medicine in 1955 for his discoveries concerning the nature and mode of action of oxidation enzymes.

thermophilic Describing microorganisms that require high temperatures (around 60°C) for growth. It is exhibited by certain bacteria that grow in hot springs or compost and manure. *Compare* mesophilic; psychrophilic.

thiamine (**vitamin** B_1) One of the water-soluble B-group of vitamins. Good sources of thiamine are unrefined cereal grains, liver, heart, and kidney. Thiamine deficiency predominantly affects the peripheral nervous system, the gastrointestinal tract, and the cardiovascular system. Thiamine has been shown to be of value in the treatment of beriberi. Thiamine, in the form of thiamine diphosphate, is the coenzyme for the decarboxylation of acids such as pyruvic acid. *See also* vitamin B complex.

thin-layer chromatography (TLC) A chromatographic method in which a glass plate is covered with a thin layer of inert absorbent material (e.g. cellulose or silica gel) and the materials to be analyzed are spotted near the lower edge of the plate. The base of the plate is then placed in a solvent, which rises up the plate by capillary action, separating the constituents of the mixtures. The principles involved are similar to those of paper chromatography and, like paper chromatography, two-dimensional methods can also be employed.

threonine *See* amino acids.

threshold The minimum stimulus intensity that will initiate a response in an irritable tissue, such as a muscle or nerve cell.

thrombin An enzyme that converts the soluble protein fibrinogen into the fibrous fibrin during blood clotting. It is formed from prothrombin under the influence of thromboplastin, calcium ions, and other clotting factors, which are activated when blood is removed from the circulation, usually by injury. *See also* blood clotting.

thrombocyte *See* platelet.

thymidine The NUCLEOSIDE formed when thymine is linked to D-ribose by a β-glycosidic bond.

thymine A nitrogenous base found in DNA. It has a PYRIMIDINE ring structure.

thymus gland A gland consisting of two lobes, which lie in the lower part of the neck and upper part of the chest. The lobules, which are found in each lobe, are composed of an outer cortex and an inner medulla portion. The thymus controls lymphoid tissues in the body and is a source of immunological activity. It is large in the young and involved in the production of lymphocytes. It degenerates after the animal reaches sexual maturity.

thyroid gland A gland situated in the neck, consisting of two lateral lobes on either side of the trachea and larynx, giving the gland a butterfly appearance. Its main function is the regulation of metabolic rate by production of thyroid hormones.

thyroid hormones Hormones produced by the thyroid gland, which increase cell metabolism. The most important is THYROXINE. *See also* triiodothyronine.

thyrotropin (**thyroid-stimulating hormone**) A hormone, produced by the anterior pituitary gland, that stimulates the thyroid gland to release thyroxine. The level of thyroxine controls thyrotropin release by a negative feedback mechanism.

thyroxine (**thyroid hormone**) An iodine-containing polypeptide hormone that is secreted by the thyroid gland and is essential for normal cell metabolism. Its many effects include increasing oxygen consumption and energy production. It is used therapeutically to treat hypothyroidism.

Tiselius, Arne Wilhelm Kaurin (1902–71) Swedish chemist who worked on serum proteins. He was awarded the Nobel Prize for chemistry in 1948 for his research on electrophoresis and adsorption analysis.

tissue A group of cells that is specialized for a particular function. Examples are connective tissue, muscular tissue, and nervous tissue. Several different tissues are often incorporated in the structure of each organ of the body.

tissue culture The growth of cells, tissues, or organs in suitable media *in vitro*. Such media must normally be sterile, correctly pH balanced, and contain all the necessary micro- and macronutrients, carbohydrates, vitamins, and hormones for growth. Studies of such cultures have shed light on physiological processes that would be difficult to follow in the living organism. The cytokinins were discovered through work on tobacco-pith tissue culture.

Plant tissue-culture techniques also have important practical applications, enabling, for example, large-scale multiplication of plants by micropropagation and the generation of disease-free material by meristem-tip culture. They are also vital to the development of genetic engineering.

titer A measure of the concentration of an antibody in serum. It is estimated from the highest dilution that will still produce a detectable reaction with the appropriate antigen and is expressed as the reciprocal of this dilution.

TLC *See* thin-layer chromatography.

tocopherol *See* vitamin E.

Todd, Alexander Robertus Todd, Lord Todd (1907–97) British organic chemist. He was awarded the Nobel Prize for chemistry in 1957 for his work on nucleotides and nucleotide coenzymes.

tolerance *See* immunological tolerance.

tonoplast The membrane that surrounds the large central vacuole of plant cells.

tonsils Small bodies of lymphoid tissue in tetrapods, situated near the back of the throat. They produce lymphocytes and are therefore concerned with defense against bacterial invasion.

totipotency The ability to form all the types of cells and tissues that constitute the mature organism. Embryonic stem cells are totipoent. Cells that can form a limited number of mature cells, e.g. bone marrow, are called *pluripotent*. Given the appropriate conditions, some fully differentiated cells can exhibit totipotency, e.g. the formation of adventitious embryos in carrot tissue cultures. This demonstrates that each cell retains the full genetic potential of the species. The cloning of Dolly (*see* clone) shows that theoretically mature cells are totipotent but we do not yet know how to access their potential. We can only do this for embryonic stem cells.

toxin A chemical produced by a pathogen (e.g. bacteria, fungi) that causes damage to a host cell in very low concentrations. Toxins are often similar to the enzymes of the host and interfere with the appropriate enzyme systems. *See also* endotoxins; exotoxins.

trace element An element required in trace amounts (a few parts per million of food intake) by an organism for health. *See table overleaf.*

trans- Designating an isomer with groups that are on opposite sides of a bond or structure. *See* isomerism.

transaminase An enzyme that catalyzes TRANSAMINATION.

transamination The transfer of an amino group from an amino acid to an α-keto acid, producing a new α-keto acid and a new amino acid. This is catalyzed by a *transaminase* enzyme in conjunction with the coenzyme pyridoxal phosphate. The amino group becomes attached to the coenzyme to form pyridoxamine phosphate, and is then transferred to the α-keto acid, which is usually pyruvic acid, oxaloacetic acid, or α-ketoglutaric acid. Transamination is important in amino acid synthesis and in the ornithine cycle.

transcription The process in living cells whereby RNA is synthesized according to the template embodied in the base sequence of DNA, thereby converting the cell's genetic information into a coded message (MESSENGER RNA, mRNA) for the assembly of proteins, or into the RNA components required for protein synthesis (ribosomal RNA and transfer RNA). The term is also applied to the formation of single-stranded DNA from an RNA template, as performed by the enzyme reverse transcriptase, for example in retrovirus infections. Details of DNA transcription differ between prokaryote and eukaryote cells, but essentially it involves the following steps. Firstly, the double helix of the DNA molecule is unwound in the region of the site marking the start of transcription for a particular gene. The enzyme RNA polymerase moves along one of the DNA strands, the transcribed strand (or noncoding strand, since the code is carried by the complementary base sequence of the RNA), and nucleotides are assembled to form a complementary RNA molecule. The POLYMERASE enzyme proceeds until reaching a stop signal, when formation of the RNA strand is terminated. Behind the enzyme, the DNA double helix re-forms, stripping off the newly synthesized RNA strand. In eukaryotes, transcription is initiated and regulated by a host of proteins called *transcription factors*; in prokaryotes an accessory *sigma factor* is essential for transcription.

In many eukaryotic genes the functional message is contained within discontinuous segments of the DNA strand (EXONS), interrupted by nonfunctional segments (INTRONS). Initially, both exons and introns are transcribed to form so-called *heterogeneous nuclear RNA* (*hnRNA*). Subsequently the noncoding intron sequences are spliced out to form the fully functional mature mRNA transcript, which then leaves the nucleus to direct protein assembly in the cytoplasm, in the process called TRANSLATION. *See also* RNA splicing.

transduction The transfer of part of the DNA of one bacterium to another by a bacteriophage. The process does occur naturally but is mainly known as a technique in recombinant DNA technology, and has been used in mapping the bacterial chromosome.

transferase An enzyme that catalyzes reactions in which entire groups or radicals are transferred from one molecule to another. Hexokinase catalyzes the transfer of a high-energy terminal phosphate group from ATP to glucose to give glucose–6–phosphate and ADP.

IMPORTANT TRACE ELEMENTS IN PLANTS AND ANIMALS

Trace element	*Compounds containing*	*Metabolic role*
copper	cytochrome oxidase	oxygen acceptor in respiration
	plastocyanin	electron carrier in photosynthesis
	hemocyanin	respiratory pigment in some marine invertebrates
	tyrosinase	melanin production – absence causes albinism
zinc	alcohol dehydrogenase	anaerobic respiration in plants – converts acetaldehyde to ethanol
	carbonic anhydrase	CO_2 transport in vertebrate blood
	carboxy peptidase	hydrolysis of peptide bonds
cobalt	vitamin B_{12}	red blood cell manufacture – absence causes pernicious anaemia
molybdenum	nitrate reductase	reduction of nitrate to nitrite in roots
	a nitrogen-fixing enzyme	nitrogen fixation
manganese	enzyme cofactor	oxidation of fatty acids
		bone development
fluorine	associated with calcium	tooth enamel and skeletons
boron		mobilization of food in plants?

transfer RNA (tRNA) A type of RNA that participates in protein synthesis in living cells. It attaches to a particular amino acid and imports this to the site of polypeptide assembly at the ribosome when the appropriate codon on the messenger RNA is reached. Each tRNA molecule consists of roughly 80 nucleotides; some regions of the molecule undergo base pairing and form a double helix, while in others the two strands separate to form loops. When flattened out the tRNA molecule has a characteristic 'cloverleaf' shape with three loops; the base of the 'leaf' carries the amino acid binding site, and the middle loop contains the anticodon, whose base triplet pairs with the complementary codon in the mRNA molecule. Hence, because there are 64 possible codons in the genetic code, of which about 60 or so code for amino acids, there may be up to 60 or so different tRNAs in a cell, each with a different anticodon, although most cells apparently have only about 10 which indicates there is some flexibility (or 'wobble') in the base pairing between the codon and anticodon. In experiments it is the third base of the codon that shows imprecise binding, which fits in with the degeneracy of the genetic code.

The correct amino acid is attached to a tRNA molecule by an enzyme called an *aminoacyl–tRNA transferase*. There are 20 of these, one for each type of amino acid. This attachment also involves the transfer of a high-energy bond from ATP to the amino acid, which provides the energy for peptide bond formation during translation. *See* translation.

transformation **1.** A permanent genetic recombination in a cell, in which a DNA fragment is incorporated into the chromosome of the cell. This may be demonstrated by growing bacteria in the presence of dead cells, culture filtrates, or extracts of related strains. The bacteria acquire genetic characters of these strains.
2. The conversion of normal cells in tissue culture to cells having properties of tumor cells. The change is permanent and transformed cells are often malignant. It may be induced by certain viruses or may occur spontaneously.

transgenic Describing organisms, especially eukaryotes, containing foreign genetic material. Genetic engineering has created a wide range of transgenic animals, plants, and other organisms, for both experimental and commercial purposes. Examples include dairy cows that secrete drugs in their milk, and herbicide-resistant crop plants. *See* recombinant DNA.

transition state (**activated complex**) Symbol: ‡ A short-lived high-energy molecule, radical, or ion formed during a reaction between molecules possessing the necessary activation energy. The transition state decomposes at a definite rate to yield either the reactants again or the final products. The transition state can be considered to be at the top of the energy profile.

For the reaction,

$$X + YZ = X...Y...Z^{\ddagger} \rightarrow XY + Z$$

the sequence of events is as follows. X approaches YZ and when it is close enough the electrons are rearranged producing a weakening of the bond between Y and Z. A partial bond is now formed between X and Y producing the transition state. Depending on the experimental conditions, the transition state either breaks down to form the products or reverts back to the reactants.

translation The process whereby the genetic code of messenger RNA (mRNA) is deciphered by the machinery of a cell to make proteins. Molecules of mRNA are in effect coded messages based on information in the cell's genes and created by the process of transcription. They relay this information to the sites of protein synthesis, the RIBOSOMES. In eukaryotes these are located in the cytoplasm, so the mRNA must migrate from the nucleus. The first stage in translation is *initiation*, in which the two subunits of the ribosome assemble and attach to the mRNA molecule near the *initiation codon* (AUG), which signals the beginning of the message. This also involves various proteins called *initiation factors*, and the initiator TRANSFER RNA (tRNA), which always carries the amino acid *N*-formyl methionine (in prokaryotes and mitochondria) or methionine (in eukaryotes).

The next stage is *elongation*, in which the peptide chain is built up from its component amino acids. tRNA molecules successively occupy two sites on the larger ribosome subunit in a sequence determined by consecutive codons on the mRNA. As each pair of tRNAs occupies the ribosomal sites, their amino acids are joined together by a peptide bond. As the ribosome moves along the mRNA to the next codon, the next tRNA enters the first ribosomal site, and so on, leading to elongation of the peptide chain. Having delivered its amino acid, the depleted tRNA is released from the second ribosomal site, which is then occupied by the tRNA with the growing chain. This process continues until the ribosome encounters a *termination codon*. Then the polypeptide chain is released and the ribosome complex dissociates, marking the *termination* of translation.

Following its release, the polypeptide may undergo various changes, such as the removal or addition of chemical groups, or even cleavage into two. This *post-translational modification* produces the fully functional protein. Folding of the protein is assisted by a class of molecules called *chaperones*. *See also* transcription.

translocation 1. The movement of mineral nutrients, elaborated food materials, and hormones through the plant. In vascular plants, the xylem and phloem serve to translocate such substances. Carbohydrates, amino acids, and other organic compounds are moved both upward and downward in the phloem, whereas water moves from roots to leaves through the xylem. There is evidence that mineral salts are moved in both the xylem and phloem. *See* transpiration.
2. *See* chromosome mutation.

transpiration The loss of water vapor from the surface of a plant. Most is lost through stomata when they are open for gaseous exchange. Typically, about 5% is lost directly from epidermal cells through the cuticle (*cuticular transpiration*) and a minute proportion through lenticels. A continuous flow of water, the *transpiration stream*, is thus maintained through the plant from the soil via root hairs, root cortex, xylem, and tissues such as leaf mesophyll served by xylem. Water evaporates from wet cell walls into intercellular spaces and diffuses out through stomata. Transpiration may be useful in maintaining a flow of solutes through the plant and in helping to cool leaves through evaporation, but is often detrimental under conditions of water shortage, when wilting may occur. It is favored by low humidity, high temperatures, and moving air.

transplantation The transfer of a tissue or organ from one part of an animal to another part or from one individual to another. *See also* graft.

transplantation antigen *See* histocompatibility.

transposon (**transposable element; jumping gene**) A segment of an organism's DNA that can insert at various sites in the genome, either by physically moving from place to place, or by producing a copy that inserts elsewhere. The simplest types are called *insertion sequences*; these comprise about 700–1500 base pairs. More complex ones, called *composite transposons*, have a central portion, which may contain functional genes, flanked by insertion sequences. Transposons can affect both the genotype and phenotype of the organism, e.g by disrupting gene expression, or causing deletions or inversions. In eukaryotes, transposons account for much of the repetitive DNA in the genome. *See* repetitive DNA.

triacylglycerol *See* triglyceride.

tricarboxylic acid cycle (**TCA cycle**) *See* Krebs cycle.

triglyceride (**triacylglycerol**) An ester of glycerol in which all the –OH groups are esterified; the acyl groups may be the same or different. Many LIPIDS are triglycerides in which the parent acid(s) of the acyl group(s) are long-chain fatty acids. In animals the triglycerides are more frequently saturated and have higher melting points than the triglycerides of plant origin, which are generally unsaturated. The triglycerides that are synthesized by organisms act as supporting material for internal organs, cell walls, etc., as transport mechanisms for nonpolar material, and as a food reserve. Triglycerides accumulate in adipose tissue and are the main energy store in mammals. Complete oxidation of triglycerides gives around 9 kcal/g compared with 4 kcal/g for carbohydrates and proteins. Triglycerides are hydrolyzed to glycerol and fatty acids by lipases. *See also* carboxylic acid; glyceride; lipase.

triiodothyronine A hormone secreted by the thyroid gland that has similar actions to but is metabolically more potent than thyroxine. Its levels in the blood are much lower than those of thyroxine. It is also produced in tissues, such as the kidney and liver, by conversion of thyroxine.

triploid A cell or organism containing three times the haploid number of chromosomes. Triploid organisms arise by the fusion of a haploid gamete with a diploid gamete that has resulted from the nondisjunction of chromosomes at meiosis. Triploids are usually sterile because one set of chromosomes remains unpaired at meiosis, which disrupts gamete formation. In flowering plants the endosperm tissue is usually triploid, resulting from the fusion of one of the pollen nuclei with the two polar nuclei. *Compare* diploid; haploid.

tritiated compound A compound in which one or more ^{1}H atoms have been replaced by tritium (^{3}H) atoms.

tritium Symbol: T, ^{3}H A radioactive isotope of hydrogen of mass number 3. The nucleus contains 1 proton and 2 neutrons. Tritium decays with emission of low-energy beta radiation to give ^{3}He. The half-life is 12.3 years. It is useful as a tracer in studies of chemical reactions. Compounds in which ^{3}H atoms replace the usual ^{1}H atoms are said to be *tritiated*. A positive tritium ion, T^{+}, is a *triton*.

tRNA *See* transfer RNA.

tropism (**tropic movement**) A directional growth movement of part of a plant in response to an external stimulus. Tropisms are named according to the stimulus. The organ is said to exhibit a positive or negative tropic response, depending on whether it grows towards or away from the stimulus respectively, e.g. shoots are positively phototropic but negatively gravitropic (geotropic). Growth straight towards or away from the stimulus (0° and 180° orientation respectively) is called *orthotropism*. Primary roots and shoots are orthotropic to light and gravity. By contrast, growth at any other angle to the direction of the stimulus as by branches or lateral roots is called *plagiotropism*. The mechanism involved in the latter is poorly understood. Since the receptor for the stimulus is often separate from the region of growth, tropic movements are often mediated by hormones. *See* phototropism; geotropism. *See also* taxis.

tropomyosin A protein found in skeletal muscle that regulates contraction of the muscle filaments. It forms a rodlike molecule that, in the resting stage, blocks the myosin binding sites on the actin filaments. Another protein, *troponin*, is associated with tropomyosin. A nerve impulse triggers contraction by causing an influx of calcium ions from the sarcoplasmic reticulum. The troponin molecules bind calcium ions and in so doing cause the tropomyosin molecules to shift position slightly, so exposing the myosin binding sites and allowing contraction to proceed. Tropomyosin, but not troponin, is also found in smooth muscle. *See* sarcomere.

troponin *See* tropomyosin.

trypsin An enzyme that catalyzes the partial hydrolysis of peptides. It catalyzes the hydrolysis of peptide bonds formed from the carbonyl group of lysine and arginine residues. Trypsin is found in pancreatic tissue and in pancreatic juices in an inactive form, trypsinogen.

trypsinogen The inactive form of trypsin, found in pancreatic tissues and pancreatic juices. Trypsinogen is converted into the active form by action of the enzyme enteropeptidase (enterokinase).

tryptophan *See* amino acids.

tubulin *See* microtubule.

tumor suppressor gene A gene whose normal function helps to prevent cells becoming cancerous. Like proto-oncogenes, they code for proteins involved in the regulation of cell growth but transformation requires tumor suppressor genes to be inactivated. Mutations to tumor suppressor genes causing loss of function are heavily implicated in human cancers, e.g. mutations to the tumor suppressor gene p53 are found in most cancers. *Compare* oncogene.

turgor The state, in a plant or prokaryote cell, in which the protoplast is exerting a pressure on the cell wall owing to the intake of water by osmosis. The cell wall, being slightly elastic, bulges but is rigid enough to prevent water entering to the point of bursting. The cell is then said to be turgid. Turgidity is the main means of support of herbaceous plants. *See* osmosis; plasmolysis.

tyrosine *See* amino acids.

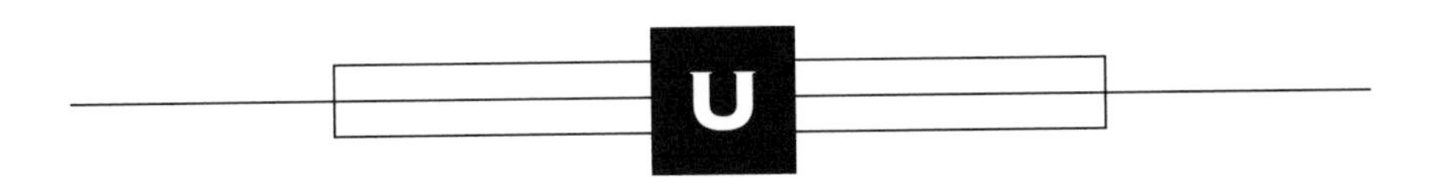

U

ubiquinone An electron-transporting coenzyme that is a component in the electron-transport chain. It was formerly called *coenzyme Q*.

ubiquitin A small protein that is involved in PROTEOLYSIS.

ultracentrifuge A high-speed centrifuge, operating at up to a million revolutions per second, that is used to sediment protein and nucleic acid molecules. Ultracentrifuges operate under refrigeration in a vacuum chamber and forces 50 million times gravity may be reached. The rate of sedimentation depends on the molecular weight of the molecule and thus the ultracentrifuge can be used to separate a mixture of large molecules, and estimate sizes.

ultramicrotome *See* microtome.

ultrastructure (**fine structure**) The detailed structure of biological material as revealed, for example, by electron microscopy, but not by light microscopy.

undulipodium A whiplike organelle that protrudes from a eukaryotic cell and is used chiefly for locomotion (e.g. sperm) or feeding (e.g. ciliate protoctists). Undulipodia include all eukaryotic cilia (*see* cilium) and flagella, which share the same essential structure, and differ markedly from bacterial flagella (*see* flagellum). The shaft comprises a cylindrical array of nine doublet microtubules surrounding a central core of two single microtubules. The outer wall of the shaft is an extension of the cell membrane. Movement of the shaft is produced by projecting arms of a protein (dynein) on the microtubules, which cause adjacent microtubule doublets to slide past each other. This requires energy from ATP. At the base of the shaft near the cell surface is a basal body, or kinetosome. This consists of a cylindrical array of nine triplet microtubules, from which the shaft grows. Its structure is similar to that of a centriole. Flagella tend to be larger than cilia, and produce successive waves of bending that are propagated to the tip of the shaft, as in a sperm. Cilia characteristically beat in a different manner, with a power stroke and a recovery stroke.

universal indicator *See* indicator.

unsaturated compound An organic compound that contains at least one double or triple bond between two of its carbon atoms. The multiple bond is relatively weak; consequently unsaturated compounds readily undergo addition reactions to form single bonds. *Compare* saturated compound.

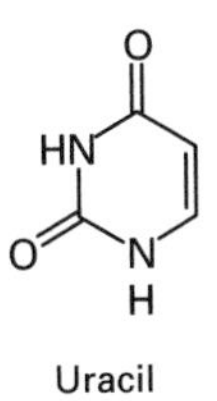

Uracil

uracil A nitrogenous base that is found in RNA, replacing the thymine of DNA. It has a pyrimidine ring structure.

urea A water-soluble nitrogen compound, $H_2N.CO.NH_2$. It is the main excretory product of catabolism of amino acids in certain animals (ureotelic animals). *See also* ornithine cycle.

urea cycle *See* ornithine cycle.

ureotelic Excreting nitrogen in the form of urea. Amphibians and mammals are ureotelic. *Compare* uricotelic.

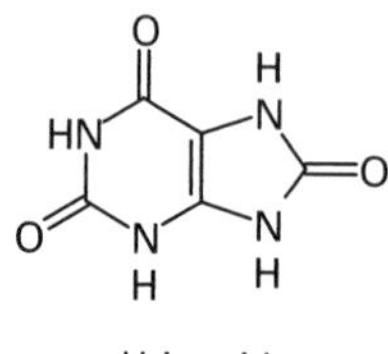

Uric acid

uric acid A nitrogen compound produced from purines. In certain animals (uricotetic animals), it is the main excretory product resulting from breakdown of amino acids. In humans, uric acid crystals in the joints are the cause of gout.

uricotelic Excreting nitrogen in the form of uric acid. Reptiles and birds are uricotelic. *Compare* ureotelic.

uridine The nucleoside formed when uracil is linked to D-ribose by a β-glycosidic bond.

urine The liquid excreted through the urethra or cloaca. It is produced in the kidneys and contains urea or uric acid, and numerous other substances in small amounts.

uronic acid *See* sugar acid.

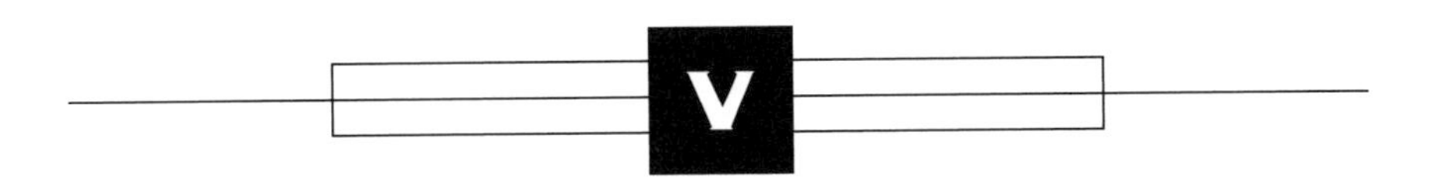

vaccination The introduction of antigens into the body to induce production of specific antibodies, either to confer immunity against subsequent infection by the same antigen or, less commonly, to treat a disease. The various types of antigens used include attenuated or dead microorganisms or harmless microorganisms that are closely related to pathogenic types. They are made into a suspension (a vaccine), which is usually injected or ingested.

vacuole A spherical fluid-filled organelle of variable size found in plant and animal cells, bounded by a single membrane and functioning as a compartment to separate a variety of materials from the cytoplasm. Vacuoles have a variety of specialized functions, for example as food vacuoles, contractile vacuoles, and autophagic vacuoles. Many mature plant cells have a single large central vacuole that confines the cytoplasm to a thin peripheral layer. It is bounded by a membrane called the tonoplast and contains cell sap. This contains substances in solution, e.g. sugars, salts, and organic acids, often in high concentrations resulting in a high osmotic pressure. Water therefore moves into the vacuole by osmosis making the cell turgid. Vacuoles may also contain crystals and waste substances.

valeric acid *See* pentanoic acid.

valine *See* amino acids.

van der Waals force An intermolecular force of attraction, considerably weaker than chemical bonds and arising from weak electrostatic interactions between molecules (the energies are often less than 1 J mol^{-1}).

The van der Waals interaction contains contributions from three effects; permanent dipole–dipole interactions found for any polar molecule; dipole–induced dipole interactions, where one dipole causes a slight charge separation in bonds that have a high polarizability; and dispersion forces, which result from temporary polarity arising from an asymmetrical distribution of electrons around the nucleus. Even atoms of the rare gases exhibit dispersion forces.

Vane, (Sir) John Robert (1927–) British pharmacologist. He was awarded the Nobel Prize for physiology or medicine in 1982 jointly with S. K. Bergström and B. I. Samuelsson for their discoveries concerning prostaglandins and related biologically active substances.

Varmus, Harold Eliot (1939–) American microbiologist and molecular biologist. He was awarded the Nobel Prize for physiology or medicine in 1989 jointly with M. J. Bishop for their discovery of the cellular origin of retroviral oncogenes.

vasoconstriction The reduction in diameter of small blood vessels due to contraction of the smooth muscle in their walls. It results from stimulation by vasoconstrictor nerve fibers or from secretion (or injection) of epinephrine in response to decreased blood pressure, low external temperature, pain, etc. *See also* vasomotor nerves.

vasodilatation (**vasodilation**) The increase in diameter of small blood vessels due to relaxation of the smooth muscle in their walls. It results from stimulation by vasodilator nerve fibers or inhibition of vasoconstrictor nerve fibers in response to increased blood pressure, exercise, high external temperature, etc. *See also* vasomotor nerves.

vasomotor nerves Nerve fibers of the autonomic nervous system that control the diameter of blood vessels. They transmit impulses from the vasomotor center in the medulla oblongata of the brain to the smooth muscle in the vessel walls, causing them to become constricted (vasoconstrictor nerve fibers) or dilated (vasodilator nerve fibers).

vasopressin (antidiuretic hormone; ADH) A peptide hormone produced by the hypothalamus and the posterior pituitary gland. It stimulates contraction of muscles around the capillaries and arterioles, raising the blood pressure. It increases peristalsis and has some effect on the uterus. It stimulates water resorption in the kidney tubules, leading to concentration of the urine. *Compare* oxytocin.

vector 1. An animal, often an insect or tick, that carries a disease-causing organism from an infected to a healthy animal or plant, causing the latter to become infected; for example, the mosquito transmits malaria and other diseases to humans.
2. (**cloning vector**) An agent used as a vehicle for introducing foreign DNA, for example a new gene, into host cells. Several types of vector are used in gene cloning, notably bacterial plasmids and bacteriophages. The segment of DNA is first spliced into the DNA of the vector, then the vector is transferred to the host cell (e.g. the bacterium, *E. coli*), where it replicates along with the host cell. The result is a clone of cells, all of which contain the foreign gene. If the vector has a promoter (*see* operon) it will lead to the foreign gene being expressed by the host cell machinery. *See* gene cloning.

vegetative Describing or relating to an involuntary function, such as digestion or the autonomic nervous activity, or to a structure or stage in development that is concerned with nutrition and growth rather than with sexual reproduction. Vegetative reproduction is asexual reproduction.

velocity constant *See* rate constant.

ventral 1. (*Zoology*) Designating the side of an animal nearest the substrate, i.e. the lower surface. However, in bipedal animals, such as humans, the ventral side is directed forwards corresponding to the anterior side of other animals.
2. (*Botany*) Designating the upper or adaxial surface of the lateral organs of plants, e.g. leaves.

Compare dorsal.

vesicle A small vacuole of variable origin, such as a Golgi vesicle or pinocytotic vesicle. *See* Golgi apparatus; pinocytosis.

vicinal positions Positions in a molecule at adjacent atoms. For example, in 1,2-dichloroethane the chlorine atoms are in vicinal positions, and this compound can thus be named *vic*-dichloroethane.

Vigneaud *See* du Vigneaud.

villus One of the microscopic finger-like projections in the lining of the small intestine. Millions of these villi give the appearance of velvet to the lining and enormously increase the surface area for absorption. Each villus is covered by a single layer of columnar cells through which the soluble products of digestion can readily pass into the blood or lymph. Each also contains a network of blood capillaries and a lacteal. Strands of muscle contract rhythmically, to shorten the villus and empty the lacteal and capillaries, so keeping up the diffusion gradient. The surface area is further increased by *microvilli*, visible only through an electron microscope, which cover the surface of each cell of the columnar epithelium.

Similar structures also occur in the chorion (*chorionic villi*), especially in the placenta where they provide a large surface area for exchange of materials between fetal and maternal blood.

violaxanthin A xanthophyll pigment found in the brown algae. *See* photosynthetic pigments.

virion The extracellular inert phase of a virus. A virion consists of a protein coat (the capsid) surrounding one or more strands of DNA or RNA. Virions may be polyhedral or helical and vary greatly in size.

viroid A tiny infectious agent found in plants that is similar to a virus but lacks a capsid, consisting simply of a circle of RNA, 300–400 nucleotides long. Viroids replicate within the plant cell and cause characteristic disease symptoms; examples are the potato spindle tuber viroid and the hop stunt viroid.

virulent phage A bacteriophage that infects a bacterial cell and immediately replicates, causing lysis of the host cell. *Compare* temperate phage.

virus An extremely small infectious agent that causes a variety of diseases in plants and animals, such as smallpox, the common cold, and tobacco mosaic disease. Viruses can only reproduce in living tissues and outside the living cell they exist as inactive particles consisting of a core of DNA or RNA surrounded by a protein coat (*capsid*). The inert extracellular form of the virus, termed a *virion*, penetrates the host membrane and liberates the viral nucleic acid into the cell. Usually, the nucleic acid is translated by the host cell ribosomes to produce enzymes necessary for the reproduction of the virus and the formation of daughter virions. The virions are released by lysis of the host cell. Other viruses remain dormant in the host cell before reproduction and lysis, their nucleic acid becoming integrated with that of the host. Some viruses are associated with the formation of tumors. *See also* oncogene; retrovirus.

Viruses that infect bacteria are called phages (or bacteriophages) and have different properties to animal and plant viruses. *See* phage.

visual purple *See* rhodopsin.

vital stains Nontoxic coloring materials that can be used in dilute concentrations to stain living material without damaging it. Examples of vital stains include Janus green, which selectively stains mitochondria and nerve cells, and trypan blue, which has an affinity for the macrophages of the reticuloendothelial system. *See also* staining.

vitamin A (**vitamin A_1; retinol**) A fat-soluble vitamin (a derivative of the yellow pigment, carotene) occurring in milk, butter, cheese, liver, and cod-liver oil. It can also be formed in the body by oxidation of carotene, which is present in fresh green vegetables and carrots. Deficiency in vitamin A can result in a reduced resistance to disease and in night blindness. *See also* rhodopsin.

vitamin B complex A group of ten or more water-soluble vitamins, which tend to occur together. They can be obtained from whole grains of cereals and from meat and liver. Since the B vitamins are present in most unprocessed food, deficiency diseases only occur in populations living on restricted diets. Many of the B vitamins act as coenzymes involved in the normal oxidation of carbohydrates during respiration.

The vitamins of the B complex include thiamine (vitamin B_1), riboflavin (vitamin B_2), nicotinic acid (niacin), pantothenic acid (vitamin B_5), pyridoxine (vitamin B_6), cyanocobalamin (vitamin B_{12}), biotin, lipoic acid, and folic acid.

vitamin C (**ascorbic acid**) A water-soluble vitamin, which is widely required in metabolism. The major sources of vitamin C are fresh fruit and vegetables and severe deficiency results in scurvy.

vitamin D A fat-soluble vitamin found in fish-liver oil, butter, milk, cheese, egg yolk, and liver. Its principal action is to increase the absorption of calcium and phosphorus from the intestine. The vitamin also has a direct effect on the calcification process in bone. Deficiency results in inadequate deposition of calcium in the bones, causing rickets in young children and osteomalacia in adults.

The term vitamin D refers, in fact, to a group of compounds, all sterols, of very similar properties. The most important are vitamin D_2 (*calciferol*) and vitamin D_3 (cholecalciferol). Precursors of these are converted to the vitamins in the body by the action of ultraviolet radiation.

vitamin E (**tocopherol**) A fat-soluble vitamin found in wheat germ, dairy products, and in meat. Severe deficiency in infants may lead to high rates of red-blood cell destruction and hence to anemia. However, there are very few deficiency effects apparent in humans.

vitamin K (**phylloquinone; menaquinone**) A fat-soluble vitamin that is required to catalyze the synthesis of prothrombin, a blood-clotting factor, in the liver. Intestinal microorganisms are capable of synthesizing considerable amounts of vitamin K in the intestine and this, together with dietary supply, insures that deficiency is unlikely to occur in any but the newborn. A newborn child may be deficient as the intestine is sterile at birth and the level supplied by the mother during gestation is limited. Thus during the first few days of life blood-clotting deficiency may be observed, but this is readily rectified by a small injection of the vitamin.

vitamins Organic compounds that are essential in small quantities for metabolism. The vitamins have no energy value; most of them seem to act as catalysts for essential chemical changes in the body, each one influencing a number of vital processes. Vitamins A, D, E, and K are the fat-soluble vitamins, occurring mainly in animal fats and oils. Vitamins B and C are the water-soluble vitamins. If a diet lacks vitamins, this results in the breakdown of normal bodily activities and produces disease symptoms. Such deficiency diseases can usually be remedied by including the necessary vitamins in the diet. Plants can synthesize vitamins from simple substances, but animals generally require them in their diet, though there are exceptions to this. These include vitamins synthesized by bacteria in the gut, and some that can be manufactured by the animal itself. A precursor of vitamin D_2 (ergosterol), for example, can be converted in the skin by ultraviolet radiation.

VLDL (**very low-density lipoprotein**) *See* lipoprotein.

voluntary muscle *See* skeletal muscle.

von Euler, Hans *See* Euler-Chelpin.

von Euler, Ulf Svante (1905–83) Son of H. von Euler. Swedish neurophysiologist and pharmacologist notable for his recognition of the importance of norepinephrine as the main neurotransmitter in mammalian adrenergic neurons. He was awarded the Nobel Prize for physiology or medicine in 1970 jointly with J. Axelrod and B. Katz.

von Hevesy *See* Hevesy.

von Laue *See* Laue.

Waksman, Selman Abraham (1888–1973) Russian-born American soil microbiologist who discovered streptomycin in 1952 (with Albert Schatz). He was awarded the Nobel Prize for physiology or medicine in 1952.

Wald, George (1906–97) American physiologist. He was awarded the Nobel Prize for physiology or medicine in 1967 jointly with R. Granit and H. K. Hartline for their discoveries concerning the primary physiological and chemical visual processes in the eye.

Walker, John E. (1941–) British biochemist who worked on the enzyme mechanism underlying the synthesis of ATP. He was awarded the 1997 Nobel Prize for chemistry jointly with P. D. Boyer. The prize was shared with J. C. Skou.

Warburg, Otto Heinrich (1883–1970) German physiologist and biochemist. He was awarded the Nobel Prize for physiology or medicine in 1931 for his discovery of the nature and mode of action of the respiratory enzyme.

water potential Symbol: ψ The chemical potential of water in a biological system compared to the chemical potential of pure water at the same temperature and pressure. It is measured in kilopascals (kPa), and pure water has the value 0 kPa; solutions with increasing concentrations of solute have more negative values of ψ, since the solute molecules interfere with the water molecules. This effect is termed the *solute potential* (or osmotic potential), denoted by ψ_s; it is measured in kPa and always has a negative value, with increasing concentrations of solute having increasingly negative values of ψ_s. In turgid plant cells there is also a pressure exerted by the walls of the cell; this is called the *pressure potential*; it is denoted by ψ_p, and has a positive value (although in xylem cells it may be negative due to water movement in the transpiration stream). Hence for a plant cell the water potential of the cell, ψ_{cell}, is given by:

$$\psi_{cell} = \psi_s + \psi_p$$

If the water potential of a cell is less (i.e. more negative) than its surroundings the cell will take up water by osmosis, until the solute potential is just balanced by the pressure potential, when the cell is described as being fully turgid. If a cell's water potential is greater (i.e. less negative) than its surroundings, the cell will lose water to its surroundings. *See* osmosis.

Watson, James Dewey (1928–) American biologist. He was awarded the Nobel Prize for physiology or medicine in 1962 jointly with F. H. C. Crick and M. H. F. Wilkins for their discoveries concerning the molecular structure of nucleic acids and its significance for information transfer in living material.

wax One of a group of water-insoluble substances with a very high molecular weight; they are esters of long-chain alcohols with fatty acids. Waxes form protective coverings to leaves, stems, fruits, seeds, animal fur, and the cuticles of insects, serving principally as waterproofing. For example, waxy deposits on some plant organs add to the efficiency of the cuticle in reducing transpiration, as well as cutting down airflow over the surface and forming a highly reflective surface, thus reducing energy available for evaporation. They may also occur in plant cell walls, e.g. leaf mesophyll. They are used in varnishes, polishes, and candles.

Western blotting A technique analogous to Southern blotting used to separate and identify proteins instead of nucleic acids. The protein mixture is separated by electrophoresis and blotted onto a nitrocellulose filter. Antibodies specific to the protein of interest are applied and bind to their target proteins. A second antibody, specific to the first antibody, is then applied. This carries a radioactive label, so enabling it to be located by autoradiography wherever it binds to the antibody–protein complex. *See* Southern blotting.

white blood cell (**white blood corpuscle**) *See* leukocyte.

Wieland, Heinrich Otto (1877–1957) German chemist who worked on the constitution of the bile acids and related substances. He was awarded the Nobel Prize for chemistry in 1927.

Wieschaus, Eric F. (1947–) American biochemist who was jointly awarded the 1995 Nobel Prize for physiology or medicine with E. B. Lewis and C. Nüsslein-Volhard for work on the genetic control of early embryonic development.

wild type The most commonly found form of a given gene in wild populations. Wild-type alleles, often designated +, are usually dominant and produce the 'normal' phenotype.

Wilkins, Maurice Hugh Frederick (1916–) New Zealand-born British biophysicist. He was awarded the Nobel Prize for physiology or medicine in 1962 jointly with F. H. C. Crick and J. D. Watson for their discoveries concerning the molecular structure of nucleic acids and its significance for information transfer in living material.

Willstätter, Richard Martin (1872–1942) German organic chemist who worked on plant pigments, especially chlorophyll. He was awarded the Nobel Prize for chemistry in 1915.

Windaus, Adolf Otto Reinhold (1876–1959) German chemist. He was awarded the Nobel Prize for chemistry in 1928 for his research into the constitution of the sterols and their connection with the vitamins.

Wöhler's synthesis A synthesis of urea from inorganic compounds (Friedrich Wöhler, 1828). The urea was produced by evaporating NH_4NCO (ammonium cyanate). The experiment was important as it demonstrated that 'organic' chemicals (i.e. ones produced in living organisms) could be made from inorganic starting materials.

wood The hard fibrous structure found in woody perennials such as trees and shrubs. It is formed from the secondary xylem and thus only found in plants that show secondary thickening, namely the gymnosperms and dicotyledons. Water and nutrients are only transported in the outermost youngest wood, termed the sapwood. The nonfunctional compacted wood of previous seasons' growth is called the heartwood and it is this that is important commercially. Wood is classified as hardwood or softwood depending on whether it is derived from dicotyledons (e.g. oak) or conifers (e.g. pine). Hardwood is generally harder than softwood but the distinction is actually based on whether or not the wood contains fibers and vessels in addition to tracheids and parenchyma. *See* xylem.

Woodward, Robert Burns (1917–79) American organic chemist. He was awarded the Nobel Prize for chemistry in 1965 for his outstanding achievements in the art of organic synthesis.

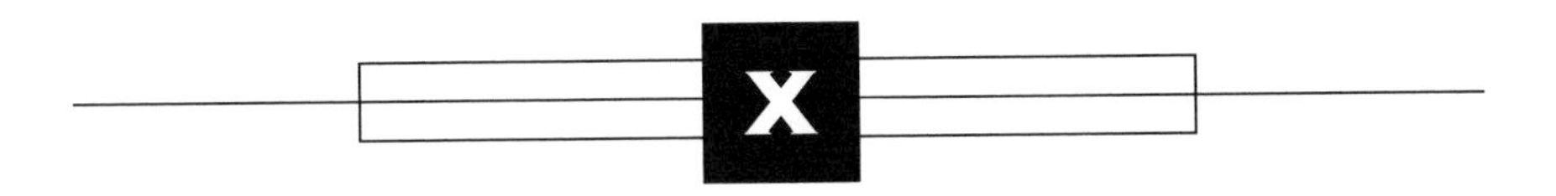

xanthine A poisonous colorless crystalline organic compound that occurs in blood, coffee beans, potatoes, and urine. It is used as a chemical intermediate.

xanthone A colorless crystalline organic compound found as a pigment in gentians and other flowers. It is used as an insecticide and in making dyestuffs.

xanthophyll One of a class of yellow to orange pigments derived from carotene, the commonest being lutein. *See* carotenoids; photosynthetic pigments.

xanthoproteic test A standard test for proteins. Concentrated nitric acid is added to the test solution. A yellow precipitate produced either immediately or on gentle heating indicates a positive result.

X chromosome The larger of the two types of sex chromosome in mammals and certain other animals. It is similar in appearance to the other chromosomes and carries many sex-linked genes. In the homogametic sex, one of the X chromosomes is inactivated in the embryo to prevent the individual having double the quantity of X chromosome gene product. The inactivation is random with respect to whether it is the maternal or paternal chromosome that is inactivated. Descendants of each cell have the same X chromosome inactive so a homogametic individual has some cells with the maternal X chromosome functional and some with the paternal.

x-ray crystallography The use of *x-ray diffraction* by crystals to give information about the 3-D arrangement of the atoms in the crystal molecules. When x-rays are passed through a crystal a *diffraction pattern* is obtained as x-rays, whose wavelength is comparable with the distances between atoms, are diffracted by the atoms, rather as light is diffracted by a diffraction grating. The technique has been successful in helping to determine the structures of some large biological molecules such as DNA, RNA, viruses, and a variety of proteins, e.g. hemoglobin, myoglobin, and lysozyme.

x-ray diffraction *See* x-ray crystallography.

xylem The water-conducting tissue in vascular plants. It consists of dead hollow cells (the tracheids and vessels), which are the conducting elements. It also contains additional supporting tissue in the form of fibers and sclereids and some living parenchyma. The secondary cell walls of xylem vessels and tracheids become thickened with lignin to give greater support. Movement of water from roots to leaves via the xylem is termed the transpiration stream.

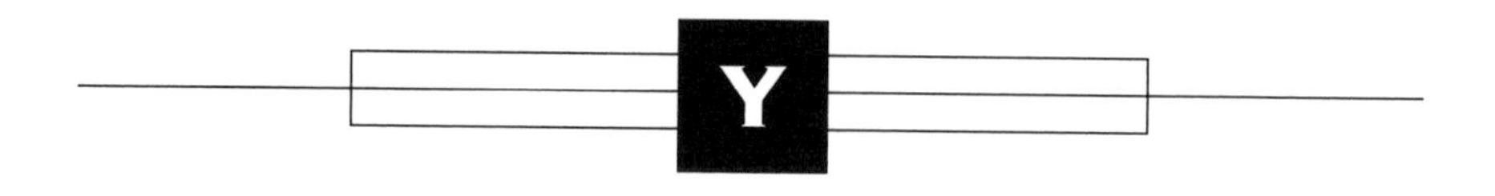

Y chromosome The smaller of the two types of sex chromosome in mammals and certain other animals. It is found only in the heterogametic sex. *See* sex chromosomes; sex determination.

Yalow, Rosalyn Sussman (1921–) American physicist. She was awarded the Nobel Prize for physiology or medicine in 1977 for the development of radioimmunoassays of peptide hormones. The prize was shared with R. Guillemin and A. V. Schally.

yeasts *See Saccharomyces.*

yocto- Symbol: y A prefix denoting 10^{-24}. For example, 1 yoctometer (ym) = 10^{-24} meter (m).

yolk The food store, consisting of proteins and fats, in eggs.

yolk sac The sac, connected to the gut of the embryo and developed from it, that contains the yolk in reptiles, sharks, and birds. When birds hatch from the egg, the yolk sac is drawn into the abdomen of the newly hatched chick. The yolk gives the chick food for the first few days, until it becomes able to feed itself.

yotta- Symbol: Y A prefix denoting 10^{24}. For example, 1 yottameter (Ym) = 10^{24} meter (m).

zepto- Symbol: z A prefix denoting 10^{-21}. For example, 1 zeptometer (zm) = 10^{-21} meter (m).

zeolite A member of a group of hydrated aluminosilicate minerals, which occur in nature and are also manufactured for their ion-exchange and selective-absorption properties. They are used for water softening and for sugar refining. The zeolites have an open crystal structure and can be used as molecular sieves.

See also ion exchange, molecular sieve.

zetta- Symbol: Z A prefix denoting 10^{21}. For example, 1 zettameter (Zm) = 10^{21} meter (m).

zinc *See* trace elements.

Zinkernagel, Rolf M. (1944–) Swiss biochemist who was jointly awarded the 1996 Nobel Prize for physiology or medicine with P. C. Doherty for work on the specificity of the cell-mediated immune defense.

zona pellucida The thick clear membrane surrounding the mammalian egg. It is surrounded by *cumulus cells* in the freshly ovulated egg, but these disperse as the sperms pass between them and penetrate the zona by enzyme action.

zwitterion (**ampholyte ion**) An ion that has both a positive and negative charge on the same species. Zwitterions occur when a molecule contains both a basic group and an acidic group; formation of the ion can be regarded as an internal acid-base reaction. For example, amino-ethanoic acid (glycine) has the formula $H_{2+}N.CH_2$-COOH. Under neutral conditions it exists as the zwitterion $H_3N.CH_2$-COO^-, which can be formed by transfer of a proton from the carboxyl group to the amine group. At low pH (acidic conditions) the ion ^+H_3N-CH_2-COOH is formed; at high pH (basic conditions) $H_2N.CH_2.COO^-$ is formed.

zygote The diploid cell resulting from the fusion of two haploid gametes. A zygote usually undergoes cleavage immediately. *See also* embryo; gamete.

zygotene In meiosis, the stage in mid-prophase I that is characterized by the active and specific pairing (synapsis) of homologous chromosomes leading to the formation of a haploid number of bivalents.

zymogen granule A secretory granule found in large numbers in enzyme-secreting cells. The granules are vesicles containing an inactive precursor of the enzyme, the zymogen, e.g. trypsinogen in exocrine cells of the pancreas. It is activated after secretion via exocytosis at the plasma membrane. The vesicles are usually derived from the Golgi apparatus, where the enzyme is processed and concentrated after synthesis on the rough endoplasmic reticulum. *See* Golgi apparatus.

APPENDIXES

Appendix I

Chronology

Major Events in the Development of Biochemistry and Molecular Biology

1833	First discovery of an enzyme, diastase, by the French chemist Anselme Payen (1795–1871)
1836	The digestive enzyme pepsin is discovered by the German physiologist Theodor Schwann (1810–82)
1860	The French chemist Louis Pasteur (1822–95) shows that fermentation is produced by substances from yeasts and bacteria (which he called 'ferments'; later (1877) designated as enzymes)
1869	Nucleic acids discovered by the German biochemist Johann Friedrich Miescher (1844–95)
1890	The lock-and-key model of enzyme action is proposed by the German chemist Emil Fischer (1852–1919)
1901	The first hormone, epinephrine, is isolated by the Japanese biochemist Jokichi Takamine (1854–1922)
1903	The enzyme zymase is discovered by the German biologist Eduard Buchner (1860–1917)
1904	Discovery of coenzymes by the British biologist Arthur Harden (1865–1940)
1909	Ribose is identified in DNA by the Russian-born American biochemist Phoebus Levene (1869–1940)
1921	The Canadian physiologist Frederick Banting (1891–1941) and the American – physiologist Charles Best (1899–1978) discover insulin.
1925	Cytochrome is discovered by the Russian-born British scientist David Keilin (1887– 1963)
1926	The American biochemist James Sumner (1877–1955) first isolates and crystallizes an enzyme (urease)

Chronology

1929	The German chemist Hans Fischer (1881–1945) determines the structure of heme in hemoglobin
1932	The muscle protein myoglobin is isolated by the Swedish biochemist Hugo Theorell (1903–82)
1937	The German-born British biochemist Hans Krebs (1900–81) describes the Krebs cycle
1940	The German-born American biochemist Fritz Lipmann (1899–1986) puts forward the proposal that ATP is the carrier of chemical energy in cells
1943	The American biochemist Britton Chance (b. 1913) suggests that enzymes operate through an enzyme–substrate complex
1952	The American biologist Alfred Hershey (1908–97) demonstrates that genetic information is carried by DNA
1953	The British biologist Francis Crick (b. 1916) and the American biologist James Watson (b. 1928) discover the structure of DNA
1955	The complete amino acid sequence of bovine insulin is discovered by the British biochemist Frederick Sanger (b. 1918)
1956	The American biochemist Arthur Kornberg (b. 1918) discovers DNA polymerase
1956	The American molecular biologist Paul Berg (b. 1926) identifies the nucleic acid that is later called transfer DNA
1957	Interferon is discovered by the British biologist Alik Isaacs (1921–67)
1959	The structure of hemoglobin is determined by the Austrian-born British biochemist Max Perutz (b. 1914)
1960	Messenger RNA is identified by the South African-born British molecular biologist Sydney Brenner (b. 1927) and the French biochemist Francois Jacob (b. 1920)
1969	The amino acid sequence of immunoglobulin G is determined by the American biochemist Gerald Edelman (b. 1929)
1970	The enzyme reverse transcriptase is discovered by the American virologists Howard Temin (1934–94) and David Baltimore (b. 1938)

Chronology

1970	Restriction enzymes are first found by the American molecular biologist Hamilton Smith (1931)
1973	Restriction enzymes are used to produce recombinant DNA by the American biochemists Stanley Cohen (b. 1935) and Herbert Boyer (b. 1936)
1977	Frederick Sanger determines the complete base sequence of DNA in bacteriophage phiX174
1985	DNA fingerprinting is invented by the British biochemist Alec Jeffreys (b. 1950)
1989	The Human Genome Project is launched
2001	First draft of the human genome sequence is published covering 97% of the human genome

Appendix II

The Genetic Code

first base	second base: U	second base: C	second base: A	second base: G	third base
U	UUU Phe	UCU Ser	UAU Tyr	UGU Cys	U
U	UUC Phe	UCC Ser	UAC Tyr	UGC Cys	C
U	UUA Leu	UCA Ser	UAA stop codon	UGA stop codon	A
U	UUG Leu	UCG Ser	UAG stop codon	UGG Trp	G
C	CUU Leu	CCU Pro	CAU His	CGU Arg	U
C	CUC Leu	CCC Pro	CAC His	CGC Arg	C
C	CUA Leu	CCA Pro	CAA Gln	CGA Arg	A
C	CUG Leu	CCG Pro	CAG Gln	CGG Arg	G
A	AUU Ile	ACU Thr	AAU Asn	AGU Ser	U
A	AUC Ile	ACC Thr	AAC Asn	AGC Ser	C
A	AUA Ile	ACA Thr	AAA Lys	AGA Arg	A
A	AUG Met (start codon)	ACG Thr	AAG Lys	CUG Arg	G
G	GUU Val	GCU Ala	GAU Asp	GGU Gly	U
G	GUC Val	GCC Ala	GAC Asp	GGC Gly	C
G	GUA Val	GCA Ala	GAA Glu	GGA Gly	A
G	GUG Val	GCG Ala	GAG Glu	GGG Gly	G

KEY

Ala	Alanine	Leu	Leucine
Arg	Arginine	Lys	Lysine
Asn	Asparagine	Met	Methionine
Asp	Aspartic acid	Phe	Phenylalanine
Cys	Cysteine	Pro	Proline
Gln	Glutamine	Ser	Serine
Glu	Glutamic acid	Thr	Threonine
Gly	Glycine	Trp	Tryptophan
His	Histidine	Tyr	Tyrosine
Ileu	Isoleucine	Val	Valine

Appendix III

Amino Acids

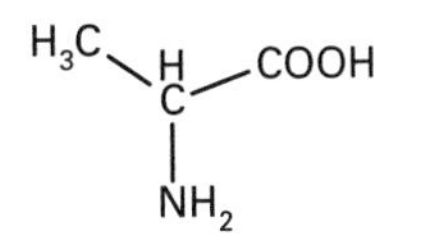

alanine

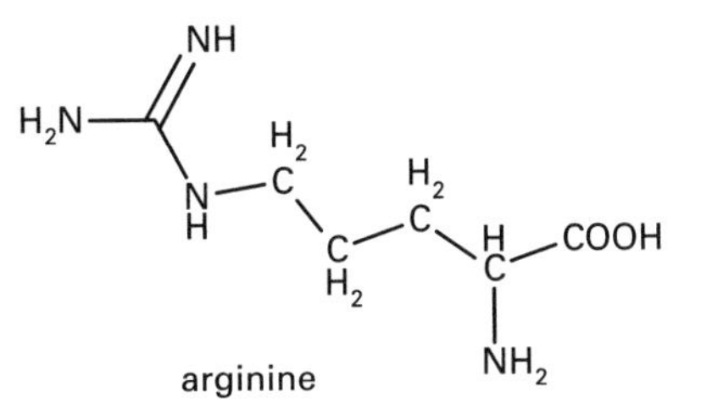

arginine

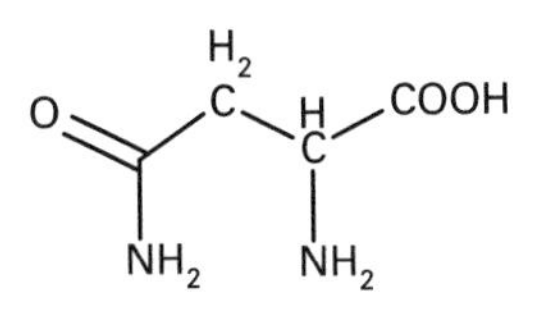

asparagine

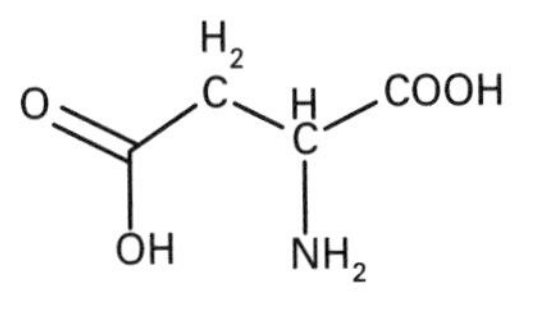

aspartine

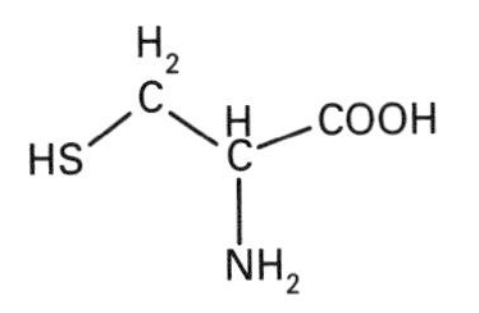

cysteine

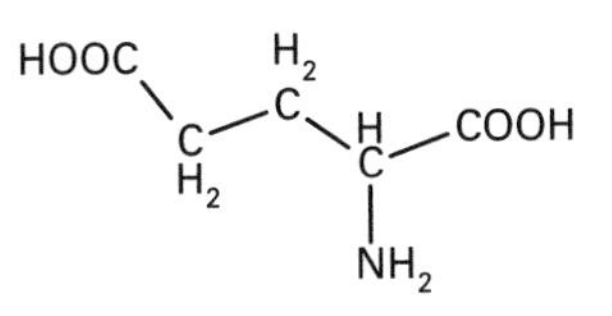

glutamic acid

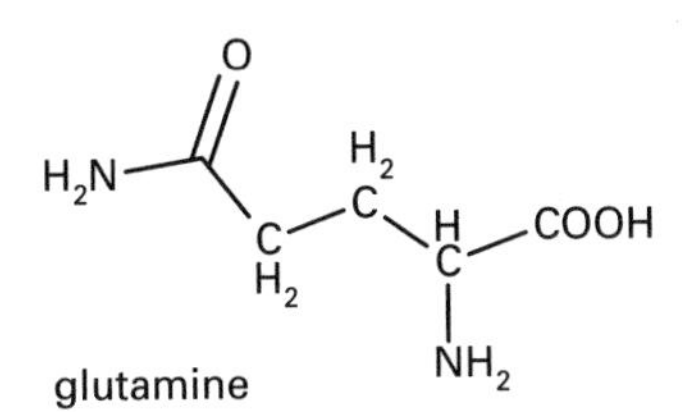

glutamine

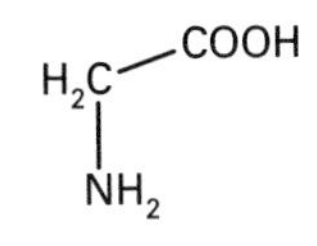

glycine

Amino Acids

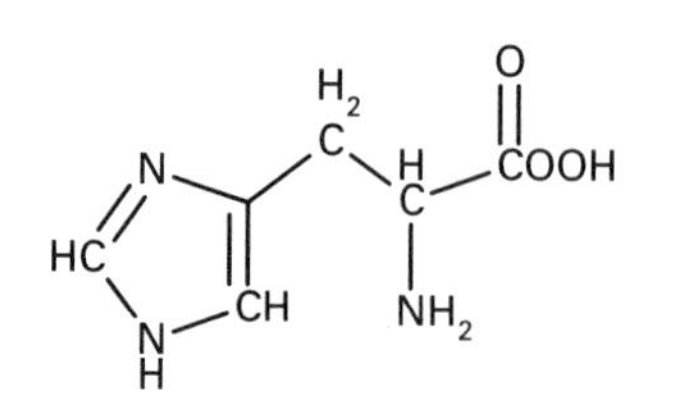

histidine

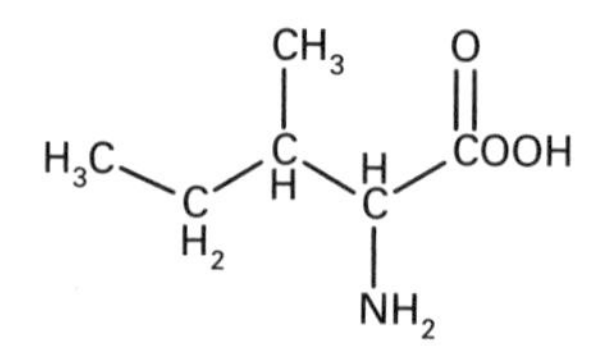

isoleucine

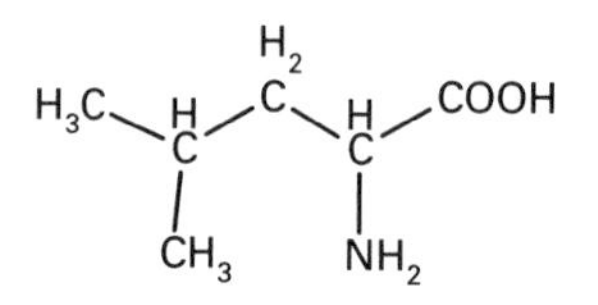

leucine

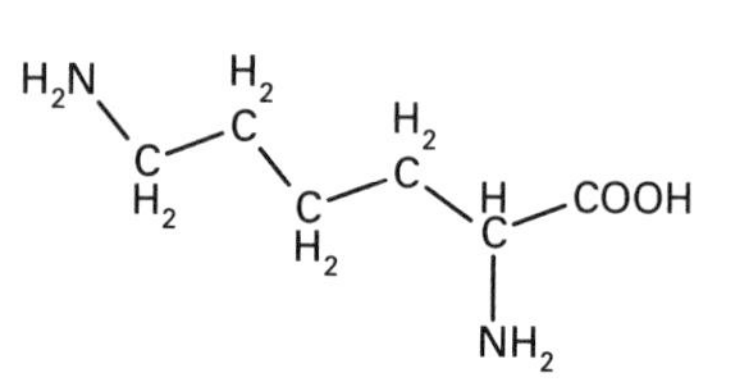

lysine

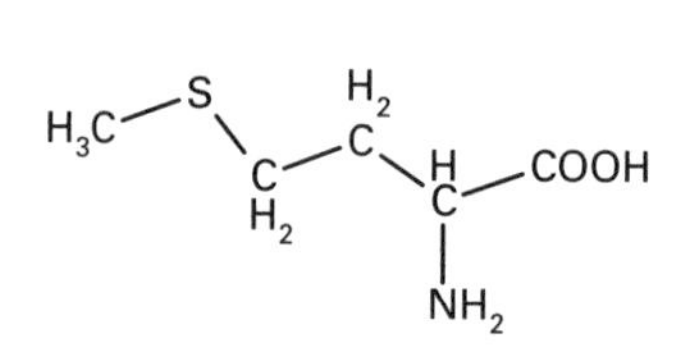

methionine

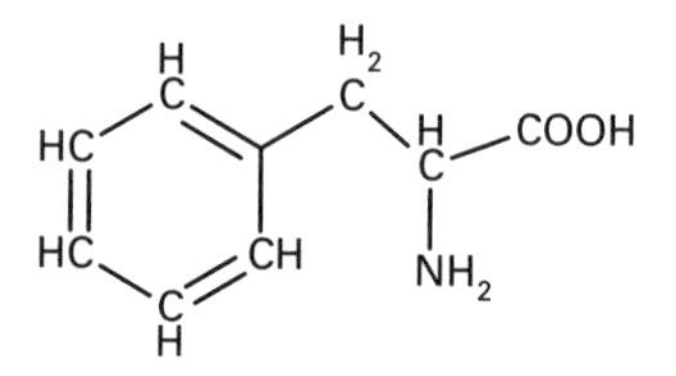

phenylalanine

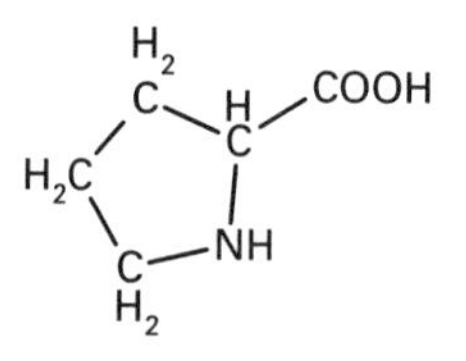

proline

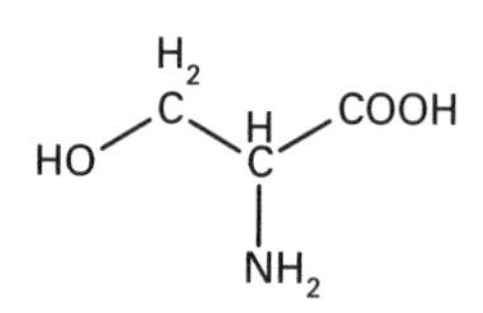

serine

Amino Acids

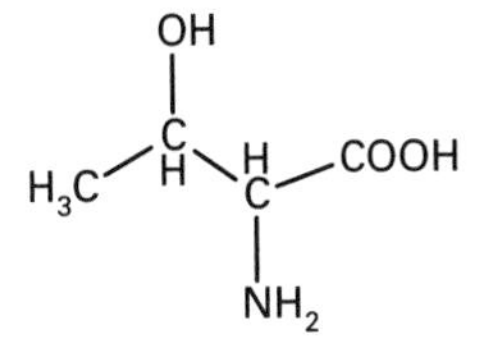

threonine

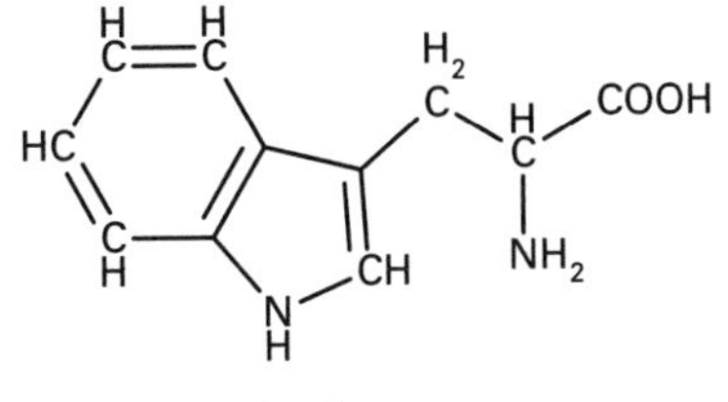

tryptophan

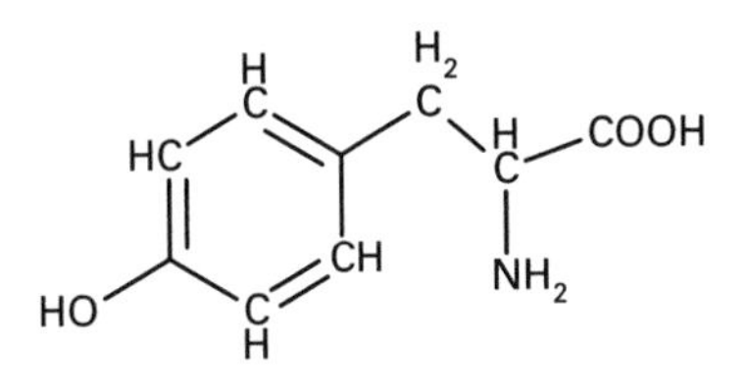

tyrosine

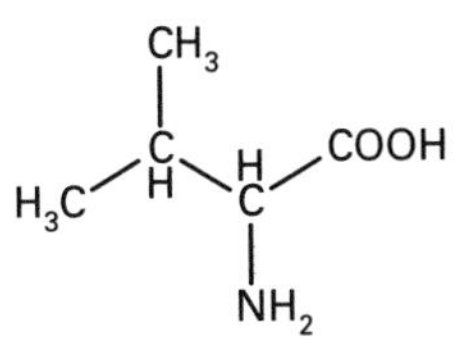

valine

Appendix IV

Periodic Table of The Elements

Periodic Table of the Elements - giving group, atomic number, and chemical symbol

Period	1	2	3	4	5	6	7	8	9	10	11	12	13	14	15	16	17	18
1	1 H																	2 He
2	3 Li	4 Be											5 B	6 C	7 N	8 O	9 F	10 Ne
3	11 Na	12 Mg											13 Al	14 Si	15 P	16 S	17 Cl	18 Ar
4	19 K	20 Ca	21 Sc	22 Ti	23 V	24 Cr	25 Mn	26 Fe	27 Co	28 Ni	29 Cu	30 Zn	31 Ga	32 Ge	33 As	34 Se	35 Br	36 Kr
5	37 Rb	38 Sr	39 Y	40 Zr	41 Nb	42 Mo	43 Tc	44 Ru	45 Rh	46 Pd	47 Ag	48 Cd	49 In	50 Sn	51 Sb	52 Te	53 I	54 Xe
6	55 Cs	56 Ba	57-71 La-Lu	72 Hf	73 Ta	74 W	75 Re	76 Os	77 Ir	78 Pt	79 Au	80 Hg	81 Tl	82 Pb	83 Bi	84 Po	85 At	86 Rn
7	87 Fr	88 Ra	89-103 Ac-Lr	104 Rf	105 Db	106 Sg	107 Bh	108 Hs	109 Mt									

Period																
6	Lanthanides	57 La	58 Ce	59 Pr	60 Nd	61 Pm	62 Sm	63 Eu	64 Gd	65 Tb	66 Dy	67 Ho	68 Er	69 Tm	70 Yb	71 Lu
7	Actinides	89 Ac	90 Th	91 Pa	92 U	93 Np	94 Pu	95 Am	96 Cm	97 Bk	98 Cf	99 Es	100 Fm	101 Md	102 No	103 Lr

The above is the modern recommended form of the table using 18 groups. Older group designations are shown below.

Modern form	1	2	3	4	5	6	7	8	9	10	11	12	13	14	15	16	17	18
European convention	IA	IIA	IIIA	IVA	VA	VIA	VIIA	VIII (or VIIIA)			IB	IIB	IIIB	IVB	VB	VIB	VIIB	0 (or VIIIB)
N. American convention	IA	IIA	IIIB	IVB	VB	VIB	VIIB	VIII (or VIIIB)			IB	IIB	IIIA	IVA	VA	VIA	VIIA	VIIIA (or 0)

Appendix V

The Chemical Elements

(* *indicates the nucleon number of the most stable isotope*)

Element	*Symbol*	*p.n.*	*r.a.m*	*Element*	*Symbol*	*p.n.*	*r.a.m*
actinium	Ac	89	227*	fermium	Fm	100	257*
aluminum	Al	13	26.982	fluorine	F	9	18.9984
americium	Am	95	243*	francium	Fr	87	223*
antimony	Sb	51	112.76	gadolinium	Gd	64	157.25
argon	Ar	18	39.948	gallium	Ga	31	69.723
arsenic	As	33	74.92	germanium	Ge	32	72.61
astatine	At	85	210	gold	Au	79	196.967
barium	Ba	56	137.327	hafnium	Hf	72	178.49
berkelium	Bk	97	247*	hassium	Hs	108	265*
beryllium	Be	4	9.012	helium	He	2	4.0026
bismuth	Bi	83	208.98	holmium	Ho	67	164.93
bohrium	Bh	107	262*	hydrogen	H	1	1.008
boron	B	5	10.811	indium	In	49	114.82
bromine	Br	35	79.904	iodine	I	53	126.904
cadmium	Cd	48	112.411	iridium	Ir	77	192.217
calcium	Ca	20	40.078	iron	Fe	26	55.845
californium	Cf	98	251*	krypton	Kr	36	83.80
carbon	C	6	12.011	lanthanum	La	57	138.91
cerium	Ce	58	140.115	lawrencium	Lr	103	262*
cesium	Cs	55	132.905	lead	Pb	82	207.19
chlorine	Cl	17	35.453	lithium	Li	3	6.941
chromium	Cr	24	51.996	lutetium	Lu	71	174.967
cobalt	Co	27	58.933	magnesium	Mg	12	24.305
copper	Cu	29	63.546	manganese	Mn	25	54.938
curium	Cm	96	247*	meitnerium	Mt	109	266*
dubnium	Db	105	262*	mendelevium	Md	101	258*
dysprosium	Dy	66	162.50	mercury	Hg	80	200.59
einsteinium	Es	99	252*	molybdenum	Mo	42	95.94
erbium	Er	68	167.26	neodymium	Nd	60	144.24
europium	Eu	63	151.965	neon	Ne	10	20.179

The Chemical Elements

Element	*Symbol*	*p.n.*	*r.a.m*	*Element*	*Symbol*	*p.n.*	*r.a.m*
neptunium	Np	93	237.048	seaborgium	Sg	106	263*
nickel	Ni	28	58.69	selenium	Se	34	78.96
niobium	Nb	41	92.91	silicon	Si	14	28.086
nitrogen	N	7	14.0067	silver	Ag	47	107.868
nobelium	No	102	259*	sodium	Na	11	22.9898
osmium	Os	76	190.23	strontium	Sr	38	87.62
oxygen	O	8	15.9994	sulfur	S	16	32.066
palladium	Pd	46	106.42	tantalum	Ta	73	180.948
phosphorus	P	15	30.9738	technetium	Tc	43	99*
platinum	Pt	78	195.08	tellurium	Te	52	127.60
plutonium	Pu	94	244*	terbium	Tb	65	158.925
polonium	Po	84	209*	thallium	Tl	81	204.38
potassium	K	19	39.098	thorium	Th	90	232.038
praseodymium	Pr	59	140.91	thulium	Tm	69	168.934
promethium	Pm	61	145*	tin	Sn	50	118.71
protactinium	Pa	91	231.036	titanium	Ti	22	47.867
radium	Ra	88	226.025	tungsten	W	74	183.84
radon	Rn	86	222*	uranium	U	92	238.03
rhenium	Re	75	186.21	vanadium	V	23	50.94
rhodium	Rh	45	102.91	xenon	Xe	54	131.29
rubidium	Rb	37	85.47	ytterbium	Yb	70	173.04
ruthenium	Ru	44	101.07	yttrium	Y	39	88.906
rutherfordium	Rf	104	261*	zinc	Zn	30	65.39
samarium	Sm	62	150.36	zirconium	Zr	40	91.22
scandium	Sc	21	44.956				

Appendix VI

The Greek Alphabet

Α	α	alpha	Ν	ν	nu
Β	β	beta	Ξ	ξ	xi
Γ	γ	gamma	Ο	ο	omikron
Δ	δ	delta	Π	π	pi
Ε	ε	epsilon	Ρ	ρ	rho
Ζ	ζ	zeta	Σ	σ	sigma
Η	η	eta	Τ	τ	tau
Θ	θ	theta	Υ	υ	upsilon
Ι	ι	iota	Φ	φ	phi
Κ	κ	kappa	Χ	χ	chi
Λ	λ	lambda	Ψ	ψ	psi
Μ	μ	mu	Ω	ω	omega

Appendix VII

Webpages

The following all have information on biochemistry and on cell and molecular biology:

European Bioinformatics Centre	ebi.ac.uk
Harvard University Department of Molecular and Cellular Biology	mcb.harvard.edu/BioLinks.html
Human Genome Project	www.ornl.gov/hgmis
Internation Union of Biochemistry and Molecular Biology (Nomenclature)	www.chem.qmul.ac.uk/iubmb
Leeds University Bioinformatics Links	www.bioinf.leeds.ac.uk/bioinformatics.html
National Biological Information Infrastructure	www.nbii.gov
Oxford University Bioinformatics Centre	www.molbiol.ox.ac.uk
Southwest Biotechnology and Informatics Center	www.swbic.org
UK Human Genome Mapping Project	www.hgmp.mrc.ac.uk

Bibliography

Basic texts covering biochemistry and cell and molecular biology:

Alberts, Bruce; Johnson, Alexander; Lewis, Julian; Raff, Martin; Roberts, Keith; & Walter, Peter *Molecular Biology of the Cell.* 4th ed. New York: Garland Science, 2002

Aldridge, Susan *Biochemistry for Advanced Biology.* Cambridge, U.K.: Cambridge University Press, 1994

Boyer, Rodney F. *Modern Experimental Biochemistry.* San Francisco, Calif.: Benjamin Cummings, 2000

Campbell, Mary K. *Biochemistry.* 3rd ed. Philadelphia: Harcourt Brace, 1999

Elliott, William H. & Elliott, Daphne C. *Biochemistry and Molecular Biology.* 2nd ed. Oxford, U.K.: Oxford University Press, 2001

Horton, H. Robert; Moran, Laurence A.; Ochs, Raymond S.; Raun, David J.; & Scrimgeour, K. Gray *Principles of Biochemistry.* 3rd ed. Upper Saddle River, NJ: Prentice Hall, 2002

Lodish, Harvey; Berk, Arnold; Zipursky, S. Lawrence; & Matsudaira, Paul *Molecular Cell Biology.* 4th ed. New York: W. H. Freeman, 2000

McKee, Trudy & McKee, James R. *Biochemistry: The Molecular Basis of Life.* 3rd ed. Boston, Mass.: McGraw Hill, 1999

Voet, Donald; Voet, Judith G.; & Pratt, Charlotte W. *Fundamentals of Biochemistry.* New York: Wiley, 1999

A more advanced text is:

Nelson, David L. & Cox, Michael M. *Lehninger Principles of Biochemistry.* 3rd. ed. New York: Worth Publishers, 2000